Advanced Information and Knowledge Processing

Series Editors
Professor Lakhmi Jain
Lakhmi.jain@unisa.edu.au

Professor Xindong Wu
Xwu@emba.uvm.edu

Also in this series

Gregoris Mentzas, Dimitris Apostolou, Andreas Abecker and Ron Young
Knowledge Asset Management
1-85233-583-1

Michalis Vazirgiannis, Maria Halkidi and Dimitrios Gunopulos
Uncertainty Handling and Quality Assessment in Data Mining
1-85233-655-2

Asunción Gómez-Pérez, Mariano Fernández-López, Oscar Corcho
Ontological Engineering
1-85233-551-3

Arno Scharl (Ed.)
Environmental Online Communication
1-85233-783-4

Shichao Zhang, Chengqi Zhang and Xindong Wu
Knowledge Discovery in Multiple Databases
1-85233-703-6

Jason T.L. Wang, Mohammed J. Zaki, Hannu T.T. Toivonen and Dennis Shasha (Eds)
Data Mining in Bioinformatics
1-85233-671-4

C.C. Ko, Ben M. Chen and Jianping Chen
Creating Web-based Laboratories
1-85233-837-7

Manuel Graña, Richard Duro, Alicia d'Anjou and Paul P. Wang (Eds)
Information Processing with Evolutionary Algorithms
1-85233-886-0

Colin Fyfe
Hebbian Learning and Negative Feedback Networks
1-85233-883-0

Yun-Heh Chen-Burger and Dave Robertson
Automating Business Modelling
1-85233-835-0

Dirk Husmeier, Richard Dybowski and Stephen Roberts (Eds)
Probabilistic Modeling in Bioinformatics and Medical Informatics
1-85233-778-8

Amit Konar and Lakhmi Jain

Cognitive Engineering

A Distributed Approach to Machine Intelligence

With 124 Figures

Foreword by Professor Witold Pedrycz

 Springer

Amit Konar, METelE, MPhil, PhD
Department of Electronics and Telecommunication Engineering
Jadavpur University, Calcutta, India

Lakhmi Jain, BE(Hons), ME, PhD, Fellow IE(Aust)
University of South Australia, Adelaide, SA 5095, Australia

British Library Cataloguing in Publication Data
A catalogue record for this book is available from the British Library

Advanced Information & Knowledge Processing ISSN 1610-3947

ISBN-13: 978-1-84996-984-0   e-ISBN-13: 978-1-84628-234-8

Printed in the United States of America   (EB)

9 8 7 6 5 4 3 2 1

Springer Science+Business Media
springeronline.com

Foreword

What we profoundly witness these days is a growing number of human-centric systems and a genuine interest in a comprehensive understanding of their underlying paradigms and the development of solid and efficient design practices. We are indeed in the midst of the next information revolution, which very likely brings us into a completely new world of ubiquitous and invisible computing, Ambient Intelligent (AMI), and wearable hardware. This requires a totally new way of thinking in which cognitive aspects of design, cognitive system engineering and distributed approach play a pivotal role. This book fully addresses these timely needs by filling a gap between the two well-established disciplines of cognitive sciences and cognitive systems engineering.

As we put succinctly in the preface, with the psychological perspective of human cognition in mind, "*the book explores the computational models of reasoning, learning, planning and multi-agent coordination and control of the human moods*". This is an excellent, up to the point description of the book. The treatise is focused on the underlying fundamentals, spans across a vast territory embracing logic perspectives of human cognition, distributed models, parallel computing, expert systems, and intelligent robotics.

The leading formal framework Professors Amit Konar and Lakhmi Jain decided to use here are Petri nets. Originally introduced by Karl Adam Petri in 1962, the concept of these nets has been re-visited and enriched in many different ways leading to new ideas, efficient modeling techniques, and interesting practical insights. Some important generalization that we witness in this book concerns fuzzy Petri nets along with their learning mechanisms (those of supervised as well as unsupervised character). Fuzziness, or granularity of information, in general, is essential to human reasoning and cognitive processes. It efficiently supports various levels of abstraction we deem of interest and relevance to a given problem. The fuzzy Petri networks come with a lot of interesting applications to abductive reasoning, belief reasoning and belief revision. The distributed character of processing is also vividly presented in the models of fuzzy cognitive maps covered in this book. Given the inherently distributed and multimodal character of the environment, multi-agent planning becomes another important topic. Here the book offers an interesting view on this subject by showing its application to mobile robots and their coordination. Context aware computing, yet another dominant subject area, is discussed in the setting of a detection of human mood in which the recognition processes are concentrated on the classification of facial expressions.

Studying this book is both a rewarding and enjoyable experience. It delivers a wealth of timely material, written in an authoritative and lucid manner. The authors are experts in their research areas who are willing to share their knowledge and practical experience with their reader. And this really shows up. Studying the book is enjoyable as it positions itself at the center of the exciting developments, which are going to form our reality in the years to come. The authors share their vision and keep us in touch with the recent developments and future directions.

Professors Konar and Jain did a superb job by bringing to the research community a timely and important volume. The book definitely offers a systematic, highly readable and authoritative material. Unquestionably, it will be equally appreciated by those advancing new frontiers of research in this area as well as the readers interested in gaining a solid and systematic knowledge about the subject matter.

January 9, 2005

Witold Pedrycz
University of Alberta
Edmonton, AB, Canada

Preface

There are books on cognitive science and cognitive systems engineering. The books on cognitive science mostly deal with the psychological aspects of human cognition. The titles on cognitive systems engineering usually focus attention on human-machine interaction models. Unfortunately, there is no book on cognitive engineering that can bridge the gap between cognitive science and cognitive systems engineering. The book *Cognitive Engineering: A Distributed Approach to Machine Intelligence* fills this gap.

Beginning with the psychological perspectives of the human cognition, *Cognitive Engineering* gradually explores the computational models of reasoning, learning, planning, and multi-agent coordination and control of the human moods. Humans usually perform the above cognitive tasks by activating distributed modules in their brain. To incorporate the humanlike ability of distributed processing, a specialized distributed framework similar to Petri nets has been selected.

Chapter 1 of the book introduces the basic psychological processes, such as memory and attention, perception, and pattern recognition using the classical theories proposed by the philosophers over a century. The chapter finally proposes Petri nets as a distributed framework for cognitive modeling.

Chapter 2 presents a distributed model of logic programming using extended Petri nets. The model facilitates concurrent resolution of program clauses. The extended Petri net, which is designed to support the model, provides a massive parallelism without sacrificing resource utilization rate.

Chapter 3 examines the scope of fuzziness in Petri nets. It reviews the existing literature on fuzzy reasoning, fuzzy learning, and consistency analysis of fuzzy production rules using Petri nets. The chapter also assesses the scope of fuzzy Petri nets in abductive reasoning, reciprocity, duality, and nonmonotonicity.

Chapter 4 proposes forward reasoning in both acyclic and cyclic fuzzy Petri nets. The acyclic model propagates fuzzy beliefs of propositions from the axioms (staring events) to the terminal (query) propositions in the network. The cyclic model is concerned with belief revision through local computation in the network. A reachability analysis for the acyclic model to prove its deadlock freedom and a stability analysis of the cyclic model to prove its conditional stability are presented.

Chapter 5 is an extension of Chapter 4 to demonstrate the application of the belief-revision model in an illustrative expert system for criminal investigation. A detailed analysis of time-complexity of the algorithms used to build up the proposed expert system is also included.

Chapter 6 is concerned with designing learning models for causal networks. The stability analysis for the learning dynamics is also given. Application of the proposed learning models in an illustrative weather prediction problem has been undertaken.

Chapter 7 deals with unsupervised learning in a fuzzy Petri net. Two distinct models of unsupervised learning have been proposed. Stability analysis of the models has been included. Application of the models in knowledge acquisition problem of expert systems has also been studied. The relative merits of the models have been compared in the concluding section.

Chapter 8 is concerned with supervised learning in a fuzzy Petri net. The convergence analysis of the model is presented. The application of the model in object recognition from noisy training instances has been demonstrated.

Chapter 9 is an extension of the fuzzy Petri net model presented in Chapter 4. The extension is needed to represent discrete membership functions instead of singleton membership (belief) at the places of the network. Further extension of the model for abductive reasoning, bi-directional iff type reasoning, reciprocity, and duality have been studied.

Chapter 10 proposes a cybernetic approach to the modeling of human mood detection and control. The mood detection has been performed through analysis of facial expressions of the subjects using Mamdani-type fuzzy relational model. The mood control is accomplished by presenting appropriate music, video clips, and audio messages to the subject. A fuzzy relational model is employed to determine the appropriateness of the above items in a given context.

Chapter 11 deals with multi-agent planning and coordination of mobile robots. Principles of multi-agent planning have been briefly introduced. A case study on material transportation problem has been undertaken to familiarize the readers with the design aspects of a typical multi-agent robotic system. Relative merits of the proposed multi-agent system with respect to its single-agent implementation have been studied using a timing analysis.

The book will serve as a unique resource to the students and researchers of cognitive science and computer science. Graduate students of mathematics, psychology, and philosophy having a keen interest to pursue their research in cognitive engineering will also find this book useful.

Most of the chapters in the book are self-contained. To avoid cross-referencing, terms defined in a chapter are sometimes redefined in a later chapter, as and when needed. For convenience of the readers, we provide a traversal graph, where an arrow from chapter i to chapter j indicates that chapter j should be read after completion of chapter i.

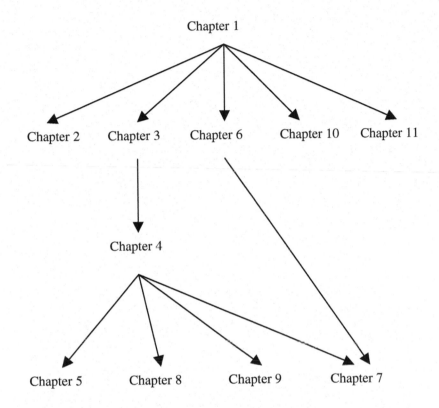

November 3, 2004

Amit Konar
and Lakhmi C. Jain

Acknowledgments

First and foremost, the first author wishes to thank Professor A.N. Basu, Vice-Chancellor, Jadavpur University, for providing him the necessary support to write the book. He is equally indebted to Professor M. Mitra, Dean, faculty of Engineering and Technology, Jadavpur University, for encouraging him to write the book. The first author also wishes to thank Prof. S.K. Sanyal, the Deputy Vice-Chancellor, Jadavpur University, for his constant mental support in writing the book. He gladly acknowledges the support extended by the current and the two former HODS of his department: Professor A.K. Bandyopadhyay, Professor H. Saha, and Professor R. Nandi for the successful completion of the book.

The first author gratefully acknowledges the research funding provided by UGC-sponsored Projects on i) AI and Image Processing and ii) University with Potential for Excellence Program in Cognitive Science. He would like to thank his research scholars, Parbati Saha, Aruna Chakraborty, and Alakananda Bhattacharya, for several hours of discussions on the subject undertaken in this book. He is equally grateful to Professor A.K. Mandal of the department of Electronics and Tele-Communication Engineering, Jadavpur University, and Profesor Amita Chatterjee of the Center for Cognitive Science, Jadavpur University, for their advice and support. The first author also wishes to thank Swagatam Das, an M.E. student of his department, for helping him in many ways in connection with the completion of the book.

Lastly, the first author wishes to express his deep gratitude to his parents, Mr. Sailen Konar and Mrs. Minati Konar, who always stood by him throughout his life and guided him in his time of crisis. He would like to thank his brother Sanjoy and sister-in-law Mala, who always inspired him to fulfill his ambitions. He also wishes to thank his wife, Srilekha, for relieving the author from the ins and outs of the household responsibilities, without which the book could not be completed.

The first author mourns the sudden death of his father-in-law, Mr. Sahadeb Samanta, who would have been very happy to see the book in print.

The first author finally remembers the charming faces of his son Dipanjan, his nephew Arpan, and niece Samiparna, whose love and imagination created inspiration and encouragement for the successful completion of the book.

The authors would like to thank Professor Witold Pedrycz, University of Alberta, Edmonton, Canada, for writing a foreword of this book.

Amit Konar
and Lakhmi C. Jain

Contents

Chapter 3: Distributed Reasoning by Fuzzy Petri Nets: A Review

Chapter 4: Belief Propagation and Belief Revision Models in Fuzzy Petri Nets

Chapter 5: Building Expert Systems Using Fuzzy Petri Nets

Chapter 6: Distributed Learning Using Fuzzy Cognitive Maps

Chapter 7: Unsupervised Learning by Fuzzy Petri Nets

Chapter 8: Supervised Learning by a Fuzzy Petri Net

Chapter 9: Distributed Modeling of Abduction, Reciprocity, and Duality by Fuzzy Petri Nets

Chapter 10: Human Mood Detection and Control: A Cybernetic Approach

Chapter 11: Distributed Planning and Multi-agent Coordination of Robots

Chapter 1

The Psychological Basis of Cognitive Modeling

This chapter examines the scope of the psychological models of the human cognitive processes in designing a computational framework of machine intelligence. Special emphasis is given to the representation of sensory information on the human mind and understanding instances of the real world by matching the received stimuli with the pre-stored information. The construction of mental imagery from the visual scenes and interpretation of the unknown scenes with the imagery have also been presented. The chapter also proposes a cybernetic "model of cognition" that mimics the mental states and their inter-relationship through reasoning and automated machine learning. It concludes with a discussion on the importance of distributed modeling of cognitive processes and proposes a specialized framework similar with Petri nets for the realization of these processes.

1.1 Introduction

A *process* in engineering sciences is represented by two attributes: i) its input(s)/output(s) and ii) the principles or techniques by which the given input(s) is transformed to the desired output(s). A heating coil, for example, may be defined as a process, where the current passing through the coil for a given time duration denotes its input and the heat generated by the coil represents its output. A simple engineering system, such as a heater, usually consists of a single process. A complex engineering system such as a distillation column may, however, include more than one process.

In designing an engineering system, we first need to model the physical/chemical processes involved in the system. The models are then analyzed and the behavior of the processes is compared to that of the models. If the model includes one or more parameters, judicious selection of their range becomes an important issue to control the behavior of the model. System identification techniques are usually employed to determine the parameters of the model.

The psychological (cognitive) processes of the human beings are much more complex than the engineering processes. Some of the well-known cognitive processes include the act of i) memorizing (encoding) and recall; ii) recognizing visual, auditory, touch, taste, and smell stimuli; iii) translating and rotating visual images in the human mind; and iv) forming ideas about size and shape of the objects. These elementary processes are part and parcel of relatively complex human thought processes such as reasoning, learning, planning, perception building, understanding, and coordinating multiple tasks. The aim of the book is to construct models of cognitive processes so as to automate the complex tasks including reasoning, learning, planning, and controlling human emotions. To model these systems, we would prefer to use conventional tools and techniques as used in modeling engineering systems.

To model an engineering system, we should be familiar with the behavioral characteristics of the system. Similarly, to model a cognitive system, we should have familiarity with the cognitive processes. There are two approaches to understand the behavior of the cognitive processes. First, we can develop a questionnaire and collect the response of the individuals to study each cognitive process. This is definitely time-consuming, and if we start it right now, we need several years to complete the study. An alternative to this is to consider the classical models of cognitive processes developed by the philosophers and the psychologists over the past few decades. There are controversies among the models, but we can accept only those models, which are free from any controversies. The rest of our work then lies with building a computational framework and automation for the models.

Sections 1.2 through 1.5 of this chapter introduce the psychological models of various cognitive processes, including memory, pattern recognition, understanding, imaging, and their mental transformations. Section 1.6 provides a cybernetic model of the mental states and their interactions. Classical models of the cognitive processes usually presume a centralized organization of the processes in the human brain. Unfortunately, these processes are distributed in the human brain, and naturally for realization of a complex task/ goal, we should construct distributed models of cognitive systems. Section 1.7 emphasizes the need for a distributed framework for cognitive modeling. In Section 1.8, we introduce a distributed framework called Petri nets. The modifications of the basic Petri net models for our applications are suggested in Section 1.9. The scope of the book is covered in Section 1.10. Conclusions are listed in Section 1.11.

1.2 Cognitive Models of Pattern Recognition

The process of recognizing a pattern involves "identifying a complex arrangement of sensory stimuli" [20], such as a character, a facial image, or a signature. Four distinct techniques of pattern recognition with reference to both contexts and experience will be examined in this section.

1.2.1 Template-Matching Theory

A "template" is part or whole of a pattern that one saves in his/ her memory for subsequent matching. For instance, in template matching of images, one may search the template in the image. If the template is part of an image, then matching requires identifying the portion (block) of the image that closely resembles the template. If the template is a whole image, such as the facial image of one's friend, then matching requires identifying the template among a set of images [4]. Template matching is useful in contexts, where pattern shape does not change with time. Signature matching or printed character matching may be categorized under this head, where the size of the template is equal to the font size of the patterns.

Example 1.1: This example illustrates the principle of the template-matching theory. Figure 1.1 (a) is the template, searched in the image of a boy in Figure 1.1 (b). Here, the image is partitioned into blocks [5] equal to the size of the template and the objective is to identify the block in the image (Fig. 1.1 (b)) that best matches with the template (Fig. 1.1 (a)).

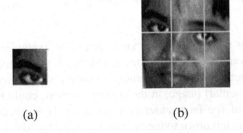

(a)   (b)

Fig. 1.1: Matching of the template (a) with the blocks in (b).

The template-matching theory suffers from the following counts.

i) **Restriction in font size and type:** Template-matching theory is not applicable to cases when the search domain does not include the template of the same size and font type. For instance, if someone wants to match a large-sized character, say Z, with an image containing a different font or size of letter Z, the template-matching theory fails to serve the purpose.

ii) **Restriction due to rotational variance:** In case the search space of the template contains a slightly rotated version of the template, the theory is unable to detect it in the space. Thus, the template-matching theory is sensitive to rotational variance of images.

It may be added here that the template-matching theory was framed for exact matching and the theory as such, therefore, should not be blamed for the reason for which it was meant. However, in case one wants to overcome the above limitations, he/she may be advised to use the prototype-matching theory as outlined below.

1.2.2 Prototype-Matching Theory

"Prototypes are idealized/abstract patterns" [20] in memory, which is compared with the stimulus that people receive through their sensory organs. For instance, the prototype of stars could be an asterisk (*). The prototype of the letter A is a symbol that one stores in his memory for matching with any of the patterns in Figure 1.2 (a) or the like.

Fig. 1.2 (a): Various fonts and size of A.

Prototype-matching theory works well for images also. For example, if one has to identify his friend among many people, he should match the prototype of the facial image and his structure, saved in memory, with the visual images of individuals. The prototype (mental) image, in the present context, could include an approximate impression of the face under consideration. How exactly the prototype is kept in memory is unknown to the researchers until this date.

1.2.3   Feature-Based Approach for Pattern Recognition

The main consideration of this approach is to extract a set of primitive features, describing the object and to compare it with similar features in the sensory patterns to be identified. For example, suppose we are interested to identify whether character H is present in the following list of characters (Fig. 1.2 (b)).

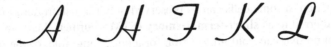

Fig. 1.2 (b): A list of characters including H.

Now, first the elementary features of H such as two parallel lines and one line intersecting the parallel lines roughly at half of their lengths are detected. These features together are searched in each of the characters in Figure 1.2 (b). Fortunately, the second character in the figure approximately contains similar features and consequently it is the matched pattern.

For matching facial images by the feature-based approach, the features like the shape of eyes, the distance from the nose tip to the center of each eye, etc. are first identified from the reference image. These features are then matched with the corresponding features of the unknown set of images. The image with the best-matched features is then identified. The detailed scheme for image matching by specialized feature descriptors such as **fuzzy moments** [5] is available in many technical literatures [19], [27].

1.2.4 The Computational Approach

Though there exist quite a large number of literature on the computational approach for pattern recognition, the main credit in this field goes to David Marr. Marr [19] pioneered a new concept of recognizing 3-dimensional objects. He stressed the need for determining the edges of an object and constructed a 2 ½-D model that carries more information than a 2-D but less than a 3-D image. An approximate guess about the 3-D object, thus, can be framed from its 2 ½- D images.

Currently, computer scientists are in favor of a neural model of perception. According to them, an electrical analogue of the biological neural net can be trained to recognize 3-D objects from their feature space. A number of training algorithms have been devised during the last two decades to study the behavioral properties of perception. The most popular among them is the well-known back-propagation algorithm, designed after Rumelhart in connection with their research on Parallel Distributed Processing (PDP) [30]. The possible realization of the neural algorithms in pattern recognition will be covered in the latter part of the book.

1.3 Cognitive Models of Memory

Sensory information is stored in the human brain at closely linked neuronal cells. Information in some cells can be preserved only for a short duration. Such memory is referred to as **short-term memory** (STM). Further, there are cells in the human brain that can hold information for a quite long time, of the order of years. Such memory is called **long-term memory** (LTM). STMs and LTMs can also be of two basic varieties, namely iconic memory and echoic memory. The **iconic memories** can store visual information, while the **echoic memories** participate in storing audio information. These two types of memories together are generally called sensory memory. Tulving alternatively classified human memory into three classes, namely **episodic, semantic,** and **procedural memory**. Episodic memory saves facts on their happening, the semantic memory constructs knowledge in structural form, while the procedural ones help in taking decisions for actions. In this section, a brief overview of memory systems will be undertaken. Irrespective of the type/variety of memory; these memory systems together are referred to as cognitive memory. Three distinct classes of cognitive memory models such as Atkinson-Shiffrin's model, Tulving's model, and the PDP model will be outlined in this section.

1.3.1 Atkinson-Shiffrin's Model

The Atkinson-Shifrin's model consists of a three-layered structure of memory (Fig. 1.3). Sensory information (signals) such as scene, sound, touch, and smell are received by receptors for temporary storage in sensory registers (memory). The **sensory registers** (Reg.) are large-capacity storage that can save information with high accuracy. Each type of sensory information is stored in separate (sensory) registers. For example, visual information is saved in iconic registers, while audio information is recorded in echoic registers. The sensory registers decay at a fast rate to keep provisions for entry of new information. Information from the sensory registers is copied into short-term memory (STM). STMs are fragile but less volatile than sensory registers. Typically STMs can hold information with significant strength for around 30 seconds, while sensory registers can hold it for just a fraction of a second. Part of the information stored in STM is copied into long-term memory (LTM). LTMs have large capacity and can hold information for several years.

STMs have faster access time [1] than LTMs. Therefore, for the purpose of generating inferences, useful information from LTMs are copied into STMs. This has been shown in Figure 1.3 by a feedback path from LTM to STM. Because of its active participation in reasoning, STMs are sometimes called **active memory**.

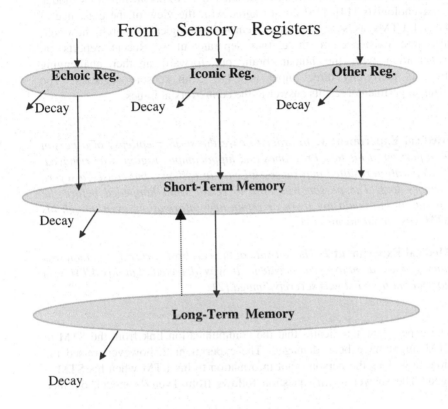

Fig. 1.3: Three-level hierarchical model of cognitive memory.

The hierarchical structure of Atkinson-Shiffrin's cognitive memory model can be compared with the memory systems in computers. The STM is similar with cache, while the LTM could be compared with main (RAM) memory. The reasoning system in the human brain is analogous with the central processing unit (CPU) of the computer. The reasoning system fetches information from the STM, as CPU receives information from cache, when the addressed words are available in the cache. In case the addressed words are not available in cache, they are fetched from the main memory and transferred to the cache and the CPU as well. Analogously, when the information required for reasoning are not found in the STM, they could be transferred from the LTM to the STM and then to the reasoning system.

1.3.2 Debates on Atkinson-Shiffrin's Model

Researchers of various domains have debated the Atkinson-Shiffrin's model. Many psychologists [14], [15] do not agree with the view of the existence of STMs and LTMs as separate units. Neuro-physiological research, however, supports the existence of these two separate units. Recent reports on experimentation with the human brain put forward another challenging question: Is there any direct input to LTM from sensory registers? The following experimental results answer to the controversial issues.

Medical Experiment 1: *In order to cure the serious epilepsy of a person X, a portion of his temporal lobes and hippocampus region was removed. The operation resulted in a successful cure in epilepsy, but caused a severe memory loss. The person was able to recall what happened before the operation, but could not learn or retain new information, even though his STM was found normal [21].*

Medical Experiment 2: *The left side of the cerebral cortex of a person was damaged by a motorcycle accident. It was detected that his LTM was normal but his STM was severely limited [2].*

The experiment 1 indicates that the communication link from the STM to the LTM might have been damaged. The experiment 2, however, raised the question: how does the person input information to his LTM when his STM is damaged? The answer to this question follows from Tveter's model, outlined below.

Without referring to the Atkinson-Shiffrin's model, Tveter in his recent book [34] considered an alternative form of memory hierarchy (Fig. 1.4). The sensory information, here, directly enters into the LTM and can be passed on to the STM from the LTM. The STM has two outputs leading to the LTM. One output of the STM helps in taking decisions, while the other is for permanent storage in the LTM.

1.3.3 Tulving's Model

The Atkinson-Shiffrin's model discussed a flow of control among the various units of the memory system. Tulving's model, on the other hand, stresses the significance of abstracting meaningful information from the environment by cognitive memory and its utilization in problem solving. The model comprises of three distinct units namely **episodic, semantic, and procedural memory.**

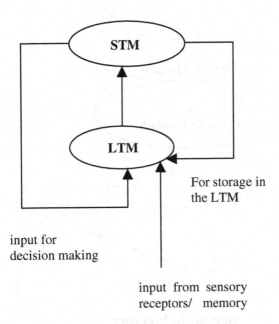

Fig. 1.4: Tveter's model showing direct entry to the LTM.

Episodic memory stores information about happened events and their relationship. The semantic memory, on the other hand, represents knowledge and does not change frequently. The procedural memory saves procedures for execution of a task [33]. A schematic architecture of Tulving's model is presented in Figure 1.5.

The episodic memory in Figure 1.5 receives an information "the sky is cloudy" and saves it for providing the necessary information to the semantic memory. The semantic memory stores knowledge in an antecedent-consequent form. The nodes in the graph denote information and the arcs denote the causal relationship. Thus the graph represents two rules: Rule 1 and Rule 2, given below.

Rule 1: *If the sky is cloudy*
Then it will rain.

Rule 2: *If it rains*
Then the roads will be flooded.

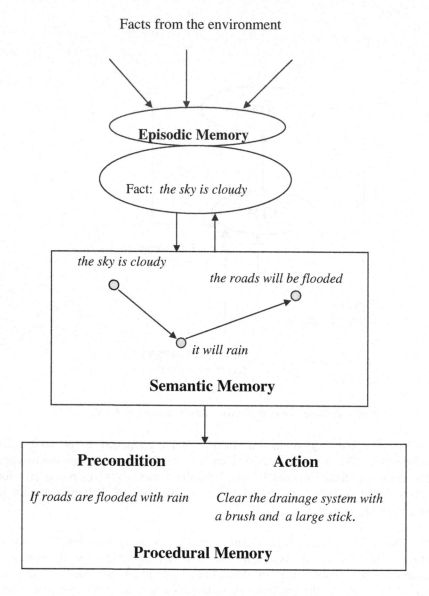

Fig. 1.5: Schematic architecture of Tulving's cognitive memory.

After execution of these two sequential rules, the semantic memory derives that "the roads will be flooded". The procedural memory first checks the pre-condition: "the road will be flooded" and then implements the action of cleaning the drainage system.

Tulving's model bridges a potential gap between the Atkinson-Shiffrin's model and the modern aspects of knowledge representation. For instance, the episodic memory contains only facts like data clauses in a PROLOG program. The semantic memory is similar with "semantic nets" used for structured knowledge representation and reasoning. The procedural memory may be compared with a frame that provides methods to solve a problem.

1.3.4   The Parallel Distributed Processing Approach

The outcome of the research on Parallel Distributed Processing (PDP) by Rumelhart, McClelland, and their associates opened up a new frontier in machine learning. Unlike the other models of cognitive memory, the PDP approach rests on the behavioral characteristics of single cellular neurons. The PDP team considered the cognitive system as an organized structure of neurons, which together forms a neural net. Such a network has an immense capability to learn and save the learned information/knowledge for subsequent usage. The PDP approach thus supports the behavioral features of cognitive memory but cannot explain the true psychological perspectives of cognition. For instance, it cannot differentiate the STM with the LTM, but experimental evidences support their co-existence. However, irrespective of any such issues, the PDP approach undoubtedly has significance from the point of view of realization of cognition on machines. The fundamental characteristics of this approach, which gave it a unique status in cognitive science, are presented below.

♦ It is one of the pioneering works on cognition that resembled the biological memory as a distributed collection of single cellular neurons that could be trained in a time parallel manner.

♦ It demonstrated a possible realization of cognition on machines.

♦ For similar input patterns, the neural net should respond similarly, while for distinct input patterns with sufficient difference, the resulting responses of the neural net will also be different sufficiently.

In fact, this is a significant observation that led to the foundation of a completely new class of pattern recognition by **supervised learning**. In a supervised learning scheme there is a trainer that can provide the desired output for the given input patterns. An example of supervised learning by a distributed approach will be covered in a separate chapter on neural nets.

♦ The PDP approach supports the content addressable features of memory, rather than addressable memory.

In conventional computers we use random access memory, where we can access the contents, when the address is supplied. But in our biological brain, we may sometimes remember part of an information and retrieve the whole after

a while. Retrieving the whole information from its part is usually done by
content addressable memory (CAM).

1.4 Mental Imagery

How do people remember scenes? Perhaps they represent scenes in some form
of image in their brain. The mental representation of scenes is informally called
mental imagery [17] in cognitive science. This section explores two important
areas in cognitive science. First, it covers mental imagery, its rotation and part-
whole relationship. Next it presents **cognitive maps**[1] that denote the spatial
relationship among objects. People form an idea of distances between any two
places by their cognitive map.

1.4.1 Mental Representation of Imagery

Psychologists have a decade-long controversy on the mental representation of
images of physical objects. One school of psychologists [17] believes that the
images are stored in human memory in analog patterns, i.e., a similar prototype
image of the object is recorded in memory. The other school [29] argues that we
store images by symbolic logic-based codes. Symbolic logic is currently used in
Artificial Intelligence to represent spatial, temporal and relational data and
knowledge but are inadequate to describe complex shape of imagery. We are,
however, trained to remember the shape of objects/animals with high accuracy,
which probably could not be reproduced (decoded) from symbolic codes.
Therefore, without going into the controversy, we may favor the opinion of the
first school of psychologists.

1.4.2 Rotation of Mental Imagery

Psychologists are of the view that people can rotate their mental imagery, as we
physically rotate the objects around a point or an axis. As experimental
evidence, let us consider the character images in Figure 1.6.

(a)   (b)

Fig. 1.6: The character A and its $180°$ rotated view around its top vertex point.

---

[1] Cognitive maps in Artificial Intelligence, however, have a more general meaning. It
stands for networks capable of acquiring, learning, encoding, and decoding information/
knowledge.

It is clear from Figure 1.6 that (b) is the inverted view of that in (a). Based on the experimental evidences on rotation of mental imagery, the following points can be envisaged.

- More complex is the shape of the original image, more (**reaction**) **time** [20] is required to identify its rotated view [31].

- More is the angle of rotation, more is the reaction time to identify the uniqueness of the mental imagery [31].

- Nonidentical but more close images and their rotated view require large reaction time to identify their nonuniqueness [32].

- Familiar figures can be rotated more quickly than the unfamiliar ones [8].

- With practice one can reduce his/ her reaction time to rotate a mental image [13].

1.4.3 Imagery and Size

In this section, the different views on imagery and size will be outlined briefly.

Kosslyn's view: Stephen Kosslyn was the pioneering researcher to study whether people make faster judgements with large images than smaller images. Kosslyn has shown that a mental image of an elephant beside that of a rabbit would force people to think of a relatively small rabbit. Again, the same image of the rabbit seems to be larger than that of a fly. Thus, people have their own relative judgement about mental imageries. Another significant contribution of Kosslyn is the experimental observation that people require more time to create larger mental imageries than smaller ones [18]. The results, though argued by the contemporary psychologists, however, follow directly from intuition.

Moyer's view: Robert Moyer provided additional information on the correspondence between the relative size of the objects and the relative size of their mental imagery. Moyer's results were based on psychophysics, the branch of psychology engaged in measuring peoples' reactions to perceptual stimuli [24]. In psychophysics, people take a longer time to determine which of the two almost equal straight lines is larger. Moyer thus stated that the reaction time to identify a larger mental image between two closely equal images is quite large.

Peterson's view: Unlike visual imagery, Intons-Peterson [28] experimented with auditory signals (also called images). She asked her students to first create a mental imagery of a cat's purring and then the ringing tone of a telephone set. She then advised her students to move the pitch of the first mental imagery up and compare it with the second imagery. After many hours of experimentation, she arrived at the conclusion that people require quite a large time to compare two mental imageries, when they are significantly different. But they require less time to traverse the mental imagery when they are very close. For instance, the imagery of purring, being close enough to the ticking of a clock, requires less time to compare them.

1.4.4 Imagery and Shape

How can people compare two similarly shaped imageries? Obviously the reasoning process looks at the boundaries and compares the closeness of the two imageries. It is evident from common-sense reasoning that two imageries of an almost similar boundary require a longer time to determine whether they are identical. Two dissimilarly shaped imageries, however, require a little reaction time to arrive at a decision about their nonuniqueness.

Paivio [26] made a pioneering contribution in this regard. He established the principle, stated above, by experiments with mental clock imagery. When the angle between the two arm positions in a clock is comparable with the same in another imagery, obviously the reaction time becomes large to determine which angle is larger. The credit to Paivio lies in extending the principle in a generic sense.

1.4.5 Part-Whole Relationship in Mental Imagery

Reed was interested in studying whether people could determine a part-whole relationship of their mental image [20]. For instance, suppose one has saved his friend's facial image in memory. If he is now shown a portion of his friend's face, would he be able to identify him? The answer in many cases was in the affirmative.

Reed experimented with geometric figures. For instance, he first showed a Star of David (vide Fig. 1.7) and then a parallelogram to a group of people and asked them to save the figures in their memory. Consequently, he asked them whether there exists a part-whole relationship in the two mental imageries. Only 14% of the people could answer it correctly. Thus determining part-whole relationship in mental imagery is difficult. But we do it easily through practicing.

Fig. 1.7: The Star of David.

1.4.6 Ambiguity in Mental Imagery

Most psychologists are of the opinion that people can hardly identify ambiguity in mental imagery, though they can do it easily with paper and pencil [20]. For instance consider the following imagery (vide Fig. 1.8). Peterson and her colleagues, however, pointed out that after some help, people can identify ambiguity also in mental imagery [28], [20].

Fig. 1.8: The letter X topped by the letter H is difficult to extract from the mental imagery but not impossible by paper and pencil.

1.4.7 Neurophysiological Similarity between Imagery and Perception

The word "perception" refers to the construction of knowledge from the sensory data for subsequent usage in reasoning. Animals including lowe- class mammals generally form perception from visual data. Psychologists, therefore, have long wondered: does mental imagery and perception have any neurophysiological similarity? An answer to this was given by Farah [12] in 1988, which earned her the Troland award in experimental psychology. Goldenbarg and his colleagues [9] noted through a series of experiments that there exists a correlation between accessing of mental imagery and increased blood flow in the visual cortex. For example, when people make judgments with visual information, the blood flow in the visual cortex increases.

1.4.8 Cognitive Maps of Mental Imagery

Cognitive maps are the internal representation of real-world spatial information. Their exact form of representation is not clearly known to date. However, most psychologists believe that such maps include both propositional codes as well as imagery for internal representation. For example, to encode the structural map of a city, one stores the important places by their imagery and the relationship among these by some logical codes. The relationship in the present context refers to the distance between two places or their directional relevance, such as place A is north to place B and at a distance of ½ Km.

How exactly people represent distance in their cognitive map is yet a mystery. McNamara and his colleagues [22] made several experiments to understand the process of encoding distance in the cognitive maps. They observed that after the process of encoding the road maps of cities is over, people can quickly remember the cities closely connected by roads to a city under consideration. But the cities far away by mileage from a given city do not appear quickly in our brain. This implicates that there must be some mechanisms to store the relative distances between elements in a cognitive map.

Besides representing distance and geographical relationship among objects, the cognitive maps also encode shapes of the pathways connecting the objects. For example, if the road includes curvilinear paths with large straight line segments, vide Figure 1.9, the same could be stored in our cognitive map easily [7]. However, experimental evidences show that people cannot easily remember complex curvilinear road trajectories (Fig. 1.10). Recently, Moar and Bower [23] studied the encoding of angle formation by two noncollinear road trajectories. They observed experimentally that people have a general tendency to approximate near right angles as right angles in their cognitive map. For example, three streets that form a triangle in reality may not appear so in the cognitive map. This is due to the fact that the sum of the internal angles of a triangle in any physical system is 180 degrees; however, with the angles close to 90 degrees being set exactly to 90 degrees in the cognitive map, the sum need not be 180 degrees. Thus a triangular path appears distorted in the cognitive map.

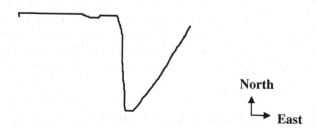

Fig. 1.9: A path with large straight-line segments.

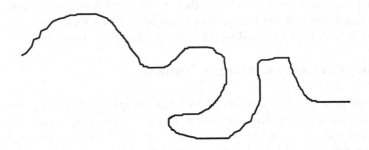

Fig. 1.10: A complex curvilinear path, difficult for encoding in a cognitive map.

1.5 Understanding a Problem

According to Greeno [10], understanding a problem involves constructing an internal representation of the problem statement. Thus to understand a sentence we must have some representation of the words and phrases and some semantic links between the words, so that the construct resembles the original sentence in meaning. The understanding of a problem, therefore, calls for understanding the meaning of the words in the problem in more elementary forms. Greeno stressed the need for three issues: *coherence, correspondence,* and *relationship to background knowledge* in connection with understanding a problem.

Coherence: A coherent representation is a pattern, so that all its components (fragments) make sense. Readers with a background of wave propagation theory, of course, will wonder: why the term "coherence"? Coherence in wave propagation theory corresponds to wavelets (small waves) with the same phase. In the present context, coherence stands for equal emphasis on each component of the pattern, so that it is not biased to one or more of its components. For example, to create a mental representation of the sentence "Tree trunks are straws for thirsty leaves and branches," one should not pay more emphasis on straws than trunks (stems of the trees). Formally, coherence calls for a mental representation of the sentence with equal emphasis on each word/ fact/ concept.

Correspondence: Correspondence refers to one-to-one mapping from the problem statement to the mental representation. If the mapping is incorrect or incomplete, then a proper understanding is not feasible. Correspondence, in most cases, however, is determined by the third issue, presented below.

Relationship to background knowledge: Background knowledge is essential to map components of the problem statement to mental representation. Without it people fail to understand the meaning of the words in a sentence and thus lose the interconnection of that word with others in the same sentence. Students can feel the significance of the background knowledge, when they attend the next class on a subject without attending the previous classes on the same topic.

1.5.1 Steps in Understanding a Problem

Understanding a problem consists of two main steps: i) identifying pertinent information from the problem description, deleting many unnecessary ones and ii) a well-organized scheme to represent the problem. Both the steps are crucial in understanding a problem. The first step reduces the scope of generality and thus pinpoints the important features of the problem. It should, of course, be added here that too much specialization of the problem features may sometimes yield a subset of the original problem and thus the original problem is lost. Determining the problem features, thus, is a main task in understanding the problem. Once the first step is over, the next step is to represent the problem features by an internal representation that truly describes the problem. The significance of the second step lies in the exact encoding of the features into the mental representation, so that the semantics of the problem and the representation have no apparent difference. The second step depends solely on the type of the problem itself and thus differs for each problem. In most cases, the problem is represented by a specialized data structure such as matrices, trees, graphs, etc. The choice of the structure and organization of the data/ information by that structure, therefore, should be given priority for understanding a problem. It should be mentioned here that the time efficiency in understanding a problem depends largely on its representation and consequently on the selection of the appropriate data structures. A few examples are presented below to give the readers some idea about understanding and solving a problem.

Example 1.2: This example demonstrates how a graphic representation can help in solving a complex problem. The problem is with a monk. He started climbing up a tall mountain on one sunny morning and reached the top on the same evening. Then he started meditating for several days in a temple at the hilltop. After several days, on another sunny morning, he left the temple and started climbing down the hill through the same road surrounding the mountain. The road is too narrow and can accommodate only one passenger at one time.

The problem is to prove that *there must be a point on the hill that the monk will visit at the same time of the day both in his upward and downward journey*, irrespective of his speed. This problem can be best solved by assuming that there are two monks, one moving up, while the other is climbing down the hill. They started moving at the same time of the day. Since the road is narrow, they must meet at some spot on the road (vide Fig. 1.11).

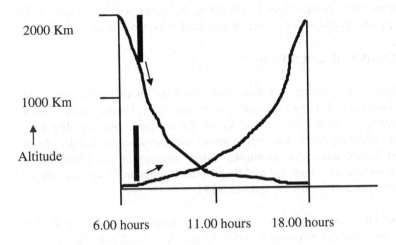

2000 Km

1000 Km

↑

Altitude

6.00 hours   11.00 hours   18.00 hours

The vertical bars denote the monks. The upward (downward) arrow indicates the upward (downward) journey of the monks. Note that the two monks must meet at some point on the road.

Fig. 1.11: Representation of the monk problem.

It may be noted that the main stress of the problem should be given to the meeting of the monks only and should not be confused with their meeting time. The solution of the given problem is a simple extension of the modified problem with two monks, which the reader can guess easily.

There exist quite a large number of interesting problems (see Exercises) that can efficiently be represented by specialized data structures. For instance, the four-puzzle problem can be described by a matrix; the water-jug problem by a graph and the missionaries-cannibals problem by a tree. There exist also a different variety of problems that could be formulated by propositional codes or other knowledge representation techniques. Identifying the best representation of a problem is an art and one can learn it through intuition only. No hard and fast rules can be framed for identifying the appropriate representation of the problem space.

1.6 A Cybernetic View of Cognition

An elementary model of cognition (vide Fig. 1.12) is proposed in this section based on its foundation in cognitive psychology [7]. The model consists of a set of five mental states, denoted by ellipses, and their activation through various physiological and psychological processes. The model includes feedback of states and is thus cyclic. The model [16] in Figure 1.12, for instance, contains three cycles, namely *the perception-acquisition cycle, the sensing-action cycle,* and the last one that passes through all states, including sensing, acquisition, perception, planning, and action, is hereafter called *the cognition cycle [16].*

1.6.1 The States of Cognition

Sensing: Apparently sensing in engineering sciences refers to reception and transformation of signal into a measurable form. However, sensing, which has a wider perspective in cognitive science, stands for all the above together with preprocessing (filtering from stray information) and extraction of features from the received information. For example, visual information on reception is filtered from undesirable noise [5] and the elementary features like size, shape, color, etc. are extracted for storing into the STM.

Acquisition: The acquisition state compares the response of the STM with already acquired and permanently stored information of the LTM. The content of the LTM, however, changes occasionally, through feedback from the perception state. This process, often called refinement of knowledge, is generally carried out by a process of unsupervised learning. The learning is unsupervised since such refinement of knowledge is an autonomous process and requires no trainer for its adaptation.

Perception: This state constructs high-level knowledge from acquired information of relatively lower level and organizes it, generally, in a structural form for the efficient access of knowledge in subsequent phases. The construction of knowledge and its organization is carried out through a process of automated reasoning that analyzes the semantic (meaningful) behavior of the low-level knowledge and their association. The state of perception itself is autonomous, as the adaptation of its internal parameters continues for years long until death. It can be best modeled by a semantic net [3], [6] .

Planning: The state of planning engages itself to determine the steps of action involved in deriving the required goal state from known initial states of the problem. The main task of this state is to identify the appropriate piece of knowledge for application at a given instance of solving a problem. It executes the above task through matching the problem states with its perceptual model, saved in the semantic memory.

It may be added here that planning and reasoning, although sharing much common formalism, have a fundamental difference that originates from their nomenclature. The reasoning may be continued with the concurrent execution of the actions, while in planning, the schedule of actions are derived and executed in a later phase. In our model of cognition, we, thus, separated the action state from the planning state.

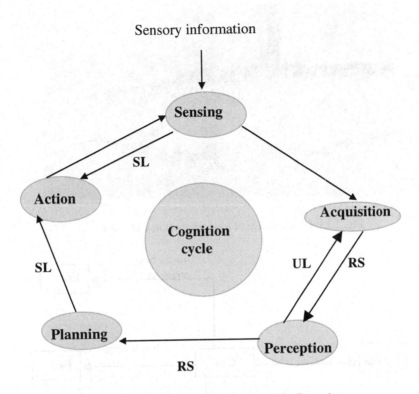

SL = Supervised learning, UL = Unsupervised learning, RS = Reasoning

Fig. 1.12: The different mental states of cognition and their relationship.

Action: This state determines the control commands for actuation of the motor limbs in order to execute the schedule of the action-plan for a given problem. It is generally carried out through a process of supervised learning, with the required action as input stimulus and the strength of the control signals as the response.

Example 1.3: This example demonstrates the various states of cognition with reference to a visual image of a sleeping cat in a corridor (Fig. 1.13).

Fig. 1.13: Digital image of a sleeping cat in a corridor.

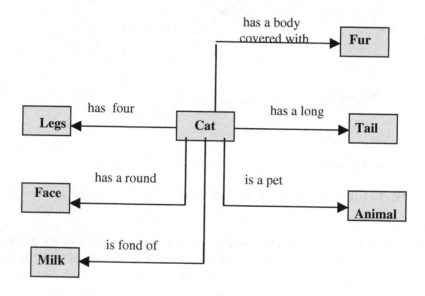

Fig. 1.14: The state of perception about a cat in the semantic memory.

Here, the sensing unit is a video camera, which received the digital image of the cat in the corridor. The image was then preprocessed and its bit-map was saved in a magnetic media, which here acts as an acquisition unit. The Acquisition State of the model of cognition, thus, contains only a pixel-wise intensity map of the scene. Human beings, however, never store the bit-map of a scene; rather, they extract some elementary features, for instance, the shape of the face (round/oval-shaped), length of the tail (long/too long/short), texture of the fur and the posture of the creature. The extracted features of a scene may vary depending on age and experience of the person. For example, a baby of 10 months only extracts the (partial) boundary edges of the image, while a child of 3 years old can extract that "the face of the creature is round and has a long tail." An adult, on the other hand, gives priority to postures, saves it in STM and uses it as the key information for subsequent search in the LTM model of perception. The LTM in the present context is a semantic net, which keeps record of different creatures with their attributes. A typical organization of a semantic net, representing a cat (Fig. 1.14) and a corridor (Fig. 1.15) is presented below.

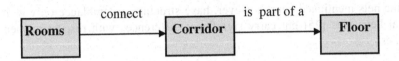

Fig. 1.15: The state of perception about a corridor in the semantic memory.

Now, for illustrating the utility of the perception, planning and action states, let us consider the semantic net for the following sentence in Figure 1.16.

"The milkman lays packets full of milk in the corridor."

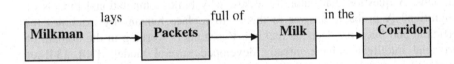

Fig. 1.16: Semantic net of a typical fact in the state of perception.

Combining all the above semantic nets together, we form a composite model of the entire scene, which together with the additional piece of knowledge *"If a living creature is fond of something and it is kept away from him, then he cannot access it,"* helps someone to generate the following schedule of plans through a process of backward reasoning [16].

Plan: *Keep the packets full of milk away from the cat.*

Further, for execution of the above plan, one has to prepare the following schedule of actions:

1. Move to the corridor.
2. Pick up the packets of milk.
3. Keep them in a safe place beyond the reach of the cat.

It may be noted that the generation of the above schedule of actions for a given plan by human beings requires almost no time. This, perhaps, is due to the supervised learning scheme that helps speeding up the generation of such a schedule.

The semantic net that serves as a significant tool for knowledge representation and reasoning requires further extension for handling various states of cognition efficiently. A specialized Petri-like net has been employed in this book, for building models of cognition for applications in inexact reasoning, learning, refinement of knowledge and control and co-ordination of tasks. The Petri-like nets mentioned here, however, have structural resemblance only with classical Petri nets [25] but carry significant differences with respect to their properties.

1.7 Computational Modeling of Cognitive Systems

The models of cognition introduced so far are based on the psychological perspectives of the human beings. These models are good enough to explain the psychological behavior of people, but are inappropriate for handling real-world problems. In this book, we would like to concentrate on the computational models of cognition, which should be capable of i) dealing with large and dynamic databases and ii) answering queries by a systematic approach like humans. A question that naturally arises: why build computational models of cognition? A probable answer to this is to replace humans by machines for efficient and bias-free decision making. Researchers working in the domain of artificial intelligence have already developed several models [30], [34] of humanlike reasoning, learning, planning and perception building. These models have their own merits and shortcomings. A notable weakness of many of the models lies in their nonparallel and nondistributed representation that logically fails to resemble true humanlike thinking.

The book takes a modest approach to cover various distributed models of cognitive processes including reasoning and machine learning. It also attempts to model multi-agent systems, where the cognitive processes of the individual agents are distributed geographically. The book finally takes into account of the models of human emotions and their control by a parallel and distributed framework.

The efficiency of distributed models to a large extent depends on the framework used for distributed computing. There exists an extensive literature on the framework for classical distributed computing. Unfortunately, there are only few papers on distributed computational frameworks for cognitive systems. In the present book, we used Petri nets as a basic framework for cognitive computation. Definitely, we cannot directly use the classical Petri nets for our framework, but need significant extension of the basic Petri net models to suit them to appropriate applications. In the next section, we briefly outline the features of ideal Petri net models.

1.8 Petri Nets: A Brief Review

Petri nets are directed bipartite graphs consisting of two types of nodes called places and transitions. Directed arcs (arrows) connect the places and transitions, with some arcs directed from the places to the transitions and the remaining arcs directed from the transitions to the places. An arc directed from a place p_i to a transition tr_j defines the place to be an input of the transition. On the other hand, an arc directed from a transition tr_k to place p_l indicates it to be an output place of tr_k. Arcs are usually labeled with weights (positive integers), where a k-weighted arc can be interpreted as the set of k-parallel arcs. A marking (state) assigns to each place a non-negative integer. If a marking assigns to place p_i a integer k (denoted by k-dots at the place), we say that p_i is marked with k tokens. A marking is denoted by a vector M, the p_i-th component of which, denoted by $M(p_i)$, is the number of tokens at place p_i. Formally, a Petri net is a 5-tuple, given by

$$PN= (P, Tr, A, W, M_0)$$

where

$P = \{p_1, p_2,...., p_m\}$ is a finite set of places,

$Tr = \{tr_1, tr_2,...., tr_n\}$ is a finite set of transitions,

$A \subseteq (P \times Tr) \cup (Tr \times P)$ is a set of arcs,

$W: A \rightarrow \{1, 2, 3, ...\}$ is a weight function,

$M_0: P \rightarrow \{0, 1, 2, 3,...\}$ is the initial marking,

$P \cap Tr = \varnothing$ and $P \cup Tr \neq \varnothing$.

Dynamic behavior of many systems can be described as transition of system states. In order to simulate the dynamic behavior of a system, a state or marking in a Petri net is changed according to the following *transition firing* rules:

1) A transition tr_j is enabled if each input place p_k of the transition is marked with at least $w(p_k, tr_j)$ tokens, where $w(p_k, tr_j)$ denotes the weight of the arc from p_k to tr_j.

2) An enabled transition fires if the event described by the transition and its input/ output places actually takes place.

3) A firing of an enabled transition tr_j removes $w(p_k, tr_j)$ tokens from each input place p_k of tr_j, and adds $w(tr_j, p_l)$ tokens to each output place p_l of tr_j, where $w(tr_j, p_l)$ is the weight of the arc from tr_j to p_l.

Example 1.4: Consider the well-known chemical reaction: $2H_2 + O_2 = 2H_2O$. We represent the above equation by a small Petri net (Fig. 1.17). Suppose two molecules of H_2 and O_2 are available. We assign two tokens to the places p_2 and p_1 representing H_2 and O_2 molecules respectively. The place p_3 representing H_2O is initially empty (Fig (1.17(a)). Weights of the arcs have been selected from the given chemical equation. Let the tokens residing at place H_2 and O_2 be denoted by $M(p_2)$ and $M(p_1)$, respectively. Then we note that

$$M(p_2) = W(p_2, tr_1) \text{ and } M(p_1) > W(p_1, tr_1).$$

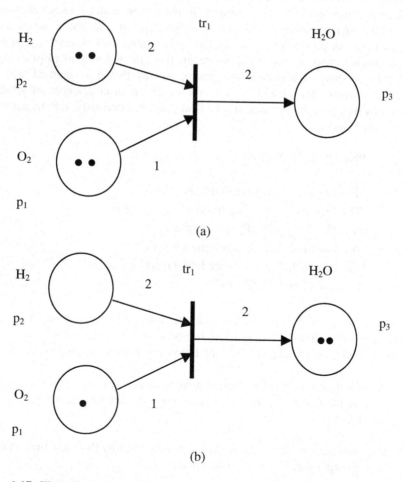

(a)

(b)

Fig. 1.17: Illustration of transition firing rule in a Petri net. The marking a) before the transition firing and b) after the transition firing.

Consequently, the transition tr_1 is enabled, and it fires by removing two tokens from the place p_2 and one token from place p_1. Since the weight $W(tr_1, p_3)$ is 2, two molecules of water will be produced, and thus after firing of the transition, the place p_3 contains 2 tokens. Further, after firing of the transition tr_1, two molecules of H_2 and one molecule of O_2 have been consumed and only one molecule of O_2 remains in place p_1.

The dynamic behavior of Petri nets is usually analyzed by a state equation, where the tokens at all the places after firing of one or mores transitions can be visualized by the marking vector M. Given a Petri net consisting of n transitions and m place.

Let

$A = [a_{ij}]$ be an (n × m) matrix of integers, called the incidence matrix, with entries

$a_{ij} = a_{ij}^+ - a_{ij}^-$ where

$a_{ij}^+ = w(tr_i, p_j)$ is the weight of the arc from transition tr_i to place p_j, and

$a_{ij}^- = w(p_j, tr_i)$ is the weight of the arc to transition tr_i from its input place p_j.

It is clear from the transition firing rule described above that a_{ij}^-, a_{ij}^+ and a_{ij}, respectively, represent the number of tokens removed, added, and changed in place j when transition tr_i fires once. Let M be a marking vector, whose j-th element denotes the number of tokens at place p_j. The transition tr_i is then enabled at marking M if

$$a_{ij}^- \leq M(j), \text{ for } j = 1, 2, \ldots, m. \tag{1.1}$$

In writing matrix equations, we write a marking M_k as an (m ×1) vector, the j-th entry of which denotes the number of tokens in place j immediately after the k-th firing in some firing sequence. Let u_k be a control vector of (n ×1) dimension consisting of (n − 1) zeroes and a single 1 at the *i*th position, indicating that transition tr_i fires at the *k*th firing. Since the ith row of the incidence matrix A represents the change of the marking as the result of firing transition tr_i, we can write the following state equation for a Petri net:

$$M_k = M_{k-1} + A^T u_k, k = 1, 2, \ldots \tag{1.2}$$

Suppose we need to reach a destination marking M_d from M_0 through a firing sequence $\{u_1, u_2, \ldots, u_d\}$. Iterating k = 0 to d in incremental steps of 1, we can then write:

$$
\left.
\begin{aligned}
M_1 &= M_0 + A^T u_1 \\
M_2 &= M_1 + A^T u_2 \\
&\cdots \cdots \cdots \cdots \\
&\cdots \cdots \cdots \cdots \\
M_{d-1} &= M_{d-2} + A^T u_{d-1} \\
M_d &= M_{d-1} + A^T u_d
\end{aligned}
\right\}
\tag{1.3}
$$

Equating the left-hand sum with the right-hand sum of the above equations we have:

$$
M_d = M_0 + \sum_{k=1}^{d} A^T u_k
\tag{1.4}
$$

$$
\text{or, } M_d - M_0 = A^T \sum_{k=1}^{d} u_k
\tag{1.5}
$$

$$
\text{or, } \Delta M = A^T x,
\tag{1.6}
$$

where

$$
\Delta M = M_k - M_0,
\tag{1.7}
$$

and

$$
x = \sum_{k=1}^{d} u_k.
\tag{1.8}
$$

Here, x is a $(n \times 1)$ column vector of non-negative integers, and is called the *firing count vector* [25]. The ith entry of x denotes the number of times that transition tr_i must fire to transform M_0 to M_d.

Example 1.5: The state equation (1.2) is illustrated below with Figure 1.18. It is clear from the figure that $M_0 = [\ 2\ \ 0\ \ 1\ \ 0\ ]^T$. After firing of transition tr_3, we obtain the resulting marking M_1 by using the state equation as follows:

$$
M_1 = M_0 + A^T u_1
$$

$$
= [\ 2\ 0\ 1\ 0\ ]^T +
\begin{pmatrix}
-2 & 1 & 1 \\
1 & -1 & 0 \\
1 & 0 & -1 \\
0 & -2 & 2
\end{pmatrix}
[\ 0\ 0\ 1\ ]^T
$$

$$= [3 \ 0 \ 0 \ 2]^{T}.$$

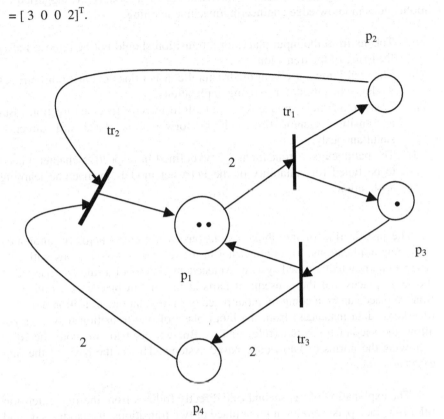

Fig. 1.18: A Petri net used to illustrate the state equation.

1.9 Extension of Petri Net Models for Distributed Modeling of Cognition

The book makes a humble attempt to model cognitive processes using a framework similar with Petri nets. However, some modifications of the basic Petri net model are needed to extend its application in the respective domains. In this section, we briefly present the extensions of the basic Petri net models and their justification. Questions may be raised on the justice of the extended models as Petri nets. In this regard, we would like to add that we need not essentially call the extended models Petri nets, we can call them specialized structures or networks or anything that the readers feel appropriate

The following amendments of the basic Petri net models are suggested for applications in knowledge engineering/machine learning.

a) Tokens from the input places of a transition should not be removed after the firing of the transition.

b) The well-known problem of structural conflict does not arise in Petri net models when used in reasoning applications.

c) Transition firing can take place in both forward or reverse directions. But a transition cannot fire both in forward and backward direction simultaneously.

d) The parameters of transitions (to be defined in respective chapters) need to be tuned for suitability of the Petri net models in machine learning applications.

The justification of the above extensions is presented here. In automated reasoning applications, transition firing in a Petri net is synonymous with rule firing of a knowledge-based system. Antecedent clauses of a rule are mapped at the input places and the consequent parts at the output places of a transition. Since a place can be a common input place of more than one transition, removal of tokens (data instances) from the input place of one transition thus does not allow the second transition (rule) to fire. But we need to fire both the rules following the norms of knowledge based systems. This is the basis of the first extension rule.

The explanation of the second rule directly follows from the first extension rule. If a place p_i is a common input place of two transitions: tr_1 and tr_2, say, and suppose both the transitions are enabled, but if one fires then the other cannot fire for removal of tokens from the input place p_i. This problem in classical Petri net theory is well known as a conflict. Since we do not remove tokens from the input place after firing of a transition, both the transitions can concurrently fire in the present context. This is the justification of the second extension rule.

In classical Petri net theory, transitions can fire only in forward directions. But in abductive reasoning problems we have to determine the premise of a rule from its consequent part if it is available as a fact. Formally, given a fact Q and a rule: if P then Q, we can infer P. This calls for backward firing of a transition to derive P from given Q, and thus justifies the extension rule (c).

Lastly, to model machine learning systems specially supervised systems, we need to adapt some parameters of the network so as to generate the desired output instance from the supplied input instance. When Petri nets are used for modeling supervised learning systems we adapt the parameters associated with transitions to train the network with the supplied instances. The extension rule (d) is required for extending Petri net models in machine learning applications.

1.10 Scope of the Book

The book attempts to model the cognitive processes involved in reasoning, learning, human mood detection and control and multi-agent co-ordination among the cognitive processes at both individual and group levels. Recent survey on experimental psychology [20] reveals that the neurons responsible for the cognitive processes are located in a distributed fashion in the cerebral cortex, the outer layer of the human brain. Unfortunately, the existing models of cognitive information processing are centralized, and therefore they do not resemble the human cognitive processes in a true sense. In this book, we present distributed models of cognitive processes with an ultimate aim to build intelligent systems. The main advantage of the distributed modeling of cognition are i) concurrent computing, ii) high fault tolerance, iii) speedy recovery of information, and, above all, iv) humanlike execution of complex tasks.

The book stresses the need for a specialized framework-like Petri nets for the modeling of distributed cognitive systems. It begins with parallel and distributed models of logic programming realized with Petri-like nets, and gradually extends the structural and behavioral characteristics of Petri nets to support reasoning with inexact data and knowledge bases. The book also discusses the scope of probabilistic and, in particular, Bayesian models for reasoning on a causal network.

The latter part of the book focuses on the learning aspects of Petri-like nets. Both supervised and unsupervised learning models on Petri-like nets have been taken into account. The book ends with discussions on neuro-fuzzy modeling of human mood detection and control, and distributed planning and coordination of multi-agent robotic systems.

1.11 Summary

Cognitive science has emerged as a new discipline of knowledge that deals with the mental aspects of the human beings. The chapter attempts to explain the psychological processes involved in human reasoning, learning, understanding, and planning. It elucidated various models of human memory and representation of imagery and cognitive maps on memory. The mechanism of understanding a complex problem is also presented here with special reference to the focus of attention in the problem.

Human beings usually perform a complex task by activating various distributed modules of their brain. The chapter takes a preparatory attempt to model distributed processes involved in human reasoning, learning, planning, etc. by a specialized distributed model similar with Petri nets. It ends with a discussion on the elementary properties of classical Petri nets and suggests the necessary extensions as required to model the brain processes.

Exercises

1. A template image of dimension of (m × m) pixels is to be searched in a digital image of dimension (n × n). Assume that mod (n / m) = 0 and n >> m. If the matching of the template block with equal sized image blocks is carried out at an interleaving of (m / 2) pixels, both row and column-wise, determine the number of times the template is compared with the blocks in the image [5].

    [**Hints:** The number of comparison per row of the image = (2n / m − 1).
              The number of comparison per column of the image = (2n/ m − 1).
              Total number of comparison = $(2n / m - 1)^2$.]

2. For matching a template image of dimension (m × m) pixels with a given image of dimension (n × n), where n >> m and mod (n / m) = 0, one uses the above interleaving.

    Further, instead of comparing pixel-wise intensity, we estimate the following five parameters (features) of the template with the same parameters in each block of the image [4] :

$$\text{Mean intensity } M_i = \sum_{b=0}^{L-1} b\, P(b),$$

$$\text{Variance of intensity } V_i^2 = \sum_{b=0}^{L-1} \{ (b - M_i)^2 P(b)\},$$

$$\text{Skewness of intensity } Sk_i = (1/ V_i^3) \sum_{b=0}^{L-1} \{ (b - M_i)^3 P(b)\},$$

$$\text{Kurtosis of intensity } Ku_i = (1 / V_i^4) \sum_{b=0}^{L-1} \{ (b - M_i)^4 P(b)\}, \text{ and}$$

$$\text{Energy of intensity } E_i = \sum_{b=0}^{L-1} \{P(b)\}^2,$$

where b represents the gray level of each pixel in a block i and L is the number of gray levels in the image.

The square of absolute deviation of these features of the i-th block from the corresponding features of the template are denoted by m_i^2, v_i^2, sk_i^2, ku_i^2, e_i^2, respectively.

Let d_i denote a measure of distance between the features of the i-th block to that of the template. Show (logically) that the *weakest feature match model* identifies the j-th block, by estimating

$$d_i = \text{Max} \{ m_i^2, v_i^2, sk_i^2, ku_i^2, e_i^2 \}, \quad 1 \le \forall \; i \le (2n / m - 1)^2$$

such that $d_j = \text{Min} \{ d_i \mid 1 \le \forall \; i \le (2n / m - 1)^2 \}$ [4].

Also show that *the strongest feature match model* identifies the j-th block by estimating

$$d_i = \text{Min} \{ m_i^2, v_i^2, sk_i^2, ku_i^2, e_i^2 \}, \quad 1 \le \forall \; i \le (2n / m - 1)^2$$

such that $d_j = \text{Min} \{ d_i \mid 1 \le \forall \; i \le (2n / m - 1)^2 \}$ [4].

Further, show that *the Euclidean least square model* identifies the j-th block by estimating

$$d_i = \{ m_i^2 + v_i^2 + sk_i^2 + ku_i^2 + e_i^2 \}^{1/2}, \quad 1 \le \forall \; i \le (2n / m - 1)^2$$

such that $d_j = \text{Min} \{ d_i \mid 1 \le \forall \; i \le (2n / m - 1)^2 \}$ [4].

3. Write a program in your favorite language to match a given template with the blocks in an image, using the above guidelines. Mark the actual block in the image that you want to identify and measure the shift of the matched blocks by executing your program. Perform the experiments with different templates on the same image and then conclude which of the three measures is the best estimator of template matching.

4. Given two jugs, one 4 liter and the other 3 liter, with no markings in them. Also given a water supply with a large storage. Using these two jugs, how can you separate 2 liters of water? Also draw the tree representing the transition of states.

[**Hints:** Define operators such as filling up a jug, evacuating a jug, transferring water from one jug to the other, etc. and construct a tree representing change of state of each jug due to application of these operators. Stop when you reach a state of 2 liters water in either of the two jugs. Do not use repeated states, since repeated states will result in a graph with loops. If you cannot solve it yourself, search for the solution in the rest of the book.]

5.   Redraw the state-space for the water jug problem, by allowing repetition of the states. Should you call it a tree yet?

6.   Show by Tulving's model which of the following information/knowledge should be kept in the episodic, semantic and procedural memories.

    a)   There was a huge rain last evening.
    b)   The sky was cloudy.
    c)   The moon was invisible.
    d)   The sea was covered with darkness.
    e)   The tides in the sea had a large swinging.
    f)   The boatmen could not control the direction of their boats.
    g)   It was a terrific day for the fishermen sailing in the sea by boats.
    h)   Since the sea was covered with darkness, the boatmen used a battery driven lamp to catch fish.
    i)   Because of the large swinging of the tides, the fishermen got a large number of fish caught in the net.
    j)   The net was too heavy to be brought to the seashore.
    k)   The fishermen used large sticks to control the motion of their boats towards the shore.

7.   The Atkinson-Shiffrin's model can be represented by first ordered transfer functions, vide Figure 1.19, presented below:

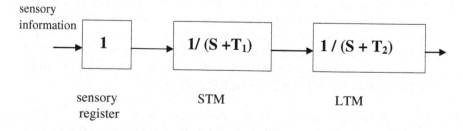

Fig. 1.19: Schematic representation of the Atkinson-Shiffrin's model.

Given $T_1 = 10$ seconds and $T_2 = 30$ minutes, find the time response of the STM and the LTM, when a unit impulse is used as an excitation signal for the sensory register. Also given that the Laplace inverse of

$1 / (S + a)$ is e^{-at}.

Now suppose we replace the excitation signal by a 0.5 unit impulse. Find the response of the STM and the LTM. Can we distinguish the current responses from the last ones? If the answer is yes, what does the result imply?

References

[1] Atkinson, R. C. and Shiffrin, R. M., "Human memory: A proposed system and its control process," In *The Psychology of Learning and Motivation: Advances in Research and Theory*, Spence, K. W. and Spence, J. T. (Eds.), vol. 2, Academic Press, New York, 1968.

[2] Baddeley, A. D., "The fractionation of human memory," *Psychological Medicine*, vol. 14, pp. 259-264, 1984.

[3] Bharick, H. P., "Semantic memory content in permastore: Fifty years of memory for Spanish learned in school," *Journal of Experimental Psychology: General*, vol. 120, pp. 20-33, 1984.

[4] Biswas, B., Mukherjee, A. K. and Konar, A., "Matching of digital images using fuzzy logic," *Advances in Modeling & Analysis*, B, vol. 35, no. 2, 1995.

[5] Biswas, B., Konar, A. and Mukherjee, A. K., "Fuzzy moments for digital image matching," Communicated to *Engineering Applications of Artificial Intelligence*, Elsevier, North-Holland, Amsterdam; also appeared in the *Proc. of Int. Conf. on Control, Automation, Robotics and Computer Vision*, ICARCV, '98, 1998.

[6] Chang, T. M., "Semantic memory: facts and models," *Psychological Bulletin*, vol. 99, pp. 199-220, 1986.

[7] Downs, R. M. and Davis, S., *Cognitive Maps and Spatial Behavior: Process and Products*, Aldine Publishing Co., 1973.

[8] Duncan, E. M. and Bourg, T., "An examination of the effects of encoding and decision processes on the rate of mental rotation," *Journal of Mental Imagery*, vol. 7, pp. 33-56, 1983.

[9] Goldenberg, G., Podreka, I., Steiner, M., Suess, E., Deecke, L. and Willmes, K., "Pattern of regional cerebral blood flow related to visual and motor imagery, Results of Emission Computerized Tomography," In *Cognitive and Neuropsychological Approaches to Mental Imagery*, Denis, M, Engelkamp, J. and Richardson, J. T. E. (Eds.), Martinus Nijhoff Publishers, Dordrecht, The Netherlands, pp. 363-373, 1988.

[10] Greeno, J. G., "Process of understanding in problem solving," In *Cognitive Theory*, Castellan, N. J., Pisoni, Jr. D. B. and Potts, G. R, Hillsdale, N. J. (Eds.), Erlbaum, vol. 2, pp. 43-84,1977.

[11] Gross, C. G. and Zeigler, H. P., *Readings in Physiological Psychology: Learning and Memory*, Harper & Row, New York, 1969.

[12] Farah, M. J., "Is visual imagery really visual? Overlooked evidence from neuropsychology," *Psychological Review*, Vol. 95, pp. 307-317, 1988.

[13] Jolicoeur, P., "The time to name disoriented natural objects," *Memory and Cognition*, vol. 13, pp. 289-303, 1985.

[14] Kintsch, W., "The role of knowledge in discourse comprehension: A construction-integration model," *Psychological Review*, vol. 95, pp. 163-182, 1988.

[15] Kintsch, W. and Buschke, H., "Homophones and synonyms in short term memory," *Journal of Experimental Psychology*, vol. 80, pp. 403-407, 1985.

[16] Konar, A. and Pal, S., "Modeling cognition with fuzzy neural nets," In *Fuzzy Logic Theory: Techniques and Applications*, Leondes, C. T. (Ed.), Academic Press, New York, 1999.

[17] Kosslyn, S. M., "Mental imagery," In *Visual Cognition and Action: An Invitation to Cognitive Science*, Osherson, D. N. and Hollerback, J. M. (Eds.) , vol. 2, pp. 73-97, 1990.

[18] Kosslyn, S. M., "Aspects of cognitive neuroscience of mental imagery," *Science*, vol. 240, pp. 1621-1626, 1988.

[19] Marr, D., *Vision*, W. H. Freeman, San Francisco, 1981.

[20] Matlin, M. W., *Cognition*, Harcourt Brace Pub., Reprinted by Prism Books Pvt. Ltd., Bangalore, India, 1994.

[21] Milner, B., "Amnesia following operation on the temporal lobe," In *Amnesia Following Operation on the Temporal Lobes*, Whitty, C. W. M. and Zangwill O. L. (Eds.), Butterworth, London, pp. 109-133,1966.

[22] McNamara, T. P., Ratcliff, R. and McKoon, G., "The mental representation of knowledge acquired from maps," *Journal of Experimental Psychology: Learning, Memory and Cognition*, pp. 723-732, 1984.

[23] Moar, I. and Bower, G. II., "Inconsistency in spatial knowledge," *Memory and Cognition*, vol. 11, pp.107-113, 1983.

[24] Moyer, R. S. and Dumais, S. T., "Mental comparison," In *The Psychology of Learning and Motivation*, Bower, G. H. (Ed.), vol. 12, pp. 117-156, 1978.

[25] Murata, T., "Petri nets: Properties, analysis and applications," *Proc. of the IEEE*, vol. 77, no. 4, pp. 541-580, 1989.

[26] Paivio, A., "On exploring visual knowledge," In *Visual Learning, Thinking and Communication*, Randhawa, B. S. and Coffman, W. E. (Eds.), Academic Press, New York, pp. 113-132, 1978.

[27] Pal, S. and Konar, A., "Cognitive reasoning with fuzzy neural nets," *IEEE Trans. on Systems, Man and Cybernetics*, Part - B, August, 1996.

[28] Peterson, M. A., Kihlstrom, J. F., Rose, P. M. and Glisky, M. L., "Mental images can be ambiguous: Reconstruals and reference-frame reversals," *Memory and Cognition*, vol. 20, pp. 107-123, 1991.

[29] Pylyshyn, Z. W., "Imagery and artificial intelligence," In Minnesota Studies in the Philosophy of Science, vol. 9: *Perception and Cognition Issues in the Foundations of Psychology*, University of Minnesota Press, Minneapolis, pp. 19-56, 1978.

[30] Rumelhart, D. E., McClelland, J. L. and the PDP research group, *Parallel Distributed Processing: Explorations in the Microstructure of Cognition*, vol. 1 and 2, MIT Press, Cambridge, MA, 1968.

[31] Shepard, R. N. and Cooper, L.A. (Eds.), *Mental Images and Their Transformations*, MIT Press, Cambridge, MA, 1981.

[32] Shepherd, R. N. and Metzler, Z., "Mental rotation of three dimensional objects," *Science*, vol. 171, pp. 701-703, 1971.

[33] Tulving, E., "Multiple memory systems and consciousness," *Human Neurobiology*, vol. 6, pp. 67-80, 1987.

[34] Tveter, D. R., *The Pattern Recognition Basis of Artificial Intelligence*, IEEE Computer Society Press, Los Alamitos, pp. 168-169, 1998.

[35] Washerman, P. D., *Neural Computing: Theory and Practice*, Van Nostrand Reinhold, New York, 1989.

Chapter 2

Parallel and Distributed Logic Programming

This chapter provides a new approach to reason with logic programs using a specialized data structure similar to Petri nets. Typical logic programs include a number of concurrently resolvable clauses. A priori detection of these clauses indeed is useful for their subsequent participation in the concurrent resolution process. The chapter explores the scope of distributed mapping of program clause components onto Petri nets so as to automatically select the participant clauses for concurrent resolution. An algorithm for concurrent resolution of clauses on Petri nets has been undertaken with a motive to improve the speed-up factor for the program without sacrificing the resource utilization rate. Examples have been introduced to illustrate all new concepts. The exercises at the end of the chapter include a number of interesting problems with sufficient hints to each so as to enable the readers to test their level of understanding.

2.1 Introduction

Logic programming has already earned much importance for its increasing applications in data and knowledge engineering. A logic program usually consists of a special type of program clauses known as **horn clauses**. Programs built with horn clauses only are called normal logic programs. Complex knowledge having multiple consequent literals cannot be represented by normal logic programs because of its structural restriction imposed by horn clauses.

In spite of limitations in knowledge representation, normal logic programs are still prevalent in relational languages like PROLOG and DATALOG for simplicity in designing their compilers. Generally, compilers for logic programming employ linear resolution that resolves each pair of program clauses at a time. Execution of a program by linear resolution thus requires a considerable amount of time. Most logic programs usually include a number of concurrently resolvable clauses; unfortunately there is hardly any literature on parallel and distributed models of logic programming, capable of resolving multiple program clauses concurrently.

In the early 1990s, Patt [12] examined a nonconventional execution model of a uniprocessor micro-engine for a PROLOG program and measured its performance with a set of 14 benchmarks. Yan [15] provided a novel scheme for concurrent realization of PROLOG on a RAP-WAM machine.[1] The WAM generates an intermediate code during the compilation phase of a PROLOG program that exploits the fullest degree of parallelism in the program itself. The RAP, on the other hand, reduces the overhead associated with the run-time management of *variable binding conflict* between goals. Hermenegildo and Tick proposed an alternative model [3] for concurrent execution of PROLOG programs by RAP-WAM combinations by representing the dependency relationship of the program clauses by a graph, where the take-off arcs at the vertex in the graph denotes parallel tasks in the program. They employed goal stacks with each processing element. When a parallel call is invoked, the concurrent tasks are mapped autonomously to the stacks of the less busy or idle processing elements. A synchronization and co-ordination for parallel calls was also implemented in their scheme.

In recent times, Ganguly, Silberschatz, and Tsur [2] presented a new algorithm for automated mapping of logic programs onto a parallel processing architecture. In their first scheme, they considered the mapping of program clauses having shared variables onto adjacent processors. This reduces the communication overhead among the program clauses. In the latter part of their work, they eliminated the above constraints at the cost of extra network management time. Takeuchi [14] presented a new language for AND-OR parallelism. Kale [6] discussed the scope of an alternative formulation of problem-solving using parallel AND-OR trees. Among the existing speed-up schemes of logic programming machines, the content addressable memory (CAM)–based architecture of PROLOG machines by Naganuma et al. [11] needs special mention. To speed up the execution performance of PROLOG programs, they employed *hierarchical pipelining and garbage collection mechanism* of a CAM for efficient backtracking on a SLD tree.

Though a number of techniques are prevalent for the realization of logic programs on a parallel architecture, none of these are capable of representing the theoretically possible maximum parallelism in a program. For realization of all types of parallelism in a logic program, a specialized data structure appropriate for representing the possible parallelism is needed. Petri net has already proved

---

[1] RAP is an acronym for **R**estricted **AND-P**arallelism, and WAM is an acronym for **W**arren **A**bstract **M**achine.

itself as a successful data structure for reasoning with complex rules. For instance, rules having more than one antecedent and consequent clause with each clause containing a number of variables can be represented by a Petri net structure [8]. Murata [10] proposed the scope of Petri net models for knowledge representation and reasoning under the framework of predicate logic.

A number of researchers are working on complex reasoning problems using Petri nets [5], [9], [13]. Because of the distributed structure of a Petri net, pieces of knowledge fragmented into components can easily be mapped onto this structure. For example, a transition firing in a Petri net may synonymously be used as rule firing in an expert system. The transitions may be regarded as the implication operator of a rule, while its input and output places may, respectively, be regarded as the antecedent and consequent clauses of a rule. The arguments associated with the predicates of a clause are also assigned at the arcs connecting the place containing the predicate and the transition describing the implication rule. Such fragmentation and mapping of the program components onto different modules of a Petri net enhances the scope of parallelism in a logic program. The objective of the chapter is to fragment a given logic program to smallest possible units, and map them onto a Petri net to fully exploit its parallelism.

The methodology of reasoning presented in the chapter is an extension of Murata's classical models on Petri nets [10], [13], applied to logic programming. Murata defined a set of rules to synonymously describe the resolution of horn clauses in a normal logic program with the firing of transitions in a Petri net. The chapter attempts to extend Murata's scheme for automated reasoning to non-Horn clause based programs as well.

Section 2.2 provides related definitions of important terminologies used in this chapter. The concept of concurrency in resolution process is introduced in Section 2.3. A new model for concurrent resolution in Petri nets is presented in Section 2.4. Performance analysis of the Petri net-based model is covered in Section 2.5. Conclusions are listed in Section 2.6. A set of numerical problems has been undertaken in the exercises of the chapter.

2.2 Formal Definitions

2.2.1 Preliminary Definitions

Definition 2.1: *A clause* cl_i *is represented by*

$$A_i \leftarrow B_i. \tag{2.1}$$

B_i *denotes the body,* A_i *denotes the head, and* '←' *denotes the implication operator. The body* B_i *usually is a conjunction of literals* $B_{jl}\,\exists j$, *i.e.,*

$$B_i = B_{i1} \wedge B_{i2} \wedge \ldots \wedge B_{il} \tag{2.2}$$

The head A_i *usually is a disjunction of literals* $A_{ik}\,\exists k$, *i.e.,*

$$A_i = A_{i1} \vee A_{i2} \vee \ldots \vee A_{ij} \tag{2.3}$$

The literals A_{ik} *and* B_{il} *have arguments containing terms that may include variables (denoted by capital letters), constants (denoted by small letters), function of variables, and function of function of variables (in a recursive form).*

Example 2.1: This example illustrates the constituents of a clause. For instance,

$$\text{Father(Y, Z), Uncle(Y, Z)} \leftarrow \text{Father(X, Y), Grandfather(X, Z).} \tag{2.4}$$

is an example of a general clause, where the body consists of Father(X, Y) and Grandfather(X, Z) and the head consists of Father(Y, Z) and Uncle(Y, Z). The clause states that if X is the father of Y and X is the grandfather of Z, then either Y is the father of Z or Y is the uncle of Z.

In case all the terms are bound variables or constants, the literals A_{ij} or B_{il} are called **ground literals**.

Example 2.2: The following is an example of a clause with all variables bound by constants, thereby resulting in ground literals Father (n, a) and Son (a, n).

$$\text{Father(n, a)} \leftarrow \text{Son(a, n).} \tag{2.5}$$

The above clause states that *if a is son of n then n is the father of a.*

Special cases:
   i)   In case of a **goal clause (query)** the consequent part A_i is absent.

       The clause (2.6) presented below contains no consequent part, and hence it is a query.

$$\leftarrow \text{Grandfather(X, Z).} \tag{2.6}$$

       Given that X is the Grandfather of Z, the clause (2.6) questions the value of X and Z.

ii) A clause with an empty body and consisting of ground literals in the head is regarded as a *fact*.

The clause (2.7) below contains no body part and the variable arguments are bound by constants. Thus, it is a fact.

$$\text{Grandfather}(r, a) \leftarrow. \tag{2.7}$$

It states that r is Grandfather of a.

iii) When the consequent part A_i includes a single literal, the resulting clause is called a **horn clause**. The details about horn clauses are given in Definition 2.2.

iv) When A_{ik} and B_{ij} do not include arguments, we call them propositions and the clause "$A_i \leftarrow B_i$." is then called a *propositional clause*.

The clause (2.8) below for instance is a propositional clause as it does not contain any arguments.

$$P \leftarrow Q, R. \tag{2.8}$$

Definition 2.2: *A **horn clause** contains a head and a body with at most one literal in its head.*

Example 2.3: The clause (2.9) is an example of a horn clause.

$$P \leftarrow Q_1, Q_2, \ldots, Q_n. \tag{2.9}$$

It represents a horn clause where P and the Q_i are literals or atomic formulas. It means if all the Q_is are true, then P is also true. Q_i is the body part and P is the head in this horn clause.

Definition 2.3: *A clause containing more than one literal in its head is known as a **non-horn clause**.*

Example 2.4: The clause (2.10) for instance is a non-horn clause.

$$P_1, P_2, \ldots, P_m \leftarrow Q_1, Q_2, \ldots, Q_n. \tag{2.10}$$

Definition 2.4: *An **extended horn clause** (EHC) contains a head and a body with at least one clause in the body and zero or more clauses in its head.*

Commas are used to denote conjunction of the literals in the body and disjunction of literals in the head.

Example 2.5: The general format of an EHC is

$$A_1, A_2, \ldots, A_n \leftarrow B_1, B_2, \ldots, B_m. \qquad (2.11)$$

Here the head and the body contain n and m number of literals, respectively.

It is important to note that an extended horn clause includes both horn clause and its extension as well.

Definition 2.5: *A program that contains extended horn clauses, as defined above, is called an* **extended logic program**.

Example 2.6: The clauses (2.12–2.15) together represent an extended logic program. It includes extended horn clause (2.12) with facts (2.13–2.15).

$$\text{Father}(Y, Z), \text{Uncle}(Y, Z) \leftarrow \text{Father}(X, Y), \text{Grandfather}(X, Z). \quad (2.12)$$
$$\text{Father}(r, d) \leftarrow . \qquad (2.13)$$
$$\neg \text{Father}(d, a) \leftarrow . \qquad (2.14)$$
$$\text{Grandfather}(r, a) \leftarrow . \qquad (2.15)$$

To represent the query "whether d is uncle of a?" the goal clause of the following form may be constructed.

$$\text{Goal:} \leftarrow \text{Uncle}(d, a). \qquad (2.16)$$

The answer to the query can be obtained by taking negation of the goal and then resolving it with the supplied clauses (2.12–2.15). In the present context, the answer to the query will be *true*.

To explain this, we need to introduce the principles of *resolvability of two clauses*. In order to explain resolvability of clauses we further need to introduce *substitution sets* and *most general unifier*.

Definition 2.6: *A substitution represented by a set of ordered pairs* $s\{t_1/v_1, t_2/v_2, \ldots, t_n/v_n\}$ *is called the* **substitution set**. The pair t_i/v_i means that the term t_i is substituted for every occurrence of the variable v_i throughout.

Example 2.7: There exist four substitution sets for the predicate $P(a, Y, f(Z))$ in the following instances:

P(a, X, f(W))
P(a, Y, f(b))
P(a, g(X), f(b))
P(a, c, f(b))

Substitution sets for the above examples are

$s_1 = \{X/Y, W/Z\}$
$s_2 = \{b/Z\}$
$s_3 = \{g(X)/Y, b/Z\}$
$s_4 = \{c/Y, b/Z\}$

To denote a substitution instance of an expression w, using a substitution s we write ws.

Example 2.8: Let the expression w = P(a, Y, f(Z)), the substitution set s = {X/Y, W/Z}. Then the substitution instance ws = P(a, X, f(W)).

2.2.2 Properties of the Substitution Set

Property 1: *(ws₁)s₂ = w(s₁s₂) where w is an expression and s₁s₂ are two substitutions.*
Example 2.9: To illustrate the above property,
let
$\quad$ w = P(X, Y),
$\quad$ $s_1 = \{f(Y)/X\}$
and $\quad$ $s_2 = \{a/Y\}$.

Now, $(ws_1)s_2 = (P(f(Y), Y))\{a/Y\}$
$\qquad\qquad = (P(f(a), a))$.

Again, $w(s_1s_2) = (P(X, Y))\{f(a)/X, a/Y\}$
$\qquad\qquad = P(f(a), a)$.

Therefore, $(ws_1)s_2 = w(s_1s_2)$.

The composition of two substitutions s_1, s_2 is hereafter denoted by $s_1 \Delta s_2$, which is the substitution obtained by first applying s_2 to the terms of s_1 and then adding the ordered pairs from s_2 not occurring in s_1. Example 2.10 below illustrates this said concept.

Example 2.10: Let $s_1 = \{f(X, Y)/Z\}$ and $s_2 = \{a/X, b/Y, c/W, d/Z\}$.

Then $s_1 \Delta s_2 = \{f(a, b)/Z, a/X, b/Y, c/W\}$.

Property 2: *Composition of substitutions is associative, i.e.,*

$$(s_1 \Delta s_2) \Delta s_3 = s_1 \Delta (s_2 \Delta s_3).$$

Example 2.11: To illustrate the associative property of composition in substitutions, let

$$s_1 = \{f(Y)/X\}$$
$$s_2 = \{a/Y\}$$
$$s_3 = \{c/Z\}$$
and $w = P(X, Y, Z)$.

Here, $(s_1 \Delta s_2) = \{f(a)/X, a/Y\}$
     $(s_1 \Delta s_2) \Delta s_3 = \{f(a)/X, a/Y, c/Z\}$

Again, $(s_2 \Delta s_3) = \{a/Y, c/Z\}$
     $s_1 \Delta (s_2 \Delta s_3) = \{f(a)/X, a/y, c/Z\}$

Therefore, $(s_1 \Delta s_2) \Delta s_3 = s_1 \Delta (s_2 \Delta s_3)$.

Property 3: *Commutability fails in case of the substitutions, i.e.,*

$$s_1 \Delta s_2 \neq s_2 \Delta s_1.$$

Example 2.12: We can illustrate the third property following the substitution sets used in example 2.11. Here,

$(s_1 \Delta s_2) = \{f(a)/X, a/Y\}$ and $(s_2 \Delta s_1) = \{a/Y, f(Y)/X\}$
So, $s_1 \Delta s_2 \neq s_2 \Delta s_1$.

Definition 2.7: *Given two predicates $P(t_1, t_2, \ldots, t_n)$ and $P(r_1, r_2, \ldots, r_n)$ and $s = \{t_i/r_i\}$ is a substitution set that on substitution in the predicates makes them identical (unifies them). Then the substitution set s is called a **unifier**. The **most general unifier (mgu)** is the simplest unifier g, so that any other unifier g' satisfies $g' = g \Delta s'$ for some substitution s'.*

Example 2.13: For the predicates $P(X, f(Y), b)$ and $P(X, f(b), b)$, $g' = \{a/X, b/Y\}$ definitely is a unifier as it unifies the predicates to $P(a, f(b), b)$, but the mgu in this case is $g = \{b/Y\}$. It is to be noted further that a substitution $s' = \{a/X\}$ satisfies $g' = g \Delta s'$.

Definition 2.8: *Let cl_i and cl_j be two clauses of the following form:*

$$Cl_i \equiv A_{i1} \vee A_{i2} \vee \ldots \vee A_{ix} \vee \ldots \vee A_{im} \leftarrow B_{i1} \wedge B_{i2} \wedge \ldots \wedge B_{il} \qquad (2.17)$$

$$Cl_j \equiv A_{j1} \lor A_{j2} \lor \ldots \lor A_{jm}' \leftarrow B_{j1} \land B_{j2} \land \ldots \land B_{jy} \land \ldots \land B_{jl}' \qquad (2.18)$$

and P be the common literal present in the head of cl_i and body of cl_j. For instance, let for a substitution s

$$[A_{ix}]_s = [B_{jy}]_s = [P]_s$$

The **resolvent** of cl_i and cl_j, denoted by $R(cl_i, cl_j) = cl_{ij}$ (say), is computed as follows under the substitution $s = s_{ij}$, say:

$$Cl_{ij} =$$
$$[(A_{i1} \lor A_{i2} \lor \ldots \lor A_{i(x-1)} \lor A_{i(x+1)} \lor \ldots \lor A_{im}) \lor (A_{j1} \lor A_{j2} \lor \ldots \lor A_{jm}') \leftarrow$$
$$(B_{i1} \land B_{i2} \land \ldots \land B_{il}) \land (B_{j1} \land B_{j2} \land \ldots \land B_{j(y-1)} \land B_{j(y+1)} \land \ldots \land B_{jl}')] \quad (2.19)$$

If cl_{ij} can be computed from the given cl_i and cl_j, we say that the clauses cl_i and cl_j are **resolvable**

Example 2.14: Given the following clauses

Cl_i: Fly(X)←Bird(X).   (2.20)
Cl_j: Bird(parrot) ←.   (2.21)
Cl_{ij}: Fly(parrot)←.   where s_{ij}={parrot/X}   (2.22)

The clauses cl_i and cl_j in this example are resolvable with the substitution s_{ij} yielding the resolvent cl_{ij}.

Definition 2.9: *If in a set of clauses there is at least one clause cl_j for each clause cl_i such that resolution holds on cl_i and cl_j, producing a resolvent cl_{ij}, then the set is called the **set of resolvable clauses**.*

Example 2.15: In the following set of clauses each clause is resolvable with at least one other clause:

Mother(Y, Z) ← Father(X,Z), Wife(Y,X).   (2.23)
Father(r,l) ←.   (2.24)
Wife(s,r) ←.   (2.25)

Here, clause 2.23 is resolvable with clause number 2.24 and clause 2.25. It is an example of the set of resolvable clauses.

Now, we briefly outline select linear definite (SLD) resolutions.

2.2.3 SLD Resolution

To understand SLD resolution we first have to learn a few definitions.

Definition 2.10: *A **definite program clause** is a clause of the form*

$$A \leftarrow B_1, B_2, \ldots, B_n,$$

which contains precisely one atom (viz. A) in its consequent (head) and a null, one or more literals in its body (viz. B_1 or B_2 or ... or B_n).

Definition 2.11: *A **definite program** is a finite set of definite program clauses.*

Definition 2.12: *A **definite goal** is a clause of the form*

$$\leftarrow B_1, B_2, \ldots, B_n.$$

i.e., a clause with an empty consequent.

Definition 2.13: ***SLD resolution*** *stands for **SL resolution for definite clauses**, where SL stands for **resolution with linear selection function**.*

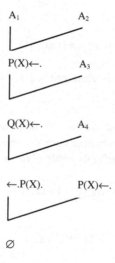

Fig. 2.1: The linear selection of clauses in the resolution tree.

Example 2.16: This example illustrates the linear resolution. Here, the following OR clauses (clauses connected by OR operator), represented by a set-like notation are considered.

Let S = {A_1, A_2, A_3, A_4},

where A_1 = P(X), Q(X)←.

A_2 = P(X)←Q(X).

$$A_3 = Q(X) \leftarrow P(X).$$
$$A_4 = \leftarrow P(X), Q(X).$$
$$\text{and goal} = \leftarrow \neg P(X).$$

By linear selection, a resolution tree can be constructed as shown in Figure 2.1. It is clear from the tree that two clauses from the set $S_1 = S \cup \{\neg \text{Goal}\}$ are first used for resolution with a third clause from the same set S_1. The process is continued until a null clause is generated. In the linear selection process, one clause, however, can be used more than once for resolution.

Definition 2.14: *Let $S = \{cl_1, cl_2, \ldots, cl_n\}$ be a set of resolvable clauses, and there exists one or more definite orders to pair-wise select the clauses for SLD resolution, without which the SLD resolution of all the n clauses fail to generate a resolvent. Under this circumstance, we say that an **orderly resolution** exists among the clauses in S.*

The resolvent of clauses cl_i and cl_j is hereafter denoted by cl_{ij} or $R(cl_i, cl_j)$, where R represents a binary resolution operator. The *order of resolution* in $R(R(cl_i, cl_j), cl_k)$ is denoted by i-j-k (or i, j, k) for brevity. It may be noted that i-j $\equiv$ j-i and i-j-k $\equiv$ k-i-j $\equiv$ k-j-i.

Example 2.17: Given the following clauses cl_1 through cl_3, we would like to illustrate the principle of orderly resolution with these clauses.

Cl_1: Paternal_uncle(X, Y), Maternal_uncle(X, Y)$\leftarrow$Uncle(X, Y).
Cl_2:$\neg$Paternal_uncle(n, a)$\leftarrow$. $\equiv$ $\leftarrow$Paternal_uncle(n, a).
Cl_3:$\neg$Maternal_uncle(n, a) $\leftarrow$. $\equiv$ $\leftarrow$Maternal_uncle(n, a).

We now demonstrate two different orders of resolution, and show that the result is unique in both the cases. One of the orders of resolution could be 1-2-3. This is computed as follows:

$R(cl_1, cl_2)$: Maternal_uncle(n, a)$\leftarrow$Uncle(n, a). with $s_{12}=\{n/X, a/Y\}$
$R(R(cl_1, cl_2), cl_3)$: $\leftarrow$Uncle(n, a). $\equiv \neg$Uncle(n, a) $\leftarrow$.

An alternative order of resolution is 3-1-2. To compute this, we proceed as follows:
$R(cl_3, cl_1)$: Paternal_uncle(n, a)$\leftarrow$Uncle(n, a). with $s_{31}=\{n/X, a/Y\}$
$R(R(cl_3, cl_1), cl_2)$: $\leftarrow$Uncle(n, a). $\equiv \neg$Uncle(n, a) $\leftarrow$.

So, both the orderly resolutions return the same resolvent: $[\neg$Uncle(n, a)$\leftarrow$.], which means n is not the uncle of a.

It is important to note that resolution of multiple clauses by different orders does not always return unique resolvents.

Definition 2.15: *If there exists only one definite order of resolution among a set of resolvable clauses, it is said to have **single sequence** of resolution. On occasions, we can obtain the same resolvent by taking a reversed order of resolution. For instance, $R(. . .(R(R(R(cl_1, cl_2), cl_3), cl_4). . . cl_{n-1}), cl_n) = R . . . (R(R(cl_n, cl_{n-1}), . . .cl_3), cl_2) ,cl_1)$ is valid if the resolvents can be computed for each resolution. We, however, consider it a single order of resolution.*

Example 2.18: To understand the single sequence of resolution, let us take the following clauses:

Cl_1: W(Z, X)←P(X, Y), Q(Y, Z).
Cl_2: P(a, b)←S(b, a).
Cl_3: S(Y, X)←T(X).

The resolution of clauses in the present context follows a definite order: 1-2-3 (or 3-2-1). It may be noted that

$R(R(cl_1, cl_2), cl_3) = R(R(cl_3, cl_2), cl_1) = W(Z, a)$ ←Q(b, Z), T(a).

Consequently, a single sequence is maintained in the process of resolution of multiple program clauses.

Definition 2.16: *When in a set of resolvable clauses, resolution takes place following different orders, **multiple sequence** is said to be present.*

Example 2.19: The following clauses are taken to illustrate the multiple sequences in a set of resolvable clauses:

Given Cl_1: S(Z, X)←P(X, Y), Q(Y, Z), R(Z, Y).
Cl_2: P(a, b)←T(c, a), U(b).
Cl_3: Q(b, c)←V(b), M(c).
Cl_4: R(Z, Y)←N(Y), O(Z).

Sequence 1: Order: 1-2-3-4.

$R(cl_1, cl_2)$:S(Z, a)←Q(b, Z), R(Z, b), T(c, a), U(b). $| s_{12=\{a/X, b/Y\}}$
$R(R(cl_1, cl_2), cl_3)$: S(c, a)←R(c, b), T(c, a), U(b), V(b), M(c). $| s_{12,3=\{c/Z\}}$
$R(R(R(cl_1, cl_2), cl_3), cl_4)$:   S(c, a)←T(c, a), U(b), V(b), M(c), N(b), O(c). $| s_{12,3,4=\{c/Z,b/Y\}}$

Sequence 2: Order: 3-1-2-4.

$R(cl_3, cl_1)$: $S(c, X) \leftarrow P(X, b), R(c, b), V(b), M(c).$ | $s_{31=\{c/Z,b/Y\}}$
$R(R(cl_3, cl_1), cl_2)$:
$S(c, a) \leftarrow T(c, a), U(b), P(a, b), R(c, b), V(b), M(c).$ | $s_{31,2=\{a/X\}}$
$R(R(R(cl_3, cl_1), cl_2), cl_4)$:
$S(c, a) \leftarrow T(c, a), U(b), V(b), M(c), N(b), O(c).$ | $s_{31,2,4=\{c/Z,b/Y\}}$

Sequence 3: Order: 1-2-4-3.

$R(R(cl_1, cl_2), cl_4)$:
$S(Z, a) \leftarrow Q(b, Z), T(c, a), U(b), N(b), O(Z).$ | $s_{12,4=\{b/Y\}}$
$R(R(R(cl_1, cl_2), cl_4), cl_3)$:
$S(c, a) \leftarrow T(c, a), U(b), V(b), M(c), N(b), O(c).$ | $s_{1,2,4,3=\{c/Z\}}$

The results of the above computation reveal that

$R(R(R(cl_1, cl_2), cl_3), cl_4)$
$= R(R(R(cl_3, cl_1), cl_2), cl_4)$
$= R(R(R(cl_1, cl_2), cl_4), cl_3)$
$= S(c, a) \leftarrow T(c, a), U(b), V(b), M(c), N(b), O(c).$

Consequently, multiple order exists in the resolution of clauses.

Readers may please note that resolution between cl_2 and cl_3, cl_3 and cl_4, and cl_2 and cl_4 are not possible.

Definition 2.17: *Let $S = \{cl_1, cl_2, \ldots, cl_n\}$ be a set of resolvable clauses and the $cl_i s$ are ordered in a manner that cl_i and cl_{i+1} are resolvable for $i = 1, 2, \ldots, (n-1)$. If cl_n is also resolvable with cl_1 we call it **circular resolution**.* Circular resolution is not allowed as it invites multiple resolution between two clauses.

Example 2.20: Consider the following propositional clauses:

Cl_1: $Q \leftarrow P.$
Cl_2: $R \leftarrow Q.$
Cl_3: $P \leftarrow R.$

Let $Cl_{12} = R(Cl_1, Cl_2)$
$= R \leftarrow P.$

However, evaluation of $R(Cl_{12}, Cl_3)$ cannot be performed as two resolutions are applicable between the clauses, which is not allowed.

Definition 2.18: *Let S be a set of resolvable clauses such that their pair-wise selection for SLD resolution from S is random. Under this condition, S is called the set of **order-less** or **order-independent** clauses.*

Example 2.21: Let S be the set of the following clauses:

Cl_1: U←P, Q, R.
Cl_2: S, M, P←V.
Cl_3: Q←W, M, T.
Cl_4: T←U, S.

In this example, we attempt to resolve each clause with others.

Cl_{12}: U, S, M←V, Q, R.
Cl_{13}:U←P, R,W, M, T.
Cl_{14}: T←P, Q, R, S.
Cl_{23}: S, P, Q←V, W, T.
Cl_{34}: Q←W, M, U, S.
Cl_{24}: M, P, T←V, U.

As each of the clauses is resolvable with each other, order-less condition for resolution holds here.

It is important to note that order-less resolution is not valid even for propositional clauses. The example below provides an insight to this problem.

2.3 Concurrency in Resolution

To speed up the execution of logic programs, we take a look at the possible parallelism/ concurrency in the resolutions involved in the program.

2.3.1 Preliminary Definitions

Definition 2.19: *If S includes multiple ordered sequence of clauses for SLD resolution and for each such sequence the final resolvent is identical then the clauses in S are **concurrently resolvable**.* Under this case, resolution of all the clauses can be done concurrently yielding the same resolvent.

Example 2.22: In this example we consider the concurrent resolution of propositional clauses.

Cl_1: R←P,Q.
Cl_2 : P←S.
Cl_3 : Q←T.

Cl_4 : T←U.

Sequence 1: Order: 1-2-3-4.

$R(Cl_1, Cl_2)$: R←Q,S.
$R(R(Cl_1, Cl_2), Cl_3)$: R←T,S.
$R(R(R(Cl_1, Cl_2), Cl_3), Cl_4)$: R←U,S.

Sequence 2: Order: 3-1-4-2.

$R(Cl_3, Cl_1)$: R←P,T.
$R(R(Cl_3, Cl_1), Cl_4)$: R←P,U.
$R(R(R(Cl_3, Cl_1), Cl_4), Cl_2)$: R←U,S.

Sequence 3: Order: 3-1-2-4.

$R(R(Cl_3, Cl_1), Cl_2)$: R←S,T.
$R(R(R(Cl_3, Cl_1), Cl_2), Cl_4)$: R←U,S.

Sequence 4: Order: 3-4-1-2.

$R(Cl_3, Cl_4)$: Q←U.
$R(R(Cl_3, Cl_4), Cl_1)$: R←P,U.
$R(R(R(Cl_3, Cl_4), Cl_1), Cl_2)$: R←U,S.

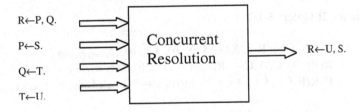

Fig. 2.2: Concurrent resolution.

Since all the four sequences yield the same resolvent, the given clauses can be resolved concurrently. Figure 2.2 thus concurrently resolves four clauses and yields a resulting clause: R←U, S. But how can we concurrently resolve these four clauses? A simple answer to this is to drop the common literals from the head of one clause and the body of another clause, and then group the remaining literals in the heads and in the body of the participating clauses. We employed this principle in Figure 2.2.

Definition 2.20: *In case multiple sequences exist in orderly resolution and the resolvent is not unique, then the substitutions used in resolution in each pair of* clauses may be propagated downstream in the process of SLD resolution. The composition of the substitution sets for every two sequential substitutions is also carried forward along the SLD tree until the final resolvent is obtained. The final substitution may now be used as the instantiation space of the resolvent and the resulting clause thus generated for each such sequence is compared. In case the instantiated resolvent generated following multiple sequences yields a unique result then the clauses in S are also called **concurrently resolvable**. The final substitution set is called the **deferred substitution set**.

Example 2.23: To illustrate the above situation
  let

  Cl_1: R(Z, X) ← P(X, Y), Q(Y, Z).
  Cl_2: P(a, b) ← S(b, a).
  Cl_3: Q(b, c) ← T(c, b).
  Cl_4: T(Z, Y) ← U(X, Y).

Sequence 1: Order: 1-2-3-4.

  $R(Cl_1, Cl_2)$: R(Z, a) ← Q(b, Z), S(b, a). $| s_{12=\{a/X, b/Y\}}$
  $R(R(Cl_1, Cl_2), Cl_3)$: R(c, a) ← T(c, b), S(b, a). $| s_{12,3=\{c/Z\}}$
  $R(R(R(Cl_1, Cl_2), Cl_3), Cl_4)$:R(c, a)←S(b, a), U(X, b). $| s_{12,3,4=\{c/Z,b/Y\}}$

∴Composition of the substitutions, $s_{12} \Delta s_{12,3} = \{a/X, b/Y, c/Z\}$ and the final composition of the substitutions, $s_{12} \Delta s_{12,3} \Delta s_{12,3,4} = \{a/X, b/Y, c/Z\}$.

Sequence 2: Order: 3-1-2-4.

  $R(Cl_3, Cl_1)$: R(c, X) ← P(X, b), T(c, b). $| s_{31=\{b/Y, c/Z\}}$
  $R(R(Cl_3, Cl_1), Cl_2)$: R(c, a) ← T(c, b), S(b, a). $| s_{31,2=\{a/X\}}$
  $R(R(R(Cl_3, Cl_1), Cl_2), Cl_4)$:R(c, a)←S(b, a), U(X, b). $| s_{31,2,4=\{c/Z,b/Y\}}$

∴Composition of the substitutions, $s_{31} \Delta s_{31,2} = \{a/X, b/Y, c/Z\}$ and the final composition of the substitutions, $s_{31} \Delta s_{31,2} \Delta s_{31,2,4} = \{a/X, b/Y, c/Z\}$.

Sequence 3: Order: 3-4-1-2

  $R(Cl_3, Cl_4)$: Q(b, c) ← U(X, b). $| s_{34=\{b/Y, c/Z\}}$
  $R(R(Cl_3, Cl_4), Cl_1)$: R(c, X) ← P(X, b), U(X, b). $| s_{34,1=\{b/Y, c/Z\}}$
  $R(R(R(Cl_3, Cl_4), Cl_1), Cl_2)$:R(c, a)←S(b, a), U(a, b). $| s_{34,1,2=\{a/X\}}$

∴Composition of the substitutions, $s_{34} \Delta s_{34,1} = \{b/Y, c/Z\}$ and the final composition of the substitutions, $s_{34} \Delta s_{34,1,2} \Delta s_{34,1,2} = \{a/X, b/Y, c/Z\}$.

Now, if we compute the deferred substitution set for the three sequences, we find it to be equal, the value of which is given by s = {a/X, b/Y, c/Z}. When the resolvents are instantiated by this deferred substitution set, they become equal and the final resolvent is given by R(c, a)←S(b, a), U(a, b).

2.3.2 Types of Concurrent Resolution

There are three types of concurrent resolution:

a) Concurrent resolution of a rule with facts,
b) Concurrent resolution of multiple rules,
c) Concurrent resolution of both multiple rules and facts.

Various well-known types of parallelisms are involved in the concurrent resolutions.

Definition 2.21: *When the predicates of a rule are attempted for matching with predicates contained in the facts, **concurrent resolution of the rule with facts** is said to take place.*

It can occur in two ways. In the first case, the literals present in the body part of a clause (AND literals) may be searched against the literals present in the heads of the available facts, which is a special case of *AND-Parallelism.*

Example 2.24: To illustrate AND-parallelism let us consider the following clauses:

$$\text{Mother(Z, Y)} \leftarrow \text{Father(X, Y), Married\_to(X, Z).} \qquad (2.26)$$
$$\text{Father(r, n)} \leftarrow. \qquad (2.27)$$
$$\text{Married\_to(r, t)} \leftarrow. \qquad (2.28)$$

Here, predicates in the heads of the facts given by the clauses (2.27) and (2.28) are concurrently resolved with the predicates in the body of the rule given by (2.26) yielding the resolvent

$$\text{Mother(t, n)} \leftarrow.$$

Again, when a literal present in the body of one rule may be searched concurrently against the literals present in the heads of more than one fact, *OR-Parallelism* is invoked.

Example 2.25: We can illustrate OR-parallelism with the help of the following clauses:

$$\text{Son(X, Y)} \leftarrow \text{Father(Y, X).} \qquad (2.29)$$

Father(r, n) ← .   (2.30)
Father(n, a) ← .   (2.31)

Here, the variables present in the body of the rule given by (2.29) can be matched concurrently with the arguments in the heads of the facts given by 2.30 and 2.31.

Definition 2.22: *When more than one rule is resolved concurrently in a given set of resolvable clauses, we say that* **concurrent resolution of multiple rules** *has taken place.*

Example 2.26: The following clauses are considered to explain the concurrent resolution of multiple rules:

Mother(Z, Y)←Father(X, Y), Wife(Z, X).   (2.32)
Wife(Y, X)←Female(Y), Married_to(Y, X).   (2.33)
Married_to(X, Y)←Marries(X, Y).   (2.34)

Here, the concurrent resolution can take place between the rules 2.32 and 2.33 in parallel with the rules 2.33 and 2.34. Moreover, the above three rules can be concurrently resolved together yielding the resolvent

Mother(Z, Y)←Father(X, Y), Female(Z), Marries(X, Z).   (2.35)

A special kind of parallelism, known as **Stream-parallelism** can be encountered while discussing concurrent resolution of multiple rules. It is explained with the following example.

Example 2.27: Let us consider the following clauses:

Integer(X+1)←Integer(X).   (2.36)
Evaluate_square(Z.Z)←Integer(Z).   (2.37)
Print(Y)←Evaluate_square(Y).   (2.38)
Integer(0)←.   (2.39)

Here, the resolvent obtained by resolving the rule (2.36) and the fact (2.39) is propagated to resolve the rule (2.37). The resolvent thus obtained is resolved further with (2.38). The process is repeated in a streamline for the integer sequence 0, 1, 2, . . . up to infinity. After one result is printed, the pipeline becomes busy rest of the time, and resolution of three pairs of clauses takes place concurrently.

Definition 2.23: *If more than one rule is resolved concurrently with more than one fact,* **concurrent resolution of both multiple rules and facts** *takes place.*

Example 2.28: To illustrate the concurrent resolution of multiple rules and facts, let us take the following clauses:

$$R(Z, X) \leftarrow P(X, Y), Q(Y, Z). \tag{2.40}$$
$$P(a, b) \leftarrow. \tag{2.41}$$
$$Q(b, c) \leftarrow. \tag{2.42}$$
$$S(U, V), T(U, V) \leftarrow R(U, V). \tag{2.43}$$
$$\leftarrow S(d, e). \tag{2.44}$$
$$\leftarrow T(d, e). \tag{2.45}$$

Here, concurrent resolution can take place in several ways. At first, the rules given by (2.40) and (2.43) can resolve generating the resolvent $S(Z, X), T(Z, X) \leftarrow P(X, Y), Q(Y, Z)$. Again, the rule (2.40) is resolved with the facts given by (2.41) and (2.42) in parallel while the rule (2.43) is resolved with the given facts (2.44) and (2.45). The resolvents in these two cases are, respectively:

$$R(c, a) \leftarrow Q(b, c). \text{ and } \leftarrow R(d, e).$$

When the predicates present in the body of one rule also occur in the head of a second rule, the latter and the former rules are said to be in pipeline. Rules in pipeline are resolvable. But in case there exists a matching ground clause for the common literals of both the rules, it is preferred to resolve the ground clause with either of (or both) the rules. The Petri-net model for extended logic programming that we would like to introduce shortly is designed based on the above concept.

The following observations can be made from the last example.

1) Resolution of a rule with one or more facts provides a scope for yielding intermediate ground inferences. For instance, the rule (2.40) in Example 2.29 when resolved with (2.41) and (2.42) yields a ground intermediate $R(c, a) \leftarrow Q(b, c)$.

2) Resolution of two or more rules yield new rules containing literals with renamed variables. The effort in doing so, on many occasions, may be fruitless. For instance, if rule (2.40) and (2.43) were resolved, a new rule would be generated with no further benefits of re-resolving the resulting rule with available facts.

3) In case there exist concurrently resolvable groups of clauses, where each group contains a rule and a few facts, then the overall computational speed of the system can be significantly improved.

Example 2.29 below provides an insight to this issue.

Example 2.29: Let us take the following clauses:

$$R(Z, X) \leftarrow P(X, Y), Q(Y, Z). \tag{2.46}$$
$$P(a, b) \leftarrow. \tag{2.47}$$
$$Q(b, c) \leftarrow. \tag{2.48}$$
$$S(U, V), T(U, V) \leftarrow R(U, V). \tag{2.49}$$
$$\leftarrow S(c, a). \tag{2.50}$$
$$\leftarrow T(c, a). \tag{2.51}$$

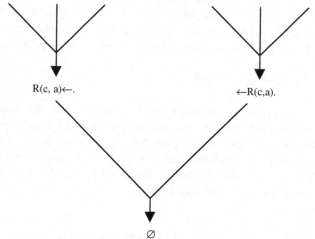

Fig 2.3: A graph illustrating concurrency in resolution.

Unlike example (2.28), where the resulting resolvents after concurrent resolution of two groups of clauses could not participate in further resolution, here the resolvents due to concurrent resolution of (2.46-2.48) and (2.49-2.51) can also take part in the resolution game, resulting in a null clause. To have an idea of speed-up, we construct a graph (Fig. 2.3) indicating the concurrency in resolution.

It is apparent from the above graph that the concurrent resolution of clauses (2.46-2.48) and (2.49-2.51) can take place in one unit time, and the resolution of the resulting clauses require one unit time. Thus, the time taken for execution of the logic program on a parallel engine is two unit times. The same problem, if solved by SLD tree, takes as many as five unit times to perform five resolutions of binary clauses (Fig. 2.4).

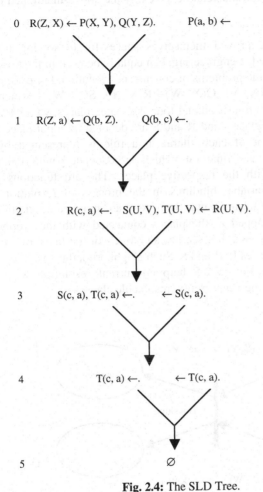

0 R(Z, X) ← P(X, Y), Q(Y, Z).   P(a, b) ←

1   R(Z, a) ← Q(b, Z).   Q(b, c) ←.

2   R(c, a) ←.   S(U, V), T(U, V) ← R(U, V).

3   S(c, a), T(c, a) ←.   ← S(c, a).

4   T(c, a) ←.   ← T(c, a).

5   ∅

Fig. 2.4: The SLD Tree.

The example shows that definitely there is a scope in speed-up due to concurrent resolution at the cost of additional expenses for hardware resources.

The most important problem in concurrent resolution is the identification of the clauses that participate in the resolution process. When there exist groups of concurrently resolvable clauses, search cost to detect the participating clauses in each group sometimes is too high to be amenable in real time. A specialized data structure, capable of performing concurrent resolution of multiple groups of clauses, is recommended to handle the problem. In fact, we are in search of a suitable structure where participating facts and rules in one group of resolvable clauses can be represented by neighborhood structural units. The search cost needed in concurrent resolution thus can be saved by the above-mentioned data structure.

Petri nets, which have proved themselves successful in solving many complex problems of knowledge engineering, can equally be used in the present context to efficiently handle the problems of concurrent resolution. For example, let us consider a clause 'P(X, Y), Q(X, W)←R(X, Y), S(Y, W).,' which is represented in a Petri net by a transition and four associated places, where P and Q are represented by output places, and R and S are denoted by input places of the transition. The argument of each literal in a rule is represented by a specialized function, called arc function, which is associated with the arc connecting the transition with the respective places. The arc functions are needed for generation of variable bindings in the process of resolution of clauses. If '¬P(a, b)←.,' 'R(a, b)←.' and 'S(b, c)←.' are supplied as additional facts then they could be mapped in the places connected with the proposed transition, and the arguments ¬<a, b>, <a, b>, and<b, c> of the facts are saved as tokens of the respective places P, R and S. Such neighborhood mapping of the rule and facts described in Figure 2.5 help concurrent resolution with no additional time for searching the concurrently resolvable clauses.

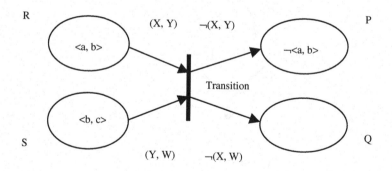

Fig. 2.5: Mapping of a rule on to a Petri net.

Reasoning in a logic program using Petri nets was first proposed by Murata [10]. In this chapter, we extended Murata's model by the following counts:

- *In Murata's model arc functions associated with the arcs of a Petri net are positive irrespective of the type[2] of the arcs. The model to be proposed shortly, however, assigns a positive sign to the arc function attached with a place-to-transition connective arc, and a negative sign*

---

[2] The directed arcs in a Petri net denote connectivity from i) places to transitions, and ii) transitions to places, and thus are of two basic types.

to the arc function attached with a transition-to-place connective arc. The attachment of sign with the arc functions facilitates the scope of matching of signed tokens of the respective places with the arc functions of the connected arcs following the formalisms of predicate logic.

♦ *Unlike Murata's model, where the arguments of body-less clauses also were represented as arc functions, in the present model these are represented as tokens of the appropriate places. Thus, in the present model, we can save additional transition firings for all those arc functions corresponding to the body-less clauses.*

♦ *The extension of Murata's model presented here allows AND-, OR-, Unification-, and Stream-parallelism in a logic program.*

2.4. Petri Net Model for Concurrent Resolution

2.4.1 Extended Petri Net

Definition 2.24: *An **extended Petri net** (**EPN**), which will be used here for reasoning with a First Order Logic (FOL) program, is a 9-tuple, given by*

$$EPN = \{P, Tr, D, f, m, A, a, I, O\}$$

where

$P = \{p_1, p_2,....., p_m\}$ *is a set of places;*
$Tr = \{tr_1, tr_2,....., tr_n\}$ *is a set of transitions;*
$D = \{d_1, d_2,....., d_m\}$ *is a set predicates;*
$P \cap Tr \cap D = \varnothing$; *Cardinality of P = Cardinality of D;*
$f: D \rightarrow P^{\infty}$ *represents a mapping from the set of predicates to the set of places;*
$m: P \rightarrow < x_i,...,y_i, X, \ldots ,Y, f, \ldots ,g>$ *is an association function, represented by the mapping from places to terms, which may include signed constant(s) like $x_i, \ldots , y_i$, variable(s) like $X, \ldots ,Y$ and function $f, \ldots ,g$ of variables; it is usually referred to as tokens of a given place;*
$A \subseteq (P \times Tr) \cup (Tr \times P)$ *is the set of arcs, representing the mapping from the places to the transitions and vice versa;*
$a: A \rightarrow (X, Y,\ldots , Z)$ *is an association function of the arcs, represented by the mapping from the arcs to terms. For arcs $A \in (P \times Tr)$ the arc functions a are positively signed, while for arcs $A \in (Tr \times P)$ the arc functions a are negatively signed;*
$I: Tr \rightarrow P^{\infty}$ *is a set of input places, represented by the mapping from the transitions to their input places;*
$O: Tr \rightarrow P^{\infty}$ *is a set of output places, represented by the mapping from the transitions to their output places.*

Example 2.30: Mapping of the above parameters onto an EPN is illustrated (Fig. 2.6) in this example with the following FOL clauses:

$$Son(Y, Z), Daughter(Y, Z) \leftarrow Father(X, Y), Wife(Z, X). \qquad (2.52)$$
$$Father(r, l) \leftarrow. \qquad (2.53)$$
$$Wife(s, r) \leftarrow. \qquad (2.54)$$
$$\neg Daughter(l, s) \leftarrow. \qquad (2.55)$$

Here, $P = \{p_1, p_2, p_3, p_4\}$;

$Tr = \{tr_1\}$;

$D = \{d_1, d_2, d_3, d_4\}$ with $d_1 =$ Father, $d_2 =$ Wife, $d_3 =$ Son and $d_4 =$ Daughter;

$f(Father) = p_1$, $f(Wife) = p_2$, $f(Son) = p_3$, $f(Daughter) = p_4$;

$m(p_1) = <r, l>$, $m(p_2) = <s, r>$, $m(p_3) = < \emptyset >$, $m(p_4) = \neg <l, s>$ initially and   can be computed subsequently through unification of predicates in the process   of resolution of clauses;

$A = \{A_1, A_2, A_3, A_4\}$, and

$a(A_1) = (X, Y)$, $a(A_2) = (Z, X)$, $a(A_3) = \neg(Y, Z)$, $a(A_4) = \neg(Y, Z)$ are the arc functions;

$I(tr_1) = \{p_1, p_2\}$, and $O(tr_1) = \{ p_3, p_4\}$.

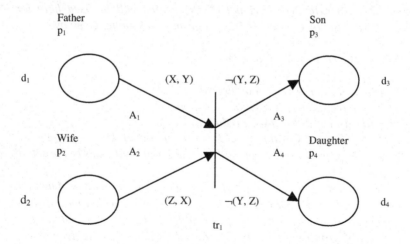

Fig. 2.6: Parameters of an EPN used to represent knowledge in FOL.

It is to be noted that if-then operator of the knowledge has been represented in the figure by tr_1 and the antecedent-consequent pairs of knowledge have been denoted by input (I)-output (O) places of tr_1. Moreover, the arguments of the predicates have been represented by arc functions.

For computing tokens at a place, the transition associated with the place needs to be enabled and fired.

Definition 2.25: *A transition tr_i is **enabled** if the conditions listed below are jointly satisfied:*

i) *all places excluding at most one place associated with the transition tr_i possess properly signed tokens, i.e., positively signed tokens for the input places and negatively signed tokens for the output places, and*

ii) *the variables in the arc functions have consistent bindings with respect to the tokens of the associated places. The consistent binding means that each variable in the arc functions should have a common value.*

Definition 2.26: *A transition, which is enabled first-time **fires**. An enabled transition that already fired can also **fire** if the value of consistent bindings of all the arc function variables with respect to the transition is different from the existing bindings obtained earlier. On firing of a transition, new tokens are generated at a place that did not participate in the generation of consistent variable bindings. The value of the token is determined by the ordered value of the variables in the arc function associated with the concerned place, and the sign of the token will be opposite to the sign of the said arc function.*

Example 2.31: The EPN shown in Fig. 2.6 is enabled because the transition tr_1 here has four input/ output places out of which three places contain properly signed tokens, and the variable in the arc functions are consistent. To test the consistency we list below the value of the variables in each arc function by matching them with their associated places.

$a(A_1) = (X, Y) = <r, l>$ on matching with tokens of place p_1.
$a(A_2) = (Z, X) = <s, r>$ on matching with tokens of place p_2.
$a(A_4) = \neg(Y, Z) = \neg<l, s>$ on matching with tokens of place p_4.

Thus, consistent binding is given by $X = r$, $Y = l$, and $Z = s$. The transition fires because it is enabled and it did not fire earlier. Since place p_3 did not participate in the construction of consistent bindings, a token is inserted in place p_3 on firing of the transition. The token is computed by taking the ordered variables in arc function A_3 and complementing the sign of the arc function. The token thus obtained at place p_3 is given by $\neg(\neg(Y, Z)) = <l, s>$.

2.4.2 Algorithm for Concurrent Resolution

Notations used in the algorithm for automated reasoning are, in order, as follows:

Current-bindings (c-b) denote the set of instantiation of all the variables associated with the transitions.

Used-bindings (u-b) denote the set of union of the current-bindings up to the last transition firing.

Properly signed token means tokens with proper signs, i.e., positive tokens for input places and negative tokens for output places of a transition.

Inactive arc functions represent the arc functions associated with a transition, which do not participate in the process of generation of consistent bindings of variables.

Inert place denotes a place that did not participate in the generation of current bindings of a transition.

The algorithm for automated reasoning to be presented shortly allows concurrent firing of multiple transitions. The algorithm in each pass checks the enabling conditions of all the transitions. If one or more transitions are found enabled, the mgus, here referred to as current-bindings for each transition is searched against the union of the preceding *mgus*, called used-bindings of the said transition. If the current-bindings are not members of the respective used-bindings, then the enabled transitions are fired concurrently. The tokens for the inert places are then computed following the current-bindings of the fired transitions.

The process of transition firing continues until no further transition is ready for firing.

Procedure Automated-reasoning
Begin
For each transition do
Par Begin
used-bindings:= Null;
Flag:= true; // transition not firable.//
Repeat
  If at least all minus one number of the input plus
  the output places of the transition possess *properly
  signed tokens*
  Then do
   Begin
    determine the set: current-bindings of all the
    variables associated with the transition;
  If (a non-null binding is available for all the
     variables) AND
     (current-bindings is not a member of used-bindings)
   Then do Begin
     Fire the transition and send tokens to the *inert place*
     using the set current-bindings and following the

inactive arc function with a presumed opposite sign;
Update used-bindings by taking union with current-bindings;
Flag:= false; //record of transition firing//
Increment no-of-firing by 1;
  End
  Else Flag:= true;
End;
Until no transition fires;
Par End;
End.

Trace of the Algorithm

The Petri net shown in Figure 2.7 is constructed with a set of rules (2.56-2.64). The reasoning algorithm presented earlier is then invoked and the trace of the algorithm thus obtained is presented in Table 2.1. It is clear from the table that the current-bindings (c-b) are not members of used bindings (u-b) in the first two firing criteria testing (FCT) cycles. Therefore, flag = 0. Thus, following the algorithm, transitions tr_1 and tr_2 both fire concurrently. In the third cycle current-bindings become members of used-bindings, and, consequently, flag = 1; so no firing takes place during the third cycle. Further, number of transition in the Petri net (vide Fig. 2.7) being 2 only, control exits the repeat-until loop in procedure Automated-reasoning after two FCT cycles.

Rules:

Father (Y, Z), Uncle (Y, Z) ← Father (X, Y), Grandfather (X, Z). (2.56)
Paternal_uncle (X, Y), Maternal_uncle (X, Y) ← Uncle (X, Y).   (2.57)
Father (r, n)←.   (2.58)
Father (r, d)←.   (2.59)
¬Father (d, a)←.   (2.60)
Grandfather (r, a)←.   (2.61)
¬ Paternal_uncle (n, a)←.   (2.62)
¬Maternal_uncle (n, a)←.   (2.63)
¬Maternal_uncle(d, a)←.   (2.64)

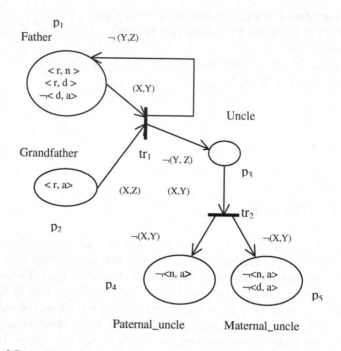

Fig. 2.7: An illustrative Petri net with initial assignment of tokens used to verify the procedure automated-reasoning.

Table 2.1: Trace of the algorithm on example net of Fig. 2.7

Time slot	Tran.	Set of c-b	Set of u-b	Flag=0, if c-b $\notin$ u-b =1, if c-b $\in$ u-b $\neq\{\phi\}$ or c-b=$\{\phi\}$
First cycle	tr_1 tr_2	$\{r/x,d/y,a/z\}$ $\{n/x,a/y\}$	$\{\{\phi\}\}$ $\{\{\phi\}\}$	0 0
Second cycle	tr_1 tr_2	$\{r/x,n/y,a/z\}$ $\{d/x,a/y\}$	$\{\{r/x,d/y,a/z\}\}$ $\{\{n/x, a/y\}\}$	0 0
Third cycle	tr_1	$\{r/x, d/y,a/z\}/$ $\{r/x,n/y,a/z\}$	$\{\{r/x,d/y,a/z\},$ $\{r/x,n/y,a/z\}\}$	1
	tr_2	$\{n/x, a/y\}/$ $\{d/x, a/y\}$	$\{\{n/x, a/y\},$ $\{d/x, a/y\}\}$	1

2.5 Performance Analysis of Petri Net-Based Models

In this section we outline two important issues: the ***speed-up factor*** and the ***resource utilization rate*** of the proposed algorithm when realized on a parallel architecture.

2.5.1 The Speed-up

A complexity analysis of a logic program of n clauses comprised of predicates of arity p reveals that the time T_u required for execution of the program by SLD-resolution on a uniprocessor architecture is given by

$$T_u = O(p.n). \tag{2.65}$$

The product (p.n) in the order of complexity appears because of SLD-resolution of n clauses with p sequential matching of arguments of predicates involved in the resolution process.

The same program comprised of m_1, m_2, ..., m_k number of concurrently resolvable clauses suppose is executed[4] on a pipelined (multiprocessor) architecture, capable of resolving max $\{m_i: 1 \le i \le k\}$ number of clauses in a unit time. Let m_i includes s_i number of supplied clauses and d_i number of derived clauses. Thus $\Sigma m_i = \Sigma s_i + \Sigma d_i$. Under this circumstance the total computational time T_m for execution of the logic program is given by

$$T_m = O\,[p\,(n - (\,s_1 + s_2 + s_3 + \dots + s_k) - 1 + 1 \times k)].$$

$$= O[p(n - \Sigma_{i=1}^{k} s_i + k-1)]$$

$$\approx O[p\,(n - \Sigma_{i=1}^{k} s_i + k)] \tag{2.66}$$

When Σs_i approaches Σm_i, T_m is maximum.

The above result presumes a k-stage pipeline of the k-sets of concurrently resolvable clauses. If the k-sets of clauses are independent, then the concurrent resolution of all the k-set of clauses can be accomplished within a unit time, and thus the computational complexity further reduces to $O[p(n - \Sigma_{i=1}^{k} s_i)]$. Thus, irrespective of a program, it can easily be ascertained that the computational time T_m of a typical logic program always lies in the interval:

---

[4] The k sets of concurrent clauses here is assumed to be resolved in sequence, i.e., resolution of one concurrent set of clause is dependent on a second set, and thus all sets of concurrent clauses cannot be resolved in parallel.

$$O[p(n - \Sigma_{i=1}^{k} s_i)] \leq T_m \leq O[p(n - \Sigma_{i=1}^{k} s_i + k)] \qquad (2.67)$$

Thus, *speed-up*[5] in the worst case is found to be

$$S = T_u/T_m$$

$$= (p.n)/[p(n - \Sigma_{i=1}^{k} s_i + k)]$$

$$= n/[n - \Sigma_{i=1}^{k} s_i + k)] \qquad (2.68)$$

In case all the n number of program clauses are exhausted by resolution, i.e., $\Sigma_{i=1}^{k} m_i$ approaches n, then $S = S_{max}$ is maximized, and the speed-up factor S_{max} reduces to

$$S_{max} = n/k. \qquad (2.69)$$

The last expression reveals that smaller is the k, larger is the S_{max}. The best case corresponds to k=1, when there is a single set of concurrently resolvable clauses. But since $\Sigma_{i=1}^{k} s_i = n$ and k=1 in the present context, $\Sigma_{i=1}^{k} s_i = \Sigma_{i=1}^{1} s_i = s_1 = n$, which means all the n set of clauses are resolvable together. Consequently, the speed-up factor is n.

On the other extreme end, when k = n, i.e., there are $s_1, s_2,..., s_n$ number of concurrently resolvable sets of clauses, then $S_{max} = n/n = 1$, and there is no speed-up. In fact, this case corresponds to typical SLD-resolution and the number of clauses $s_1 = s_2 = ... = s_n = 2$.

2.5.2 The Resource Utilization Rate

Let us assume that the number of resources available for concurrent resolution in the present context is max $\{s_k : 1 \leq k \leq n\}$. Thus, maximum degree of parallelism [4] P is given by

$$P = \max \{s_k : 1 \leq k \leq n\}. \qquad (2.70)$$

---

[5] In case of unification-parallelism, T_m reduces to $O[(n - \Sigma_{i=1}^{k} s_i + k)]$ and, consequently, the worst-case speed-up factor becomes $S_{max} = (p.n)/(n - \Sigma_{i=1}^{k} s_i + k)$.

The average degree of parallelism P_{av} is defined below following Hwang and Briggs [4] as

$$P_{av} = (\Sigma_{i=1}^{k} s_i)/ k. \tag{2.71}$$

The *Resource Utilization Rate* μ thus is found to be

$$\mu = P_{av} / P$$

$$= (\Sigma_{i=1}^{k} s_i)/ [k. \max \{s_k : 1 \le k \le n\}]. \tag{2.72}$$

When s_i for all i approaches to max $\{s_k : 1 \le k \le n\}$, $\Sigma_{i=1}^{k} s_i = k. \max\{s_k: 1 \le k \le n\}$, and consequently, μ approaches 1.

2.5.3 Resource Unlimited Speed-up and Utilization Rate

Suppose the number of resources $\ge$ n, the number of program clauses. Then the concurrent resolution of different sets of clauses may take place in parallel. Suppose, out of s_1, s_2, ..., s_k number of concurrent sets of resolvable clauses, r-sets of clauses on an average can participate in concurrent resolutions at the same time. Then the average time T_{RU} required to execute the program = O[p (n - $\Sigma_{i=1}^{k} s_i + k / r)$]. Then, *Resource Unlimited Speed-up*

$$S_{RU} = T_u / T_{RU}$$
$$=(p.n)/ [p(n - \Sigma_{i=1}^{k} s_i +k / r)]$$
$$= n / (n- \Sigma_{i=1}^{k} s_i + k / r) \tag{2.73}$$

Consequently, maximum speed-up occurs when $\Sigma_{i=1}^{k} s_i$ approaches n, and the result is

$$(S_{RU})_{max} = (n / k)r. \tag{2.74}$$

Further, maximum degree of parallelism in a resource-unlimited system is $P_{RU}= \Sigma_{i=1}^{k} s_i$, and the average degree of parallelism $P_{av} = \Sigma_{i=1}^{r} s_i$. Thus *Resource Utilization Rate* is given by

$$\mu = P_{av} / P_{RU}$$
$$= (\Sigma_{i=1}^{r} s_i)/ (\Sigma_{i=1}^{k} s_i) \tag{2.75}$$

In a special case, when $s_i=s$ for all i= 1 to k, the above ratio reduces to (r/ k).

It is to be noted that when more than one group of concurrently resolvable clauses participate in the resolution process at the same time, $\max\{s_k: 1 \le k \le n\}$ assumes the maximum of the sum of the concurrently resolvable group of clauses.

2.6 Conclusions

This chapter presented a new algorithm for automated reasoning in a logic program using extended Petri net models. Because of the structural advantage of Petri net models, the proposed algorithm is capable of handling AND-, OR-, and stream-parallelisms in a logic program. A complexity analysis of a logic program with n number of clauses and k sets of concurrently resolvable clauses reveals that the maximum speed-up factor of the proposed algorithm in the worst case is $O(n/k)$. Under no constraints on resources, the speed-up factor is improved further by an additional factor of r, where r denotes an average number of the concurrent sets of resolvable clauses. In the absence of any constraints to resources, the maximum resource utilization rate for the proposed algorithm having $s_1 = s_2 = \ldots = s_n$ is $O(r/k)$. With limited resource architecture, the proposed algorithm can execute safely at the cost of extra computational time. The selection of dimensions of such limited resource architecture depends greatly on the choice of r. Selection of r in typical logic programs in turn may be accomplished by running Monte Carlo simulations for a large set of programs. A complete study of this, however, is beyond the scope of this book.

With the increasing use of logic programs in data modeling, its utility in the next generation commercial database systems will also increase in pace. Such systems require a specialized engine that supports massive parallelisms. The proposed computational model, being capable of handling all possible parallelisms in a logic program, is an ideal choice for exploration in commercial database systems. To meet this demand, a hardware realization of the proposed algorithm is needed. We hope for the future when a specialized MIMD engine for logic programming will be commercially available for applications in the next-generation database machines.

Exercises

1. Given below a set of clauses:

    i) $r \lor s \leftarrow p \land q$.
    ii) $R(Z, f(X)) \leftarrow P(X, Y), Q(Y, Z)$.
    iii) Above $(a, c) \leftarrow$ Above (a, b), Above (b, c).
    iv) $N(f(Y), X) \leftarrow L(X, Y, Z), M(f(X), Z)$.

a) Identify the literals in the head part.
b) Identify the literals in the body part.
c) List the functions.
d) List the arguments of the head literals.
e) List the arguments of the body literals.

[**Answers:** a) The literals in the head part:
  i)   r, s;
  ii)   R(Z, f(X));
  iii)   Above(a, c);
  iv)   N(f(Y), X).

b) The literals in the body part:
  i)   p, q;
  ii)   P(X, Y), Q(Y, Z);
  iii)   Above(a, b), Above(b, c);
  iv)   L(X, Y, Z), M(f(X), Z).

c) The functions:
  i)   Nil;
  ii)   f(X);
  iii)   Nil;
  iv)   f(Y), f(X).

d) The arguments of the head literals:
  i)   Nil;
  ii)   Z, f(X);
  iii)   a, c;
  iv)   f(Y), X.

e) The arguments in the body literals:
  i)   Nil;
  ii)   X, Y & Y, Z;
  iii)   a, b & b, c;
  iv)   (X, Y, Z) & f(X), Z.]

2. Identify the following from the given set of clauses

  i)   ground literal,
  ii)   goal clause or query,
  iii)   fact,
  iv)   Horn clause,
  v)   non-horn clause,
  vi)   propositional clause.

a) Boy(X) ← Male-child(X), Non-adult(X).
b) Boy(X), Girl(X) ← Male-child(X), Non-adult(X).
c) S ← P, Q, R.
d) Boy(X) ←.
e) Boy(ram) ←.
f) ← Boy(Y).

[**Answers:**
  i)   Ground literal: Boy(ram)
  ii) Goal clause or query: ← Boy(Y).
  iii) Fact: Boy(ram) ←.
  iv) Horn clause: Boy(X) ← Male-child(X), Non-adult(X).
                    S ← P, Q, R.
                    Boy(X) ←.
                    Boy(ram) ←.
                    ← Boy(Y).
  v) Non_Horn clause: Boy(X), Girl(X) ← Male-child(X), Non-adult(X).
  vi) Propositional clause: S ← P, Q, R.]

3. Translate the following clauses into English:

   Mother (Z, Y)←Father (X, Y)∧Wife (Z, X).
   Mother (s, l)←.
   Wife (s, r)←.
   Query: ←Father (r, l).

[**Hints:** If X is the father of Y and Z is the wife of X, then Z is the mother of
   Y.
   s is the mother of l.
   s is the wife of r.
   Whether r is the father of l?]

4. Construct an Extended Logic Program from the following statements:

The animals that eat plants are herbivorous, the animals that eat animals are
carnivorous, and the animals that eat plants and animals are omnivorous.

[**Hints:** Herbivorous (X)← Animals (X)∧Plant (Y)∧Eats (X, Y).
   Carnivorous (X)← Animals (X)∧Plant (Z)∧Eats (X, Z).
   Omnivorous (X)← Animals (X)∧Plant (Y)∧Animals (Z)∧
                   Eats (X, Y)∧Eats (X, Z).]

5. Given an expression
                    w = P (X, f (Y, Z), d)
   and three substitution sets

$$s_1 = \{a/X,\ b/Y,\ c/Z\}$$
$$s_2 = \{g(Y)/X\}$$
$$s_3 = \{g(W)/X,\ b/Y,\ c/Z\}.$$

Evaluate ws_1, ws_2, and ws_3.

[**Answers:** $ws_1 = P(a,\ f(b,\ c),\ d)$,
$ws_2 = P(g(Y),\ f(Y,\ Z),\ d)$, and
$ws_3 = P(g(W),\ f(b,\ c),\ d)$.]

6. Let w be an expression and ws be an expression after substitution what is the substitution set?

$$w:\ M(Z,\ Y) \leftarrow F(X,\ Y),\ K(Z,\ X).$$
$$ws:\ M(s,\ j) \leftarrow F(r,\ j),\ K(s,\ r).$$

[**Answers:** $s:\{r/X,\ j/Y,\ s/Z\}$.]

7. Let s_1 and s_2 be two substitutions such that
s_1: $\{\ r/X\ \}$, and
s_2: $\{u(X)/Y\}$.
Evaluate the composition of the substitutions:
     a) $s_1 \Delta s_2$ and
     b) $s_2 \Delta s_1$.

[**Answers:** The composition of the substitutions:
a) $s_1 \Delta s_2 = \{r/X,\ u(X)/Y\}$,
b) $s_2 \Delta s_1 = \{u(r)/Y,\ r/X\}$.]

8. Verify the substitution set property 1, i.e., $(ws_1) \Delta s_2 = w(s_1 \Delta s_2)$ using the following items:
Let the expression w: $P(X,\ Y,\ Z)$
And the substitution sets s_1: $\{u(X)/Y,\ v(X)/Z\}$,
                      s_2: $\{r/X\}$.

[**Answers:** $(ws_1) \Delta s_2 = P(X,\ u(X),\ v(X))\{r/X\}$
            $= P(r,\ u(r),\ v(r))$.

$w(s_1 \Delta s_2) = (P(X,\ Y,\ Z))\{r/X,\ u(r)/Y,\ v(r)/Z\}$
              $= P(r,\ u(r),\ v(r))$.
$\therefore$ $(ws_1) \Delta s_2 = w(s_1 \Delta s_2)$.]

9. Verify the substitution set property 2, i.e., $(s_1 \Delta s_2) \Delta s_3 = s_1 \Delta (s_2 \Delta s_3)$ using the following items:
Let the substitution sets s_1: $\{r/X\ \}$,

s_2: $\{u(X)/Y\}$, and
s_3: $\{l/Z\}$.

[**Answers:** $s_1 \Delta s_2 = \{r/X, u(X)/Y\}$
$(s_1 \Delta s_2) \Delta s_3 = \{r/X, u(X)/Y, l/Z\}$.
$s_2 \Delta s_3 = \{u(X)/Y, l/Z\}$
$s_1 \Delta (s_2 \Delta s_3) = \{r/X, u(X)/Y, l/Z\}$.
$\therefore (s_1 \Delta s_2) \Delta s_3 = s_1(s_2 s_3)$.]

10. Verify the substitution set property 3, i.e., $s_1 \Delta s_2 \neq s_2 \Delta s_1$ using the following items:
    Let the substitution sets s_1: $\{r/X\}$,
                    and s_2: $\{u(X)/Y\}$.

[**Answers:** $s_1 \Delta s_2 = \{r/X, u(X)/Y\}$ and
$s_2 \Delta s_1 = \{u(r)/Y, r/X\}$.
$\therefore s_1 \Delta s_2 \neq s_2 \Delta s_1$.]

11. Which of the following clauses are resolvable and what is the resolvent?

    i) Cl_1: $R(Y, X) \leftarrow P(X, Y)$.
    ii) Cl_2: $P(r, n) \leftarrow$.
    iii) Cl_3: $R(n, r) \leftarrow$.

[**Hints:** As the same predicate is present in the head and the body part of the clause number cl_2 and cl_1 respectively, they are resolvable. However, due to presence of the same predicate in the head part of the clauses, cl_1 and cl_3, they are not resolvable.]

12. By definition 8, show that the following is a set of resolvable clauses.

    i) Cl_1: $R(Z, X) \leftarrow P(X, Y), Q(Y, Z)$.
    ii) Cl_2: $P(r, s) \leftarrow$.
    iii) Cl_3: $\neg R(l, r) \leftarrow$.
    iv) Cl_4: $Q(s, l) \leftarrow$.

[**Hints:** Cl_{12}: $R(Z, r) \leftarrow Q(s, Z)$.
Cl_{13}: $\leftarrow P(r, Y), Q(Y, l)$.
Cl_{14}: $R(l, X) \leftarrow P(X, s)$.

As each of the clauses is resolvable with at least one of the set of clauses, producing a resolvent, according to definition 2.9, it is a set of resolvable clauses.]

13. Identify the definite program clause with definite goal.

   Cl_1: R, S ← P, Q.
   Cl_2: R ← P, Q.
   Cl_3: ← P, Q.

   [**Hints:** According to definition 2.10, Cl_2: R ← P, Q. is a definite clause containing one atom in its head and Cl_3: ← P, Q. is a definite goal with empty consequent vide definition 2.12.]

14. Construct the resolution tree for linear selection,

   R ← P, Q.
   S ← R.
   T ← S.
   Q ←.
   P ←.
   Goal: ← T.

15. Determine whether the following are orderly or order independent clauses. If orderly, verify whether single or multiple sequence. Determine the various orders of resolution in the following set of clauses.

   Cl_1: R(Z, X) ← P(X, Y), Q(Y, Z).
   Cl_2: P(r, a) ←.
   Cl_3: Q(a, k) ←.
   Cl_4: ¬R(k, r) ←.

   [**Hints:** The clauses are to be selected pair-wise according to some definite order from the set of resolvable clauses. Otherwise they fail to generate a solution. As for example, cl_2, cl_3 or cl_3, cl_4 cannot be resolved. But, we can get results by resolving the clauses following some definite orders.
   The multiple orders are:
   Sequence 1: Order 1: 1-2-3-4/ 2-1-3-4
   $Cl_{12\text{-}3\text{-}4}$: ∅
   Sequence 2: Order 2: 1-2-4-3/ 2-1-4-3
   $Cl_{12\text{-}4\text{-}3}$: ∅
   Sequence 3: Order 3: 1-3-2-4/ 3-1-2-4
   $Cl_{13\text{-}2\text{-}4}$: ∅
   Sequence 4: Order 4: 1-3-4-2/ 3-1-4-2
   $Cl_{13\text{-}4\text{-}2}$: ∅
   Sequence 5: Order 5: 1-4-2-3/ 4-1-2-3
   $Cl_{14\text{-}2\text{-}3}$: ∅
   Sequence 6: Order 6: 1-4-3-2/ 4-1-3-2

$$Cl_{14\text{-}3\text{-}2}: \varnothing \ ]$$

16. Determine the final resolvent from the given set of clauses. If not possible, indicate why.

> Cl_1: Son(Y, X) ← Father(X, Y).
> Cl_2: Mother(i, a) ← Son(a, n), Husband(n, i).
> Cl_3: Father(Z, Y) ← Mother(X, Y), Wife(X, Z).

[**Hints:** Cl_{12}: Mother(i, a) ← Father(n, a), Husband(n, i).
  $Cl_{12\text{-}3}$ is not possible as double resolution takes place.]

17. Show that the following clauses are order independent. Explain the reason for nonvalidity.

> Cl_1: Boy(X), Girl(X) ← Child(X).
> Cl_2: Likes_to_play_indoor(X) ← Boy(X), Introvert(X).
> Cl_3: Likes_to_play_doll(X) ← Girl(X), Likes_to_play_indoor(X),
> $\qquad\qquad\qquad\qquad\qquad\qquad$ Has_doll(X).

[**Hints:** Each clause is resolvable with any other clause.
      Nonvalidity: Ultimately, double resolution takes place.]

18. Show whether the following set of clauses is concurrently resolvable:

> Cl_1: R(Z, X) ← P(X, Y), Q(Y, Z).
> Cl_2: P(r, a) ←.
> Cl_3: Q(a, k) ←.
> Cl_4: ¬R(k, r) ←.

[**Hints:** As the instantiated resolvent generated following multiple sequences yields a unique result, the set includes concurrently resolvable clauses.]

19. Test whether the following clauses are concurrently resolvable:

a) Cl_1: Mother(Z, Y) ← Father(X, Y), Son(Y, Z), Married_to(X, Z).
    Cl_2: Father(r, l) ← Has_one_son(r, l), Male(r).
    Cl_3: Female(Z) ← Mother(Z, Y).
    Cl_4: Son(l, s) ←.

b) Cl_1: Mother(Z, Y) ← Father(X, Y), Son(Y, Z), Married_to(X, Z).
    Cl_2: Father(r, l) ← Has_one_son(r, l), Male(r).
    Cl_3: Female(Z) ← Mother(Z, Y).
    Cl_4: Son(k, s) ←.

[**Answers:** (a) Concurrently resolvable, (b) not concurrently resolvable].

20. Indicate a) the AND-parallel clauses and b) the OR-parallel clauses to concurrently resolve the rule cl_1 with the rest of the clauses cl_2 through cl_5.

    Cl_1: Likes_to_play(X, Y) ← Child(X), Game(Y).
    Cl_2: Child(r) ←.
    Cl_3: Child(t) ←.
    Cl_4: Game(c) ←.
    Cl_5: Game(l) ←.

[**Answers:** a) The AND-parallel clauses:
       i) Cl_1, Cl_2, Cl_4,
      ii) Cl_1, Cl_2, Cl_5,
     iii) Cl_1, Cl_3, Cl_4,
     iv) Cl_1, Cl_3, Cl_5.

    b) The OR-parallel clauses:
     i) Cl_1, Cl_2 and Cl_1, Cl_3,
    ii) Cl_1, Cl_4 and Cl_1, Cl_5.]

21. a) Verify whether concurrent resolution is valid for the following clauses:

    Cl_1: Game(Y) ← Child(X), Likes_to_play(X, Y).
    Cl_2: Outdoor_game(Y), Indoor_game(Y) ← Game(Y).
    Cl_3: Child(X) ← Boy(X).
    Cl_4: Child(X) ← Girl(X).
    Cl_5: Boy(X) ← Likes_to_play(X, Y), Outdoor_game(Y).
    Cl_6: Girl(X) ← Likes_to_play(X, Y), Indoor_game(Y).
    Cl_7: Likes_to_play(t, c) ←.
    Cl_8: Likes_to_play(r, l) ←.
    Cl_9: Outdoor_game(c) ←.
    Cl_{10}:¬ Indoor_game(c) ←.

    b) If yes, identify the types of concurrent resolution for the following sequences:

       i) Cl_5, Cl_7, Cl_9,
      ii) Cl_1, Cl_2, Cl_5,
     iii) Cl_3, Cl_5, Cl_7, Cl_9,
     iv) Cl_1, Cl_2, Cl_6,
      v) Cl_1, Cl_2, Cl_3, Cl_7.

[**Hints:**

a) For the following ordered sequence of clauses for SLD resolution, the final resolvent is identical. So, concurrent resolution is possible in the following cases:

i)   **Sequence 1:** Order 1: 1-3-7.
     Sequence 2: Order 2: 1-7-3.
ii) **Sequence 3:** Order 3: 1-3-8.
     Sequence 4: Order 4: 1-8-3.
iii) **Sequence 5:** Order 5: 1-4-7.
     Sequence 6: Order 6: 1-7-4.
iv) **Sequence 7:** Order 7: 1-4-8.
     Sequence 8: Order 8: 1-8-4.
v)   **Sequence 9:** Order 9: 5-8-9.
     Sequence 10: Order 10: 5-9-8.

b)
   i)   Concurrent resolution between a rule and facts,
   ii)   Concurrent resolution between rules,
   iii) Concurrent resolution between rules and facts,
   iv)   Concurrent resolution between rules,
   v)   Concurrent resolution between fact and rules.]

22. Map the following FOL clauses on to an EPN and state the result after resolution.

Cl_1: Game(Y) ← Child(X), Likes_to_play(X, Y).
Cl_2: Outdoor_game(Y), Indoor_game(Y) ← Game(Y).
Cl_3: Child(t) ←.
Cl_4: Likes_to_play(t, c) ←.
Cl_5:¬ Indoor_game(c) ←.

23. Map the following program clauses onto an EPN. What result do you obtain after execution of the program by the algorithm: 'Procedure Automated Reasoning'?

The program clauses are:

Cl_1: Reproduce_by_laying_eggs(X) ← Build_nests(X), Lay_eggs(X).
Cl_2: Has_wings(X) ← Can_fly(X), Has_feather(X).
Cl_3: Bird(X) ← Reproduce_by_laying_eggs(X), Has_beaks(X),
                 Has_wings(X).
Cl_4: Build_nests(p) ←.
Cl_5: Lay_eggs(p) ←.
Cl_6: Can_fly(p) ←.
Cl_7: Has_feather(p) ←.
Cl_8: Has_beaks(p) ←.

24. Given a set of clauses (Cl_1-Cl_{13}):

Cl_1: Father(Y, Z), Uncle(Y, Z) ← Father(X, Y), Grandfather(X, Z).
Cl_2: Paternal_uncle(X, Y), Maternal_uncle(X,Y) ← Uncle(X, Y).
Cl_3: Mother(Z, Y) ← Father(X, Y), Married_to(X, Z).
Cl_4: Father(X, Y) ← Mother(Z, Y), Married_to(X, Z).
Cl_5: Father <r, n> ←.
Cl_6: Father <r, d> ←.
Cl_7: ¬ Father <d, a> ←.
Cl_8: Grandfather <r, a> ←.
Cl_9: ¬Paternal_uncle <n, a> ←.
Cl_{10}: ¬Maternal_uncle <n, a> ←.
Cl_{11}: ¬Maternal_uncle <d, a> ←.
Cl_{12}: Married_to <r, t> ←.
Cl_{13}: Married_to <n, i> ←.

a) List the possible resolutions that take place in the network.
b) Identify the concurrent resolutions among those in the above list.
c) Represent the concurrent resolutions in tabular form like Table 2.1.

[**Hints:** a) The possible resolutions are:

1-3, 1-4, 1-5, 1-6, 1-7, 1-8, 2-9, 2-10, 2-11, 3-12, 3-13, 4-12, 4-13.

b) The concurrent resolutions are:

Sequence 1: Order 1: 1-6-7-8.
$Cl_{1-6-7-8}$ = Uncle(d, a) ←.
Sequence 2: Order 2: 2-9-10.
Cl_{2-9-10} = ← Uncle(n, a). ≡ ¬ Uncle(n, a) ←.
Sequence 3: Order 3: 3-5-12.
Cl_{3-5-12} = Mother(t, n) ←.
Sequence 4: Order 4: 3-6-12.
Cl_{3-6-12} = Mother(t, d) ←.

c) The table describing concurrent resolution for problem 24 is given below.

Time slot	Transitions	Set of c-b	Set of u-b	Flag=0, if c-b ∉ u-b = 1, if c-b ∈ u-b≠{Ø} or c-b = {Ø}
First cycle	tr_1	{r/X,d/Y, a/Z}	{{Ø}}	0
	tr_2	{n/X, a/Y}	{{Ø}}	0
	tr_3	{r/X, n/Y, t/Z}/ {r/X, d/Y, t/Z}	{{Ø}}	0
Second cycle	tr_1	{r/X, n/Y, a/Z}	{{r/X, d/Y, a/Z}}	0
	tr_2	{d/X, a/Y}	{{n/X, a/Y}}	0
	tr_3	{Ø }	{{r/X, n/Y, t/Z}, {r/X, d/Y, t/Z}}	0
Third cycle	tr_1	{Ø }	{{r/X, d/Y, a/Z}}	1
	tr_2	{Ø }	{{n/X, a/Y}}	1
	tr_3	{n/X, a/Y, i/Z}	{{r/X, n/Y, t/Z}, {r/X, d/Y, t/Z}, {n/X, a/Y, i/Z}}	1

25.a) For answering the goal(← M(t, n).) with the clauses given in problem 24, draw the SLD-tree from the given set of clauses.

b) Use 'Procedure Automated_Reasoning' to verify the goal.

c) Assuming unit time to perform a resolution, determine the computational time involved for execution of the program by SLD-tree approach.

d) Compute the computational time required for execution of the program clauses on EPN using 'Procedure Automated_Reasoning'.

e) Determine the percentage time saved in the EPN approach ((a-b)/a × 100) where 'a' stands for SLD, 'b' for EPN.

26. From the given Petri nets (Fig. 2.8 and Fig. 2.9), calculate the speed-up and the Resource utilization rate from the definition given in the equations (2.68) and (2.72).

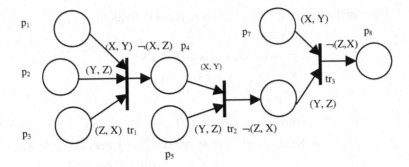

Token at the place p_1 = <a, b>, Token at the place p_2 = <b, c>, Token at the place p_3 = <c, a>, Token at the place p_5 = <c, d>, Token at the place p_7 = <e, d>.

Fig. 2.8: A Petri net.

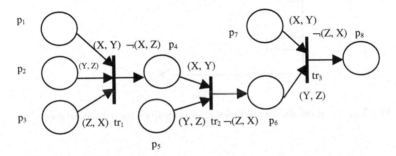

Token at the place p_1 = <a, b>, Token at the place p_2 = <b, c>, Token at the place p_3 = <c, a>, Token at the place p_5 = <c, d>, Token at the place p_7 = <e, d>, Token at the place p_8 = ¬<a, e>.

Fig. 2.9: An illustrative Petri net.

[**Hints**: Pipelining of transitions in the first and the second cases are given in Figures 2.10 and 2. 11 respectively.

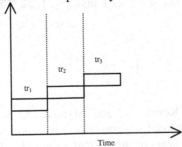

Fig. 2.10: Diagram showing pipelining of transitions for the first case.

Here, in the case of the first Petri net, as per definition:

$$n = 8,$$

$$\Sigma s_i = s_1 + s_2 + s_3 = 4+2+2 = 8,$$

$$\Sigma m_i = \Sigma s_i + \Sigma d_i = 8+3 = 11,$$

$$k = 3$$

$\therefore$ Speed-up factor $S = T_u/T_m = n /(n - \Sigma s_i + k) = 8/3 = 2 \cdot 66.$

& Resource utilization rate $\mu = \Sigma s_i /[k \cdot \max \{s_k: 1 \leq k \leq n\}]$

$$= 8/(3 \times 4) = 8/12 = 0 \cdot 66.$$

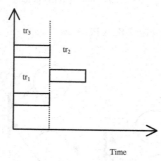

Time

Fig. 2.11: Diagram showing pipelining of transitions for the second case.

Now, for the second case, $n = 9,$

$$\Sigma s_i = 4+2+3 = 9,$$

$$k = 2.$$

$\therefore$ The speed-up factor $S = T_u/T_m = n /(n - \Sigma s_i + k)$

$$= 9/(9 - \Sigma s_i + k) = 9/2 = 4 \cdot 5.$$

& Resource utilization rate $\mu = \Sigma s_i /[k \cdot \max \{s_k: 1 \leq k \leq n\}]$

$$= 9/(2 \times 7) = 9/14 = 0 \cdot 64.]$$

References

[1]   Bhattacharya, A., Konar, A. and Mandal, A. K., "A parallel and distributed computational approach to logic programming," Proc. of *International Workshop on Distributed Computing (IWDC 2001)*, held in Calcutta, December 2001.

[2] Ganguly, S., Silberschatz, A., Tsur, S., "Mapping datalog program execution to networks of processors," *IEEE Trans. on Knowledge and Data Engg.*, vol. 7, no. 3, June 1995.

[3] Hermenegildo, M. and Tick, E., "Memory performance of AND-parallel PROLOG on shared-memory architecture," *Proc. of the 1988 Int. Conf., on Parallel Processing*, vol. II, Software, pp. 17-21, Aug. 15-19, 1988.

[4] Hwang, K. and Briggs, F. A., *Computer Architecture and Parallel Processing*, McGraw-Hill, pp. 27-35, 1986.

[5] Jeffrey, J., Lobo, J. and Murata, T., "A high–level Petri net for goal-directed semantics of Horn clause logic," *IEEE Trans. on Knowledge and Data Engineering*, vol. 8, no. 2, April 1996.

[6] Kale, M. V., Parallel Problem Solving, In *Parallel Algorithms for Machine Intelligence and Vision*, Kumar, V., Gopalakrishnan, P. S. and Kanal, L.N. (Eds.), Springer-Verlag, Heidelberg, 1990.

[7] Konar, A., *Uncertainty Management in Expert Systems Using Fuzzy Petri Nets*, Ph.D. Thesis, Jadavpur University, 1994.

[8] Konar, A. and Mandal, A. K., "Uncertainty management in expert systems using fuzzy Petri nets," *IEEE Trans. on Knowledge and Data Engg.*, vol. 8., no.1, Feb. 1996.

[9] Li, L., "High level Petri net model of logic program with negation," *IEEE Trans. on Knowledge and Data Engg.*, vol. 6, no. 3, June 1994.

[10] Murata, T. and Yamaguchi, H., "A Petri net with negative tokens and its application to automated reasoning," *Proc. of the 33rd Midwest Symp. on Circuits and Systems,* Calgary, Canada, Aug. 12-15,1990.

[11] Naganuma, J., Ogura, T., Yamada, S-I., and Kimura, T., "High-speed CAM-based Architecture for a Prolog Machine (ASCA)," *IEEE Transactions on Computers*, vol. 37, no.11, November 1988.

[12] Patt, Y. N., "Alternative Implementations of Prolog: the micro architecture perspectives," *IEEE Trans. on Systems, Man and Cybernetics,* vol. 19, no.4, July/August 1989.

[13] Peterka, G. and Murata, T., "Proof procedure and answer extraction in Petri net model of logic programs," *IEEE Trans. on Software Eng.*, vol. 15, no. 2, Feb. 1989.

[14] Takeuchi, A., *Parallel Logic Programming*, Wiley, 1992.

[15] Yan, J. C., "Towards parallel knowledge processing," In *Advanced series on Artificial Intelligence*: *Knowledge Engineering Shells, Systems and Techniques*, Bourbakis, N. G. (Ed.), vol. 2, World Scientific, 1993.

Chapter 3

Distributed Reasoning by Fuzzy Petri Nets: A Review

This chapter provides a review of the existing methodology on knowledge representation and approximate reasoning using specialized data structures called fuzzy Petri nets. Both forward and backward reasoning models of fuzzy Petri nets have been addressed with a clear emphasis on their chronological developments. The difference between acyclic and cyclic forward reasoning algorithms has been illustrated with typical examples. The scope of supervised machine learning on fuzzy Petri nets has also been briefly outlined. Lastly, the equivalence of fuzzy Petri nets from a semantic viewpoint has been addressed. The chapter ends with a discussion on the scope of fuzzy Petri nets on complex reasoning problems, many of which will be discussed in the rest of the book.

3.1 Fuzzy Logic and Approximate Reasoning

The logic of fuzzy sets is an extension of the logic of propositions/predicates by the following counts:

i) Propositional logic is concerned with (total) matching of the propositions in the antecedent part of a rule with available data/facts on the working memory (WM). Fuzzy logic in the contrary allows partial matching of the antecedent clauses with facts of the WM.

ii) The inferences derived by propositional logic thus are based on standard propositional syllogisms like *modus ponens*, *modus tollens*, and *resolution by refutation technique*. Fuzzy logic extends modus ponens and modus tollens to *fuzzy modus ponens* and *fuzzy modus tollens,* respectively. The inferences derived by fuzzy logic are based on fuzzy implication relations of the input and the output fuzzy variables and the measured membership distribution of the input variables.

For instance, given the fuzzy implication relation R(x, y) of the input variable $x \in A$, and the output variable $y \in B$, we can evaluate $\mu_B'(y)$, when $\mu_A'(x)$ for $A' \approx A$ is supplied. In fact,

$$\mu_B'(y) = [\mu_A'(x)] op \ R(x, y) \tag{3.1}$$

where "op" is a fuzzy operator, typically the *MAX-MIN composition* operator.

For a given set of fuzzy rules like

If x is A_1 then y is B_1 Else
If x is A_2 then y is B_2 Else
....
....
If x is A_n then y is B_n,

we can derive a fuzzy relation by Mamdani implication as

$$R_i(x, y) = \mu_{Ai}(x) \wedge \mu_{Bi}(y), \ 1 \le \forall i \le n. \tag{3.2}$$

Now, given the membership distribution of $\mu_A'(x)$ where $A' \approx A_1$, $A' \approx A_2$, ..., $A' \approx A_n$, we, by firing all the n rules, get the desired inference y is B_i' $1 \le i \le n$ where

$$\mu_{Bi}'(x, y) = \bigvee_{\forall x} [\mu_{Ai}'(x) \wedge R_i(x, y)] \ 1 \le \forall i \le n. \tag{3.3}$$

The composite membership distribution of $\mu_B'(y)$ can now be determined by taking the aggregation of $\mu_{Bi}'(y)$ for i=1 to n

$$\mu_B'(y) = \bigvee_{1 \le i \le n} \mu_{Bi}'(y). \tag{3.4}$$

Expression (3.4) describes a method for computation of y is B' from the known membership distribution of the antecedent clauses x is A_i'. The reasoning methodology introduced is well known as *forward reasoning* in fuzzy logic. Besides forward reasoning, there exist other types of fuzzy reasoning as well. Abductive reasoning, fuzzy bi-directional reasoning and fuzzy non-monotonic reasoning are just a few of them. Each of the above reasoning problems has its own formulation, and approach to determine its solution. The basic theme of their considerations under fuzzy sets is their unique feature of approximate reasoning using fuzzy tools and techniques.

3.2 Structured Models of Approximate Reasoning

Structured models/networks have been given prime considerations in reasoning since the origin of "semantic nets" [15]. Nilsson [11] coined the term *semantic nets*, which has later been successfully used in many AI problems, including reasoning and natural language understanding. The popularity of the structured models, including graphs, trees, or networks, was later enhanced by Pearl [13], who employed Bayesian reasoning on acyclic graphs and trees. Gradually, structured models employing Markov chain [6] and traditional logic [1] emerged. The structured approach to fuzzy reasoning, however, had remained a far cry until Professor Carl Looney in late 1980s brought forward a solution to the approximate reasoning problem using Petri nets. Because of the structural analogy of his engine with a typical Petri net, he called it fuzzy Petri net or fuzzy logic network. Like traditional Petri nets, a fuzzy Petri net (FPN) [2-5], [7-10], [14], [17-19], too, is a directed bipartite graph containing two types of nodes: places and transitions. It is usually represented by a 10-tuple

FPN=$\{P, D, Tr, A, T, Th, N, CF, I, O\}$
where
$P=\{p_1, p_2, ..., p_n\}$ is a set of places denoted by circles,
$D=\{d_1, d_2, ..., d_n\}$ is a set of association function of places,
$Tr=\{tr_1, tr_2, ..., tr_m\}$ is a set of transitions denoted by bars,
$A \subseteq (P \times Tr) \cup (Tr \times P)$ is the set of arcs,
$T=\{t_1, t_2, ..., t_m\}$ is a set of association function of transitions, such that t_i: $tr_i \rightarrow [0,1]$,
$Th=(th_1, th_2, ..., th_m\}$ is another set of association function of transitions, such that th_i: $tr_i \rightarrow [0,1]$,
$N=\{n_1, n_2, ..., n_n\}$ is a set of association function of places such that n_i: $p_i \rightarrow [0, 1]$,
$CF=\{cf_1, cf_2, ..., cf_n\}$ is a set of association function of transitions, such that cf_i: $tr_i \rightarrow [0,1]$.
I: $tr_i \rightarrow P^\infty$ denotes a mapping from a transition tr_i to its input places, and
O: $tr_i \rightarrow P^\infty$ denotes a mapping from a transition tr_i to its output places.

It is given that

$$|P| = |D| = |N| \text{ and}$$

$$|T| = |Th| = |CF| \text{ and}$$

$$P \cap T = \phi$$

where $|.|$ denotes the cardinality of a set.

Example 3.1: Figure 3.1 describes a typical FPN.

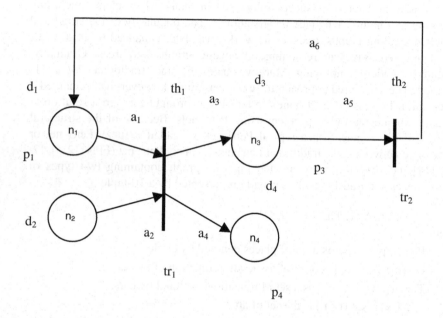

Fig 3.1: A typical fuzzy Petri net.

In Fig. 3.1,
        $P=\{p_1, p_2, p_3, p_4\}$,
        $D=\{d_1, d_2, d_3, d_4\}$
        where the propositions
            $d_1=$the sky is cloudy,
            $d_2=$ it is hot,
            $d_3=$ it will rain, and
            $d_4=$ the humidity is high;
$Tr=\{tr_1, tr_2\}$ is the set of transitions corresponding to the following rules:

      R1: If (the sky is cloudy) & (it is hot)
           Then (it will rain) or (the humidity is high), and

      R2: If (it rains) Then (the sky is cloudy), respectively.

$A \subseteq (P \times Tr) \cup (Tr \times P) = \{a_1, a_2, a_3, a_4, a_5, a_6\}$

$T = \{t_1, t_2\}$ is the set of fuzzy truth tokens (FTT) possessed by the output arcs of the transitions tr_1 and tr_2 respectively.

$Th = \{th_1, th_2\}$ is the set of thresholds associated with the transitions tr_1 and tr_2 respectively.

$N = \{n_1, n_2, n_3, n_4\}$ is the set of fuzzy beliefs/ membership distributions [18] associated with the propositions. Thus, n_i corresponds to the belief of d_i.

$CF = \{cf_1, cf_2\}$ is the set of certainty factor[1] of the rules R_1 and R_2 respectively.

It is indeed important to note that certainty factor is a measure of the degree of firing of a production rule. The larger the certainty factor, the higher is the degree of firing of the rule for a successful instantiation of its antecedent part.

3.3 Looney's Model

Looney pioneered the concept of fuzzy reasoning with Petri nets for determining the membership of a fuzzy decision variable. He considered an acyclic structure of a fuzzy Petri net and defined a recursive rule for updating FTT at the transitions and fuzzy membership at the places. His formulation can be best described as follows:

For each place p_i where $\exists_j \{tr_j \times p_i\} \subseteq A$, i.e., p_i is an output place of transition tr_j,

$$n_i(t+1) = Max [n_i(t), \{Max_{\forall j} t_j(t+1)\}] \qquad (3.5)$$

and $n_i(t+1) = n_i(t)$ when $(tr_j \times p_i) \notin A$. $\qquad (3.6)$

Further for each transition tr_j, if

$\exists k, \{p_k \times tr_j\} \subseteq A$, i.e., p_k is an input place of transition tr_j, then

$$\left. \begin{array}{l} t_j(t+1) = Min[Min_k(n_k), t_j(t)] \text{ if } Min(n_k) \geq th_j \\ = 0, \text{ otherwise.} \end{array} \right\} \qquad (3.7)$$

Looney considered updating of FTTs of the transitions first and then used the temporary results $t_j(t+1), \forall j$, to compute $n_i(t+1)$, $\forall i$. The following example describes his formulation.

Example 3.2: Consider a FPN to illustrate Looney's algorithm [9]. Given $n_1(0)=0$, $n_2(0)=0.8$, $n_3(0)=0$, $n_4(0)=0$, $n_5(0)=0.5$, $n_6(0)=0$ and $t_j(0)=1$ $\forall j$ and $th_j =0$ $\forall j$.

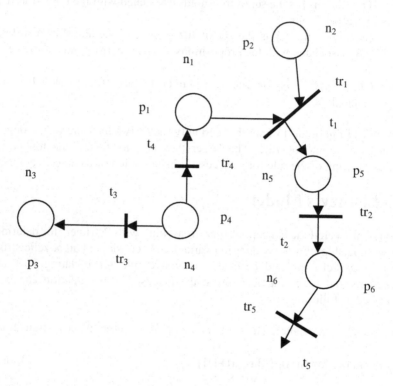

Fig. 3.2: A fuzzy Petri net used to illustrate Looney's model. Belief n_i for place p_i, for i=1 to 6 here is labeled outside the place for space-crisis inside the places.

The results of computation of all t_j's in parallel followed by n_i's in parallel are summarized in Table 3.1.

Table 3.1: Results of computation of FTT and memberships

	t_1	t_2	t_3	t_4	t_5	
	0	0	0.5	0.5	0	
t=1	n_1	n_2	n_3	n_4	n_5	n_6
	0.5	0.8	0.5	0.5	0	0
	t_1	t_2	t_3	t_4	t_5	
t=2	0	0	0.5	0.5	0	
	n_1	n_2	n_3	n_4	n_5	n_6
	0.5	0.8	0.5	0.5	0	0

Since $n_i(2) = n_i(1)$ for i = 1 to 6 have been attained in the FPN of Figure 3.2, there is no need for further updating of memberships in the given network.

3.4 The Model Proposed by Chen et al.

Chen et al. proposed an alternative model for computing belief of a proposition from the supplied belief of a set of axioms (places with no input arcs). The following definitions are needed to understand their scheme for belief computation.

Definition 3.1: *If $p_i \in I(tr_j)$ and $p_k \in O(tr_j)$, then p_k is **immediately reachable** from p_i. This is symbolically denoted by $p_k \in IRS(p_i)$.*

Definition 3.2: *If $p_k \in IRS(p_i)$ and $p_j \in IRS(p_k)$, then p_j is **reachable** from p_i. Consequently, $p_j \in RS(p_i)$, where $RS(p_i)$= reachability set of p_i. The reachability set operator is the reflexive-transitive closure of the immediate reachability set operator.*

Definition 3.3: *Let $p_x \in I(tr_k)$ and $p_y \in O(tr_k)$, then the non-empty set $AP_{xy}= I(tr_k)-\{P_x\}$ is called the set of **adjacent places**.*

For the evaluation of fuzzy singleton membership (belief) at place p_j, when the same is given at place p_s, Chen et al. represented part of the Petri net by a sprouting tree. The nodes corresponding to a place p_i in the tree is labeled with a 3-tuple: $\{p_i, n_i, IRS(p_i)\}$ where n_i denotes the belief at place p_i and $IRS(p_i)$ denotes the immediately reachable set of p_i.

In the construction of the sprouting tree, the following principles are adopted. Let p_i be a node in the sprouting tree corresponding to a place p_i in the FPN. Now,

1) If goal place $p_j \in IRS(p_i)$, then construct a node p_j and an arc p_i to p_j in the tree.
2) If $p_k \in IRS(p_i)$ and $p_j \in RS(p_k)$, then construct p_k as a node in the tree and an arc from p_i to p_k .

The algorithm thus constructs a sprouting tree, where the evaluation of membership at a node p_k is carried out by the following principle.

1. If $p_k \in IRS(p_i)$ and AP_{ik} exist, then membership n_k at place p_k is computed by
    a) temp = min (membership of the parents of p_k in the FPN);

    b) If temp< threshold th_r of the transition tr_r, where $p_i \in I(tr_r)$ and $p_k \in O(tr_r)$ then temp=0;

          c) n_k = temp*certainty factor cf_r of the transition tr_r..

2.   If $p_k \in IRS(p_i)$ and AP_{ik} does not exist then the computation of n_k is redefined as follows:

        a)   temp = membership of the parent of p_k in the FPN.

        Steps (b) and (c) remain as usual.

3.   The computation of membership at the goal place p_j having multiple inputs is performed by taking the maximum of the memberships obtained for the goal in the sprouting tree.

Example 3.3: The sprouting tree for the FPN given in Figure 3.3 is constructed below (Fig 3.4) and the membership at place p_4 is evaluated following the above principles.

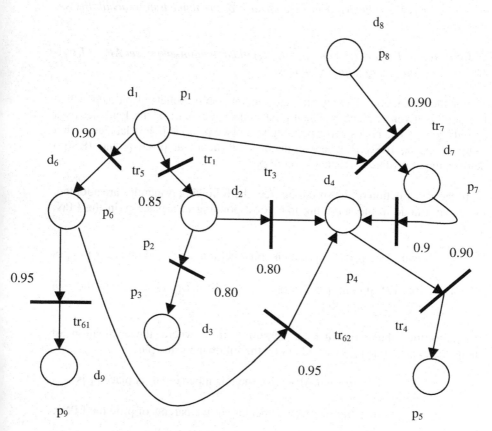

Fig. 3.3: A FPN used for illustrating the construction of the sprouting tree.

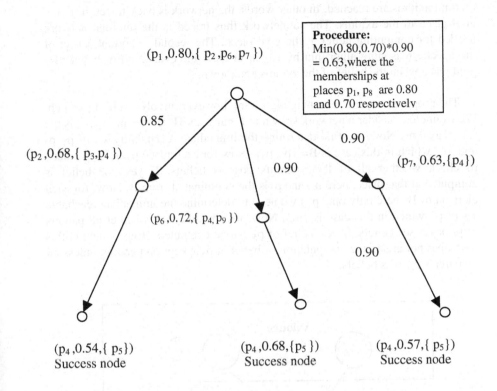

$(p_1, 0.80, \{p_2, p_6, p_7\})$

Procedure:
Min(0.80,0.70)*0.90
= 0.63,where the
memberships at
places p_1, p_8 are 0.80
and 0.70 respectively

0.85

0.90

$(p_2, 0.68, \{p_3, p_4\})$

0.90

$(p_7, 0.63, \{p_4\})$

$(p_6, 0.72, \{p_4, p_9\})$

0.90

$(p_4, 0.54, \{p_5\})$
Success node

$(p_4, 0.68, \{p_5\})$
Success node

$(p_4, 0.57, \{p_5\})$
Success node

Fig. 3.4: Membership at place p_4 of the above tree is computed by taking the max. of the success nodes at the leaves. Thus, $n_4 = \max (0.54, 0.68, 0.57) = 0.68$.

It needs mention here that the algorithm for reasoning presented by Chen et al. does not support intermediate places that belong to the output places of multiple (more than one) transitions.

3.5 Konar and Mandal's Model

Konar and Mandal [8] in a recent paper discussed the shortcomings of the algorithm for belief computation proposed by Chen et al. [4]; and proposed an alternative solution to the problem of fuzzy reasoning posed by Chen et al. The most important drawback of Chen's work is reasoning in a cyclic graph without taking care of the cycles. Konar and Mandal [8] devised two algorithms, one for reasoning in an acyclic FPN and another in a cyclic network. In an acyclic FPN, given the fuzzy beliefs of the axioms, one is interested to compute beliefs at a given goal place. To handle this problem, Konar et al. pruned a minimal sub-network, which is essential for the computation of belief at the given goal. The main theme of their algorithm is to determine the predecessors of the goal place in a given FPN, then to determine the predecessors of the predecessors and on until axioms are reached. In other words, the network is back-traced from the goal place to the axioms. The subnetwork thus traced is the minimal network needed for computing belief at the goal place. The second important aspect of the acyclic reasoning algorithm by Konar et al. is the evaluation of belief of a goal place when the belief of the axioms are known.

The proposed acyclic reasoning algorithm, however, involves a tricky search. For example, consider a network shown in Figure 3.5. Here, A={p_1, p_2, p_3} is the set of axioms. Now, we first determine the immediate reachability set of p_1, p_2 and p_3, which in this case is B= {p_4, p_5}. Now for each place p_4 and p_5 we need to check whether all of their parents possess beliefs. If yes, the belief is computed at that place, here p_4, and p_4 is then eliminated from B. Now, for each element of B (here only one, p_4) we need to determine the immediate reachable set of p_4, which in this case is {p_5}. Now, we should check whether all parents of p_5 possesses beliefs. If yes, belief of p_5 can be computed. If no (which in this case does not arise), the computation of belief at p_5 is kept on pending unless all its parents possess beliefs.

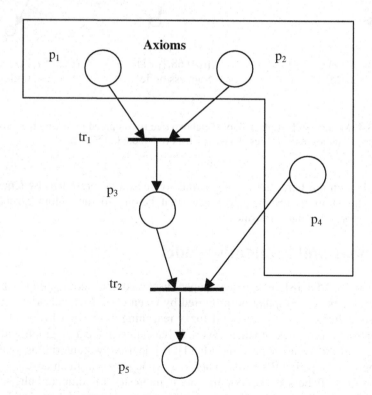

Fig. 3.5: An acyclic FPN used to illustrate the reasoning model by Konar and Mandal.

The second interesting work of Konar et al. is the formulation of belief revision model for a cyclic network. The belief revision model comprises of two major steps:

      i) Updating FTT at all transitions in parallel followed by
      ii) Updating fuzzy beliefs at all the places in parallel.

The following example illustrates the above steps.

Example 3.4: Consider the FPN shown in Figure 3.6. Here the initial beliefs of places p_1, p_2, and p_3 are given as $n_1(0) = 0.8$, $n_2(0) = 0.9$, and $n_3(0) = 0.7$. Let the thresholds associated with the transitions $th_i = 0$, say, and the certainty factors $cf_i = 1$ say for $i = 1$ to 3.

Konar et al. used the following model for belief computation.

$$n_j (t + 1) = \underset{\exists i}{\vee}\, t_i(t+1) \text{ such that place } p_j \in O(tr_i) \qquad (3.8)$$

$$= n_j (t) \text{ when } p_j \text{ has no input-arcs.}$$

and
$$t_j (t + 1) = \underset{\exists k}{\wedge}\, n_k (t) \qquad (3.9)$$

such that place $p_k \in I(tr_i)$.

    The FTT updating rule (3.9) for all i, should be applied in parallel. The belief updating rule (3.8) then needs to be invoked for all places in parallel. Table 3.2 illustrates the phenomenon.

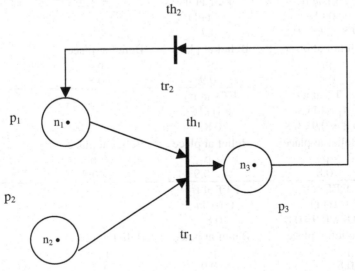

Fig. 3.6: A fuzzy Petri net used for belief revision. The dots inside the places denote the fuzzy beliefs (markings in usual Petri nets).

The process of belief revision discussed above may give rise to two possibilities. First, the beliefs at the places may reach an equilibrium value, i.e.,

$$n_i (t + 1) = n_i (t) \ \forall i. \tag{3.10}$$

Alternatively, the beliefs at the places on cycle may demonstrate stable oscillations (limit cycles). Konar et al. [8] discussed possible remedies to eliminate limit cycles from a FPN.

Table 3.2: Belief revision in Fig. 3.6

Iteration (t)	FTT at tr_1 $t_1 (t+1)$	FTT at tr_2 $t_2 (t+1)$	
1	$0.8 \wedge 0.9 = 0.8$	0.7	
	Belief at place p_1 0.7	Belief at place p_2 0.9	Belief at place p_3 0.8
2	FTT at tr_1 $t_1 (t+1)$ $0.7 \wedge 0.9 = 0.7$	FTT at tr_2 $t_2 (t+1)$ 0.8	
	Belief at place p_1 0.8	Belief at place p_2 0.9	Belief at place p_3 0.7
3	FTT at tr_1 $t_1 (t+1)$ $0.8 \wedge 0.9 = 0.8$	FTT at tr_2 $t_2 (t+1)$ 0.7	
	Belief at place p_1 0.8	Belief at place p_2 0.9	Belief at place p_3 0.8
4	FTT at tr_1 $t_1 (t+1)$ $0.8 \wedge 0.9 = 0.8$	FTT at tr_2 $t_2 (t+1)$ 0.8	
	Belief at place p_1 0.8	Belief at place p_2 0.9	Belief at place p_3 0.8
5	FTT at tr_1 $t_1 (t+1)$ $0.8 \wedge 0.9 = 0.8$	FTT at tr_2 $t_2 (t+1)$ 0.8	
	Belief at place p_1 0.8	Belief at place p_2 0.9	Belief at place p_3 0.8

3.6 Yu's Model

Yu recently extended the classical work of Murata on Petri nets for reasoning in fuzzy domain. According to him the fuzzy Petri net model can be defined as 12-tuple, given by

$$FPM = \{P, Tr, F, D, V, \pi, A_p, A_{Tr}, A_F, f, M_0, \alpha'\}$$

where

P, Tr and F (=A in previous models) have the usual definitions.
D is a non-empty finite set, called the individual set of the FPM;
V is the set of variables over D;
π is the set of dynamic predicates on D;
$A_p: P \rightarrow \pi$, Ap is a bijective mapping;
$A_{Tr}: T \rightarrow f_D$, f_D is the formula set on D;
$A_F: F \rightarrow f_s$, f_s is the symbolic sum set on D;
f: $T_r \rightarrow [0,1]$ is an association function;
M_o is a marking function, which initializes the distribution of tokens in the places of the FPM;
$\alpha': p(tok) \rightarrow [0,1]$ describes a mapping from the token "tok" resident at each place p to a membership value in [0,1]. It needs mention here that in case there exist more than one tokens at a place then for each token located at the place we have a separation.

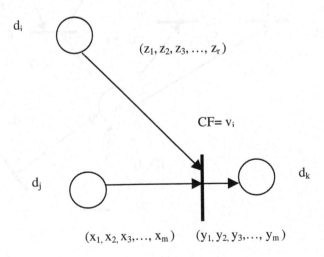

Fig. 3.7: Representation of Rule i using Petri net by Yu's method.

A transition $tr_i \in Tr$ is *enabled* if $p \in I(tr_i)$ and $\alpha'(p(tok)) \geq \lambda$, where λ is the predefined threshold. Figure 3.7 describes the representation of a typical rule using FPM.

Rule i: $d_j(x_1, x_2, x_3, ..., x_m)$ & $d_l(z_1, z_2, z_3, ..., z_r) \rightarrow d_k(y_1, y_2, y_3,...,y_m)$ (CF= v_i)

It is important to note that in Yu's original paper .tr_i (a dot prior to tr_i) was used to denote the input places of tr_i. However, to keep the uniformity in notation we prefer $I(tr_i)$ to denote it.

For the purpose of fuzzy reasoning, Yu defined an incidence matrix following Murata [10]. We illustrate the construction of the incidence matrix with reference to a given fuzzy Petri net (Fig. 3.8).

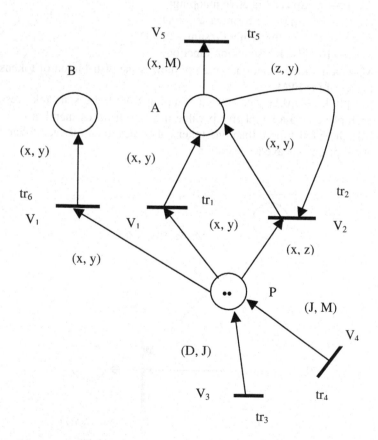

Fig. 3.8: An example FPM, where V_i denotes the CF of the transitions, the dots within a place denote the number of markings.

C=	A	B	P
tr_1	$(x, y)v_1$	0	$-(x, y)$
tr_2	$(x, y)v_2 - (z, y)$	0	$-(x, z)$
tr_3	0	0	$(D, J)v_3$
tr_4	0	0	$(J, M)v_4$
tr_5	$-(x, M)$	0	0
tr_6	0	$(x, y)v_1$	$-(x, y)$

The above incidence matrix has been constructed with the following idea. When the transition tr_i has an input place A, the negative arc function is contributed in the term $C(tr_i, A)$. Further if a place A is an output place of tr_i then a positive arc function multiplied by the CF of tr_i is contributed to $C(tr_i, A)$. Thus $C(tr_2, A)$ in the Figure 3.8 is given by

$$C(tr_2, A) = (x, y)v_2 - (z, y).$$

For reasoning with FPM, Yu defined a fuzzy certainty factor (CF) matrix as well. It is defined as

$$F_n = \begin{array}{c} tr_1 \\ tr_2 \\ \\ \\ tr_n \end{array} \begin{bmatrix} CF_1 & 0 & 0 \ldots & 0 \\ 0 & CF_2 & 0 \ldots & 0 \\ \cdot\cdot & \cdot & \cdots & \cdot \\ \cdot\cdot & \cdot & \cdots & \cdot \\ 0 & 0 & 0 & CF_n \end{bmatrix}$$

where $CF_i = f(tr_i)$ is the belief strength of some fuzzy production rule ($i = 1, 2, \ldots, n$). We now outline the major steps of Yu's fuzzy reasoning algorithm [19].

Procedure fuzzy-reasoning
Begin

1. Assign $A \leftarrow C_{n \times m}$ and $D \leftarrow F_n$.
   For i:=1 to n
   If $CF_i < \lambda$ **Then** each element in the i-th row of [D | A] should be replaced by zero;
   End For;
2. **For** i:= 1 to m **do**
   Begin
   a) Append to the matrix [D | A] every row resulting as a non-negative linear combination of unifiable row pairs from [D | A] that annual the i-th column of A, and attach the corresponding

unifier and v, the belief strength of the unifier, to each row of the pair.

b) If $v < \lambda$ then each element of the appended row should be replaced by zero.

c) Eliminate from [D | A] the rows in which the i-th column of A is non-null.

End For

End.

3.7 Chen's Model for Backward Reasoning

Given the belief of the axioms in a FPN, the forward reasoning problem is concerned with the computation of the belief of the given goal by moving stepwise forward in the given network. Thus, in the process of computing beliefs of the goal proposition, beliefs of other propositions not lying on the minimal subnet leading to the goal may also be computed. Computation of beliefs of undesired places, however, may be avoided by backward reasoning on the FPN.

Chen in a recent issue of IEEE SMC [3] provides a novel approach to backward reasoning on a FPN. He considered an equivalent AND-OR tree of the subnet terminating at the given goal position. Thus, the tree includes all possible connectivity of the FPN starting from the axioms and terminating at the goal. This undoubtedly is the minimal data structure needed for computing the belief at the goal.

In his proposed AND-OR tree Chen labeled each node by a 3-tuple given by

(Place name, {IBIS of the place}, belief).

where

place name refers to the place corresponding to the given node in the AND-OR tree,

IBIS stands immediate backward incidence set to be defined shortly,

and belief corresponds to the belief to be computed (if not known) or assigned for axioms.

Definition 3.4: *Let t_a be a transition and let p_i, p_j and p_k be three places. If $p_i \in I(t_a)$ and $p_k \in O(t_a)$, then p_i is called an immediately backward incidence place of p_k. The set of places, which contains immediate backward places of p_k, is called the **immediate backward incidence set** of p_k and is denoted by IBIS (p_k).*

In the construction of the AND-OR tree, the following points need special mention:

1) The goal place should correspond to the root of the AND-OR tree.

2) Each node in the tree should be expanded after considering its IBIS recorded in the second field of the node descriptor.

3) If the IBIS of a place p_i satisfy the following two conditions:

$p_i \in O(tr_j)$ and
$p_j, p_k, \ldots\ldots, p_l \in I(tr_j)$

then $\{p_j, p_k, \ldots\ldots, p_l\}$ should be saved within braces in the IBIS field of the tree descriptor and an AND arc must be used to label the emerging arcs from p_i to $p_j, p_k, \ldots, p_l$ is expanded.

4) When the axioms are reached in the process of expanding a node in the tree, their IBIS being empty set, the construction process of the tree ceases.

5) The certainty factor of the transition between each pair of places is labeled against the arcs in the AND-OR tree.

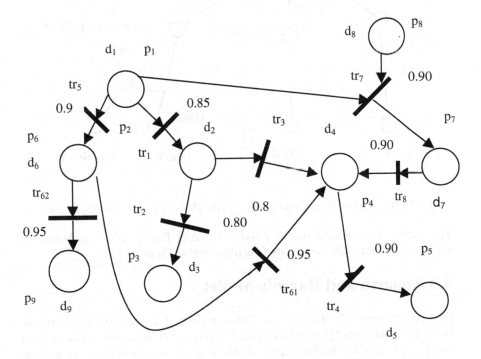

Fig. 3.9: A fuzzy Petri net used to illustrate AND-OR tree construction.

After the construction of the tree is over, the computation begins with the submission of beliefs at the leaves. If parent of a node has no children connected by AND arc, then the belief of the child multiplied by CF of the arc is placed in the belief field of the parent. In case there is an AND arc among the children of the parent of a given place, then their beliefs are ANDed and the result is multiplied by the common CF of the AND-arcs. The result, thus obtained is placed at the appropriate field of the parent. When the parent of a node has no children connected by AND arcs, then the maximum of the belief multiplied by CF of each node is transferred to the parent.

Example 3.5: Consider the FPN shown in Figure 3.9. The AND-OR tree corresponding to Figure 3.9, where the symbol "_" denotes that the truth value of the proposition d_j is unknown, is presented below:

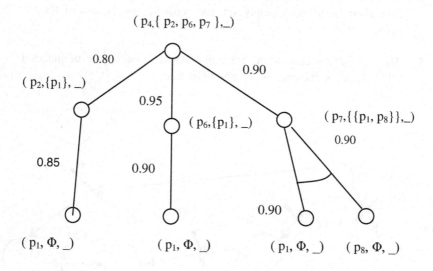

Fig. 3.10: The AND-OR tree corresponding to the FPN of Figure 3.9 with goal = p_4.

The computation process on the AND-OR tree of Figure 3.10 with beliefs of axioms as submitted in the tree itself is illustrated in Figure 3.11.

3.8 Bugarin and Barro's Model

In a recent paper [2] Bugarin and Barro, presented an alternative method of fuzzy reasoning satisfying the traditional approach for generating inferences in fuzzy logic. Before outlining their work, a brief introduction to the general scheme for fuzzy reasoning needs to be introduced. Consider for instance following r-th fuzzy production rule:

R^r: If x_1 is A_1^r and ...and x_{Mr} is A_{Mr}^r

Then x_{Mr+1} is B_1^r and...and x_{Mr+Nr} is B_{Nr}^r (τ^r)

where x_{mr}, x_{nr} for $m_r = 1, 2, \ldots, M_r$ and $n_r = M_r + 1, M_r + 2, \ldots, M_r + N_r$ are linguistic variables.

τ^r: $[0,1] \rightarrow [0,1]$

denotes the degree of fuzzy quantifier that behaves like the certainty factor of a rule based system.

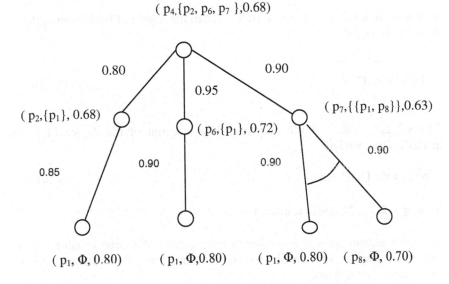

Fig. 3.11: Computation of beliefs from leafs to the root of the AND-OR tree.

Let $A_{mr}^r = \{a_{mr.i}^r\}$, for $r = 1, 2, \ldots, R$, $m_r = 1, \ldots, M_r$, $i = 1, 2, \ldots, I$ denotes the sampled distribution associated with the linguistic variable of the same name in proposition "x_{mr} is A_{mr}^r."

Similarly, $B_{nr}^r = \{b_{nr.i}^r\}$ for $r = 1, 2, \ldots, R$, $n_r = 1, \ldots, N_r$, $i = 1, 2, \ldots, I$ denotes the sampled distribution associated with the linguistic variable of the same name in proposition "x_{Mr+nr} is B_{nr}^r."

Let $AO_{mr}^r = \{ao_{mr.i}\}$ denotes the observed sampled distribution for a variable X_{mr} and $BO_{nr}^r = \{bo_{nr.i}^r\}$, the observed sampled distribution for a variable x_{Mr+nr}, where r, m_r, n_r, i are as defined earlier.

The degree of fulfillment DOF^r_{mr} between the observed distribution AO^r_{mr} and the measured distribution A^r_{mr} is given by

$$DOF^r_{mr} = \overset{I}{\underset{i=1}{\vee}} \; (a^r_{mr.i} \wedge ao^r_{mr.i}).$$

where $\wedge$ and $\vee$ denote the usual minimum and maximum operators.

The inferred membership distribution BO^r_{nr} is given by its component

$$bo_{nr.i} = DOF^r \wedge b^r_{nr.i} , \qquad\qquad (3.12)$$

for integer $n_r = 1, 2, \ldots , I$ where DOF^r, called the degree of fulfillment of the r-th rule is given by

$$DOF^r = \overset{Mr}{\underset{mr=1}{\wedge}} DOF^r_{mr}, \qquad\qquad (3.13)$$

for $r = 1, 2, \ldots , R.$ Further, taking into account of the degree of fuzzy quantification, we have

$$bo_{nr.i} = \tau^r (DOF^r \wedge b^r_{nr.i}) \qquad\qquad (3.14)$$

for $n_r = 1, 2, \ldots, N_r$ and $i = 1, 2, \ldots, I.$

If the consequent parts of a number of rules contain the same variable x (say) and R_c is the set of rules in which x occurs, then this situation can be characterized as follows:

$$\forall \, R^r \in R_c , \exists \, n_r \in \{1,2,\ldots,N_r\} / \, x^r_{Mr+nr} = x \qquad\qquad (3.15)$$

Then, the aggregation of the sampled distributions can be described by

$$\underset{R^r \in R_c}{\vee} bo^r_{nr.i} ,$$

$i = 1,2, \ldots , I, \; n_r \in \{1,2, \ldots , N_r\}$, where the aggregation operator is maximum.

The following definitions are in order for the subsequent discussion on this work.

Definition 3.5: *If $p_i \in O(tr_j)$ for $j=1, 2,..., m$, say, then each transition tr_j, $\exists j$ is called adjacent to the other, and the set $\{tr_j \mid p_i \in O(tr_j), \exists j\}$ is called the adjacent transitions set with respect to the place p_i.*

Definition 3.6: *The fulfillment function g at a place is given by*

$$g(p) = DOF(\alpha(p)),\tag{3.16}$$

where g: set of places $\rightarrow [0,1]$ and α: set of places $P = \{p_k\} \rightarrow$ set of propositions $PR = \{pr_k\}$ given by $\alpha(p_k) = pr_k$.

For computation of $g(p_i)$ where $p_i \in O(tr_j)$ and $p_k \in I(tr_j)$, we define:

$$g(p_i) = \bigwedge_{p_k \in I(tr_j)} g(p_k)\tag{3.17}$$

Definition 3.7: *A transition tr_j is enabled (active) and subsequently fired if all of its input places have a token. On firing of a transition, the markings of the tokens at its input and output places are modified by the following rule:*

$$\begin{aligned} M'(p_i) &= 0 \text{ if } p_i \in I(tr_j)\\ &= 1 \text{ if } p_i \in O(tr_j)\\ &= M(p_i), \text{ otherwise} \end{aligned}\tag{3.18}$$

where M and M' are the markings of the tokens at place p_i prior or after firing of tr_j.

Consider Figure 3.12 where p_i for $i=1$ to 3 are the places and $\alpha(p_i) \forall i$ are the values of the bijective mapping α at the respective places. Let $M(p_1) = M(p_2) = 1$ and $M(p_3) = 0$. So, by rule of enabling the transitions tr_1 is enabled (active) and hence fires.

Now, after firing

$$M'(p_1) = M'(p_2) = 0 \text{ and } M'(p_3) = 1.$$

Now, given $g(p_1)=0.8$ and $g(p_2)=0.6$, we can evaluate following expression (3.17),

$$g(p_3) = g(p_1) \wedge g(p_2)$$

$$= 0.6.$$

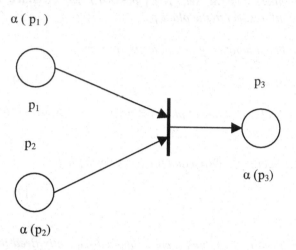

$\alpha (p_1)$

p_1

p_2

$\alpha (p_2)$

p_3

$\alpha (p_3)$

Fig. 3.12: Computation of the fulfillment function g on an illustrative Petri net.

Bugarin and Barro [2] stressed the need of a special consideration about chaining. The following definitions are in order to explain the chaining consideration.

Definition 3.8: *Two rules are* **chained** *if the consequent variables of one rule appear as the antecedent variable in another rule.*

The following pair of rules

> IF x is A then y is B
> IF y is B$'$ then z is C

for instance are chained as they use the common variable y.

The computation of $g(p_i)$ for $p_i \in O(tr_j) \exists j$, where the transition tr_j corresponds to the chaining condition, can be done by

$$g(p_i) = \vee[\tau^{rk} (g(p_k))o^{rk} \mu_{p(ki)}], \forall p_i \in O (tr_j) \qquad (3.19)$$
$$p_k \in I(tr_j)$$

where

$\tau^{rk} = f(tr_{rk}), p_k \in O (tr_{rk})$;

$o^{rk} = \wedge$, when τ^{rk} is monotonically increasing

$\quad = \vee$, when τ^{rk} is monotonically decreasing,

and $\mu_{p(ki)}$ denotes the certainty factor of the rule.

Typically, τ^{rk} can take the following forms:

$$\tau^{rk}(x) = x^2$$
$$= \sqrt{x}$$
$$= x$$

depending on the fuzzy quantifier for x to be VERY GOOD, VERY POOR, or NORMAL, respectively.

Example 3.6: Consider the following chained rules

   Rule R_1: If x_1 is A_1 and y_1 is A_2, then z_1 is B_1
   Rule R_2: If z_1 is A_3 then y_2 is B_2

where the variable z_1 is common to the consequent part of rule R_1 and antecedent part of Rule R_3. The chaining condition here may be represented by the dotted transitions in Figure 3.13.

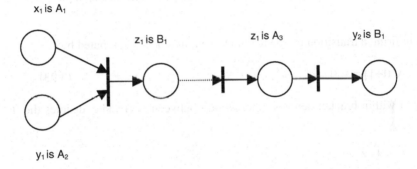

Fig. 3.13: Illustration chaining of 2 rules R_1 and R_2 having a common intermediate variable z_1.

3.9 Pedrycz and Gomide's Learning Model

Pedrycz and Gomide [14] provides a novel scheme for machine learning using fuzzy Petri nets. Their formulation is based on the usual definition of t-and s-norms. A transition tr_1 fires if its degree of firing

$$Z = \underset{i=1}{\overset{n}{T}} ((r_i \rightarrow x_i)s \, w_i) \qquad (3.20)$$

exceeds its threshold th_i.

   In the last expression " $\rightarrow$ " denotes a fuzzy implication, defined as

$$a \rightarrow b = \sup\{c \in [0,1] \,|\, a \text{ t } c \leq b \} \tag{3.21}$$

Thus, the firing condition describes a conjunctive type of firing in which all the input places of tr_i having markings (tokens/beliefs) $x_1, x_2, \ldots, x_n$ are sought as AND-wise contributions to this process. Subsequently, r_i are used to characterize the threshold level modulating the strength of the firing coming from i-th input place. It is needless to mention that

$$r_i \rightarrow x_i \equiv \neg \, r_i \vee x_i$$
$$\equiv x_i \text{ when } r_i = 1. \tag{3.22}$$

Thus, for $r_i = 1$, a propositional rule $r_i \rightarrow x_i$ is equivalent to x_i. In a fuzzy sense, if $r_i \leq x_i$ then $r_i \rightarrow x_i$ will have a truth value 1.

The weight factor w_i is used to discriminate between the input places with respect to their contributions to their overall level of firing. It is clear from (3.20) that smaller the value of w_i, the larger the influence of this input place on z.

On firing a transition tr_i, the token at its input place is computed by

$$x_i(t+1) = x_i(t) \text{ t } z' \tag{3.23}$$

where t within bracket denotes time, while t between $x_i(t)$ and z' denotes the 't' norm.

$$z' = 1 - z$$

Further on firing of the transition tr_i, the token at its output place y_j is computed by

$$y_j(t+1) = y_j(t) \text{ s } z \tag{3.24}$$

Let us denote

$$\underset{j \neq i}{\overset{n}{T}} [(r_j \rightarrow x_i) s w_j] \text{ by } z^+ \tag{3.25}$$

and

$$[(r_i \rightarrow x_i') \, s w_i] \text{ by } z^-, \tag{3.26}$$

then $z = z^+ \text{ t } z^-$ $\tag{3.27}$

includes both the excitatory component z^+ and the inhibitory component z^- respectively. To understand the inhibitory mechanism, we set $w_i = 0$ and $r_i = 1$, then $z^- = x_i$, hence $x_i = 1$ completely prohibits the transition from firing.

To understand how FPN can be used for pattern recognition, let us consider a single transition associated with several input places x and an output place y (Fig. 3.14). The training set consists of the levels of the input and the output places say $x(k)$ (input places) and $y(k)$ (output places) for $k = 1$ to N.

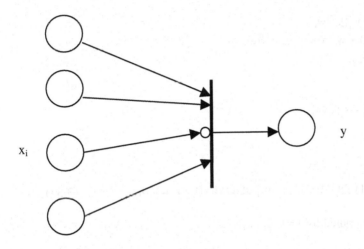

Fig. 3.14: Illustrating inhibition in a fuzzy Petri net.

Let the performance index be

$$Q = (1/2)\sum_{k=1}^{N-1} [y(k+1) - DOM(x(k); r, w)]^2 \qquad (3.28)$$

We now compute:

$$\partial Q/\partial w_i = -[y(k+1) - DOM(x(k); r, w)]\, \partial DOM(x(k); r, w)/\partial w_i \qquad (3.29)$$

where

$$\partial DOM(x(k); r, w)/\partial w_i$$

$$= \partial(y(k)OR\ z(k))/\partial w_i$$

$$= \{\partial(y(k) \text{ OR } z(k))/\partial z(k)\}.\{\partial z(k)/\partial w_i\} \qquad (3.30)$$

Since $z(k)=DOM(x(k); r, w)$

$$\partial z(k)/\partial w_i = \partial \left[\begin{array}{c} n \\ T \\ j=1 \end{array} [(r_j \rightarrow x_j)sw_j]\right]/\partial w_i \qquad (3.31)$$

with $r_j \rightarrow x_j \equiv \min(1, x_j/r_j)$, x_j, $r_j \in [0,1]$ we find (3.31) reduces to

$$\begin{aligned}
\partial z(k)/\partial w_i \\
= &\partial [A(r_j \rightarrow x_j)sw_j]]/\partial w_i \\
= &A \, \partial [(r_j \rightarrow x_j) + w_i - w_i(r_i \rightarrow x_i)]/\partial w_i \\
= &A [1 - (r_i \rightarrow x_i)] \qquad (3.32)
\end{aligned}$$

$$\begin{array}{c} n \\ \text{where } A= T \; [(r_j \rightarrow x_j)s \; w_j] \\ j \neq i \end{array}$$

Similarly,

$$\partial Q/\partial r_i = -[y(k+1) - DOM(x(k); r, w)] \, \partial DOM(x(k); r, w)/\partial r_i \qquad (3.33)$$

can easily be computed for $i = 1, 2, \ldots, n$.

A steepest descent learning algorithm then may be employed to search the optimal w_i and r_i for which Q is minimized.

3.10 Construction of Reduction Rules Using FPN

Garg et al. [5] provides a new method for consistency analysis in a fuzzy Petri net via a set of reduction rules. They derived the following three reduction rules and designed an algorithm for fuzzy consistency of knowledge in a fuzzy Petri net.

Reduction Rules

Rule 1: For a $tr_j \in Tr$, if $I(tr_j) = \phi$, $O(tr_j) = p_i$ and $(tr_j \times p_i) \in A$ (where A is the set of arcs) and FTT $t_j = \alpha$, then $\forall tr_k \in O(p_i)$, delete the arc $(p_i \times tr_k)$ and replace the FTT t_k by $\min(\alpha, t_k)$. The equivalence of FPN is presented in Figure 3.15

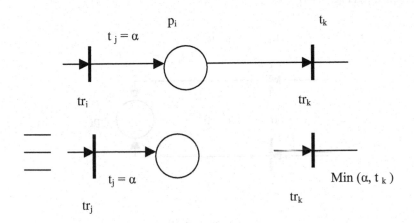

Fig. 3.15: Illustrating equivalence of 2 fuzzy Petri nets by Rule 1.

Rule 2: For $tr_j \in Tr$, if $O(tr_j) = \phi$, $I(tr_j) = p_i$ and $(tr_k \times p_i) \in A$, then for all tr_k such that $p_i \in O(tr_k)$ having FTT $t_k = \beta$, delete the arc $(tr_k \times p_i) \in A$ and replace t_k by $\min(\beta, t_j)$. The equivalence is illustrated in Figure 3.16.

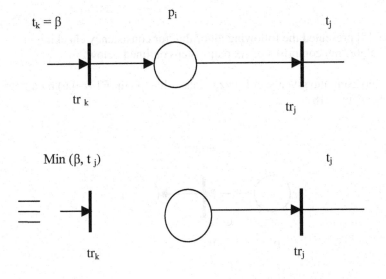

Fig. 3.16: Illustrating equivalence of two fuzzy Petri nets by Rule 2.

Rule 3: If $I(tr_i) = I(tr_j) = \phi$ and $O(tr_i) = O(tr_j) = p_k$ and $(tr_i \times p_k) \wedge (tr_j \times p_k) \in A$, $t_i= \alpha$ and $t_j= \beta$ then merge the transitions tr_i and tr_j to give one transition tr_{ij} with FTT= $max(\alpha, \beta)$ (Fig. 3.17).

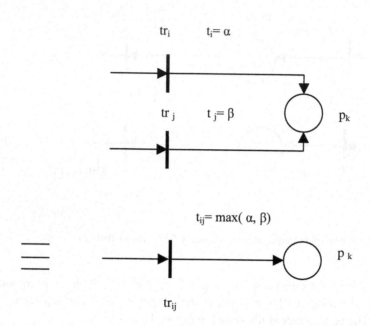

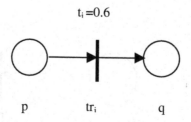

Fig. 3.17: Illustrating equivalence of two FPNs by Rule 3.

Garg et al. [5] presented the following algorithm for consistency checking in a FPN. The algorithm comprises of six major steps outlined below:

1. Represent the given set of fuzzy clauses like $p \rightarrow q$ (FTT=0.6) by a Petri net (Fig. 3.18).

$$t_i = 0.6$$

p tr_i q

Fig. 3.18: Representation of $p \rightarrow q$ (FTT=0.6) by a fuzzy Petri net.

2. Replace negative arcs with their equivalent positive arcs with reversal of directions. Since $\neg p \rightarrow q \equiv p \vee q$, the justification of the above example is meaningful. Figure 3.19 illustrates the step.

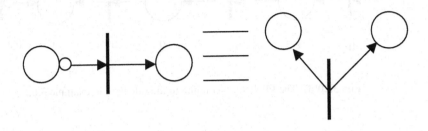

Fig. 3.19: Illustrating replacement of negative arcs by reversal of direction.

3. If a positive source transition tr_i is connected to a place, then apply reduction Rule 1.

4. If a positive sink transition tr_j is connected to a place, then apply Rule 3.

5. If two positive source or positive sink transitions are connected to one place then apply reduction Rule 3.

6. If a "null transition" i.e., an isolated transition, is obtained, then stop and declare that the given set of fuzzy clauses is inconsistent.

Example 3.7: Consider the following fuzzy rules:

$$p \rightarrow q \quad (\text{FTT}=0.9)$$
$$q \rightarrow r \quad (\text{FTT}=0.8)$$
$$p \wedge \neg q \quad (\text{FTT}=0.6)$$

The corresponding FPN is shown in Figure 3.20. Figure 3.20 (a)-(e) show the effect of applying the reduction rules in each step. Since the figures are self-explanatory, we do not provide any explanation here. In Figure 3.20(e), a transition tr_2 is isolated. So, the algorithm terminates, claiming that the rules are inconsistent.

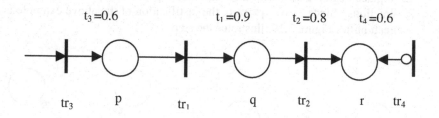

Fig. 3.20(a): The FPN corresponding to the rule base in example 3.6.

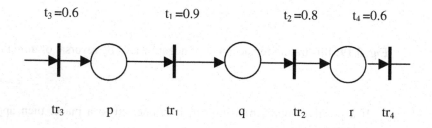

Fig. 3.20(b): Reversal of tr_4.

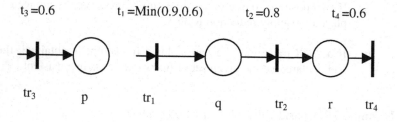

Fig. 3.20(c): Applying Rule 1 at source transition tr_3.

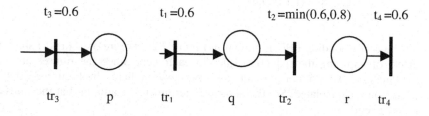

Fig. 3.20(d): Applying Rule 2 at sink transition tr_4.

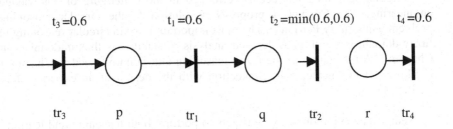

$t_3 = 0.6$ $t_1 = 0.6$ $t_2 = \min(0.6, 0.6)$ $t_4 = 0.6$

tr_3 $\quad$ p $\qquad$ tr_1 $\qquad$ q $\qquad$ tr_2 $\quad$ r $\quad$ tr_4

Fig. 3.20(e): Applying Rule 1 at source transition tr_1.

3.11 Scope of Extension of Fuzzy Reasoning on Petri Nets

Fuzzy reasoning so far introduced in this chapter deals with forward and backward reasoning only. In forward reasoning, membership distributions are assigned at the initiating places (i.e., places having no input arcs) only, and one wants to evaluate the membership of a proposition mapped at any other place of the network. Forward reasoning models propagate the memberships only in the forward direction, and search or computation is not needed to take a look at the predecessors from their respective directed successors in the network. The backward reasoning model that also computes the membership of any proposition mapped at a place of the network from the given memberships of the initiating places, however, works in a different manner. It first back-traces the network from the place where membership computation is needed and identifies that part of the network that is only sufficient for computing the membership at the pre-defined goal place. A forward pass in the identified network thus computes the membership at the desired place.

Besides forward and backward reasoning, there exist quite a good number of other reasoning problems where fuzzy Petri nets have not yet been explored. These include abductive reasoning, reasoning under bi-directional iff (if and only if) type relations, and duality. Given two atomic clauses [16], an abductive reasoning problem can be formalized as follows. If p causes q and the fact q' is given, we can infer the cause p', which is close enough to p. It may be noted that abductive reasoning does not hold in classical logic, as there may be other causes of q, so p or p' always need not be the correct answer. But in systems where there are no other causes much different from p, p or its fuzzy extensions p' may be presumed as the cause of q', when q' is said to have happened.

This book attempts to evaluate p' from the known membership of q. The fuzzy inverse [12], introduced in the latter part of the book, will be useful for the above computation.

Circular reasoning has been overlooked in most literature on knowledge engineering. The Chapter 4 proposes a solution to the circular reasoning problem using fuzzy Petri nets. The most important issue in circular reasoning in fuzzy domain is stability. A detailed analysis of stability is thus undertaken in Chapter 4. A realization of the fuzzy reasoning algorithm outlined in Chapter 4 is illustrated in Chapter 5 in connection with the design of an expert system shell.

Neural realization of fuzzy Petri nets to perform both reasoning and learning on the same framework is of great concern to the machine intelligence community. Chapters 9 and 10 propose learning algorithms for a Petri net framework and demonstrate the applications of the algorithms in knowledge acquisition/refinement and pattern classification problems. The advantage of integrating reasoning and learning on the common framework is to simulate humanlike ability of reasoning while learning and learning while reasoning.

Bi-directional reasoning introduced in the latter part of the book, is capable of computing the membership distribution of the propositions mapped at the input (output) places of the transition from the memberships of propositions mapped at the output (input) places. This holds for all transitions in the FPN. Bi-directional reasoning thus may be used for propagating memberships in either direction (forward/backward) of the transition in an acyclic FPN. One interesting characteristic of bi-directional reasoning that ensures restoration of memberships at the places after n-forward followed by n-backward passes is called *reciprocity*. The condition of reciprocity derived in a later chapter, ensures that the relational matrices associated with its transitions solely determine the topology of a reciprocal net.

The concept of duality to be addressed in this book utilizes the modus tollens property of predicate logic. Duality, when realized on FPNs, provides the users the benefit of computing the memberships of negated predicates from known memberships of the negated predicates embedded in the network. Because fuzzy complementation does not necessarily mean the one's complement of the memberships, the above computation will find massive applications, especially in criminal investigation when the suspects claim their alibi to prove them innocent.

3.12 Summary

This chapter presented principles of fuzzy reasoning by various models of fuzzy Petri nets. Broadly speaking, the models are of two basic types: forward reasoning models and backward reasoning models. The forward reasoning models are usually employed to determine the membership distribution of the concluding propositions in a fuzzy Petri net. The backward reasoning model, on the other hand, first identifies the selected axioms in the Petri net, which only can influence the membership of the goal proposition. This is implemented by back-tracing the network from the available goal (concluding) proposition until the axioms are reached. The places thus traced in the FPN are sufficient to determine the membership of the given goal proposition. The chapter also introduces the possible scheme of machine learning on a fuzzy Petri net. Because of the inherent feature of approximate reasoning, FPNs used for machine learning have a promising feature in fuzzy pattern recognition from noisy training instances.

Exercises

1. Given the rule: If height is TALL then speed is HIGH, and the membership distributions:

   $$\mu_{TALL}(height)= \{5'/0.2, 6'/0.7, 7'/0.9\}$$

   and $\mu_{HIGH}(speed)= \{8m/s/0.4, 9m/s/0.8, 10\ m/s/0.9\}$.

   a) Determine the relation R(height, speed) for the given rule using Mamdani's min implication function.
   b) Given μ_{TALL}' (height)= $\{5'/0.3, 6'/0.4, 7'/0.5\}$, evaluate μ_{HIGH}' (speed), where TALL ' $\approx$ TALL and HIGH' $\approx$ HIGH.

   [**Hints:** a) The relational matrix is given by

   $$
   \begin{array}{cccc}
    & 8 & 9 & 10 \quad \text{speed (in m/s)} \\
   \end{array}
   $$

   $$
   R(height, speed)=
   \begin{array}{c}
   \text{height} \\
   5' \\
   6' \\
   7'
   \end{array}
   \left(
   \begin{array}{ccc}
   0.2 & 0.2 & 0.2 \\
   0.4 & 0.7 & 0.7 \\
   0.4 & 0.8 & 0.9
   \end{array}
   \right).
   $$

   b) $\mu_{HIGH}'(speed)= [0.3 \quad 0.4 \quad 0.5]$ o R (height, speed)

   $$= [\ 8\ m/s/0.4,\ 9m/s/0.5,\ 10m/s/0.5],$$

   where o denotes a max-min composition operator.]

2. Consider the following two fuzzy rules:

   Rule 1: If height is TALL Then speed is HIGH
   Rule 2: If height is MODERATE Then speed is MODERATE.

   R1 and R2 below denote the relational matrices corresponding to the rules 1 and 2, respectively.

$$R_1(\text{height, speed})= \begin{array}{c} \\ \text{height} \\ 5' \\ 6' \\ 7' \end{array} \begin{array}{ccc} 8 & 9 & 10 \\ \left(\begin{array}{ccc} 0.2 & 0.3 & 0.4 \\ 0.6 & 0.8 & 0.9 \\ 0.4 & 0.5 & 0.7 \end{array} \right) \end{array} \quad \text{speed (in m/s)}$$

$$R_2(\text{height, speed})= \begin{array}{c} \\ \text{height} \\ 5' \\ 6' \\ 7' \end{array} \begin{array}{ccc} 8 & 9 & 10 \\ \left(\begin{array}{ccc} 0.2 & 0.6 & 0.4 \\ 0.5 & 0.7 & 0.3 \\ 0.8 & 0.9 & 0.2 \end{array} \right) \end{array} \quad \text{speed (in m/s)}$$

Given $\mu_{\text{MORE-OR-LESS-TALL}}(\text{height})= \{5'/0.2,\ 6'/0.4,\ 7'/0.8\}$, determine $\mu_{\text{MORE-OR-LESS-HIGH}}(\text{speed})$.

[**Hints:** $\mu_{\text{MORE-OR-LESS-HIGH}}(\text{speed})= \mu_{\text{MORE-OR-LESS-TALL}}(\text{height})\ o\ (R_1 \vee R_2)$

$$= \{8\text{m/s}/\ 0.8,\ 9\text{m/s}/\ 0.8,\ 10\text{m/s}/\ 0.7\}.]$$

3. Evaluate the beliefs at the places and FTTs at the transitions until the belief at place p_4 becomes stable. Also mention the order of firing of the transitions. Given $n_1(0) = 0.5$, $n_2(0) = 0.6$, $n_3(0) = n_4(0) = 0$ and $n_5(0) = 0.7$, and $t_1(0) = t_2(0) = t_3(0) = t_4(0) = t_5(0) = 1$. Let the thresholds for all transitions be 0.2.

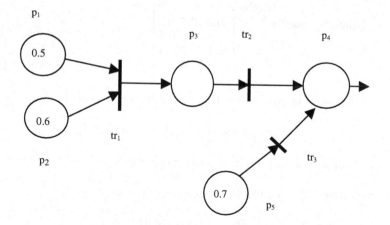

Fig. 3.21: A FPN used to clarify problem 3.

[**Hints:** Order of firing: Here transition tr_1 and tr_3 can fire concurrently, tr_2 fires after firing of tr_1. We compute the following in order.

$t_1(1) = t_1(0) \wedge (n_1(0) \wedge n_2(0))$ as $n_1(0) \wedge n_2(0) > th_1 = 0.2.$
$\qquad = 0.5.$
$t_2(1) = t_2(0) \wedge n_3(0) = 0.$
$t_3(1) = t_3(0) \wedge n_5(0) = 0.7.$

$n_1(1) = n_1(0) = 0.5$ as p_1 is an axiom.
$n_2(1) = n_2(0) = 0.5$ as p_2 is an axiom.
$n_3(1) = t_1(1) \vee n_3(0) = 0.5.$
$n_4(1) = (t_2(1) \vee t_3(1)) \vee n_4(0) = 0.7.$
$n_5(1) = n_5(0) = 0.7$ as p_5 is an axiom.

This completes the first pass of updating in the network. Similar steps are repeated in subsequent passes. The passes continue until beliefs of the places in a iteration become steady.]

4. Show that beliefs at a place by Looney's model can never decrease.

   [**Hints:** Since by Looney's model $n_i(t + 1) = $ Max $(n_i(t), \ldots)$, therefore, $n_i(t+1) \geq n_i(t)$. Hence, the statement.]

5. Show that FTT values in Looney's model never decrease.

   [**Hints:** By Looney's model $t_j(t +1) = $ Min $[t_j(t), \ldots]$, consequently, $t_j(t +1) \leq t_j(t)$. Thus, the result follows.]

6. Assuming CF= 0.9 for all the rules embedded in the FPN of Fig. 3.21, determine the fuzzy beliefs of all the places by the algorithm proposed by Chen et al.

7. The membership distributions for the necessary fuzzy linguistic variables in the rules 1 and 2 are presented below.

   Rule 1: If x is A_1 and y is B_1 then z is C_1.
   Rule 2: If x is A_2 and y is B_2 then z is C_2.

   $\mu_{A1}(x) = \{10/0.2, 20/0.3, 30/0.4\},$
   $\mu_{B1}(y) = \{50/0.6, 60/0.8, 70/0.9\},$
   $\mu_{A2}(x) = \{10/0.8, 20/0.9, 30/0.6\},$
   $\mu_{B2}(y) = \{50/0.2, 60/0.8, 70/0.7\}.$

Using Bugarin and Barro's method, determine DOFs of the rules when the observed distributions are given by

   $\mu_A'(x) = \{10/0.2, 20/0.6, 30/0.7\}$
   and $\quad \mu_B'(y) = \{50/0.8, 60/0.9, 70/0.7\}.$

[**Hints:** Computation of DOF for rule 1 is given below.

$\quad$ Let u = $\underset{\forall x}{\vee}(\mu_{A1}(x) \wedge \mu_A'(x))$

$\qquad = \vee \{10/ (0.2 \wedge 0.2), 20/ (0.3 \wedge 0.6), 30/ (0.4 \wedge 0.7)\}$
$\qquad x \in [10, 20, 30]$

$\qquad = \vee \{10/ (0.2), 20/ (0.3), 30/ (0.4)\}$
$\qquad x \in [10, 20, 30]$

$\qquad = 0.4,$

$\quad$ Let v = $\underset{\forall y}{\vee} (\mu_{B1}(y) \wedge \mu_B'(y))$

$\qquad = 0.8.$

Consequently, DOF of rule 1 = u $\wedge$ v
$\qquad\qquad\qquad\qquad = 0.4 \wedge 0.8$
$\qquad\qquad\qquad\qquad = 0.4.$

Similarly, compute the DOF of rule 2.]

8. Given the membership distribution of $\mu_{C1}(z)$ and $\mu_{C2}(z)$ as follows in problem 7.

$\quad \mu_{C1}(z) = \{10/0.4, 20/0.3, 30/0.7\}$

$\quad \mu_{C2}(z) = \{10/0.4, 20/0.4, 30/0.6\},$

compute $\mu_{C1}'(z)$ and $\mu_{C2}'(z)$.

[**Hints:** $\mu_{C1}'(z) = (\text{DOF of rule 1}) \wedge \mu_{C1}(z)$

$\qquad\qquad\qquad = 0.4 \wedge \{10/ 0.4, 20/ 0.3, 30/ 0.7\}$

$\qquad\qquad\qquad = \{10/0.4, 20/ 0.3, 30/ 0.4\}.$

Similarly, $\mu_{C2}'(z)$.]

References

[1] Bender, E. A., *Mathematical Methods to Artificial Intelligence*, IEEE Computer Society Press, Los Alamitos,1996.

[2] Bugarin, A. J. and Barro, S., "Fuzzy reasoning supported by Petri nets," *IEEE Trans. on Fuzzy Systems*, vol. 2, pp.135-145, May 1994.

[3] Chen, S. M. "Fuzzy backward reasoning using fuzzy Petri nets", *IEEE Trans. on Systems, Man and Cybernetics-Part B: Cybernetics*, vol. 30, no. 6, pp. 846-855, Dec.n2000.

[4] Chen, S. M., Ke, J. S. and Chang, J. F., "Knowledge representation using fuzzy Petri nets," *IEEE Trans. on Knowledge and Data Engineering*, vol. 2, no. 3, pp. 311-319, Sept. 1990.

[5] Garg, M. L., Ahson, S. L. and Gupta, P.V., "A fuzzy Petri net for knowledge representation and reasoning," *Information Processing Letters*, vol. 39, pp. 165-171, Aug. 1991.

[6] Kemney, G. J. and Snell, J. L., *Finite Markov Chains*, D. Van Nostrand Co., Inc., New York, 1960.

[7] Konar, A., *Uncertainty Management in Expert Systems Using Fuzzy Petri Nets*, Ph.D. Thesis, Jadavpur University, 1994.

[8] Konar, A. and Mandal, A. K., "Uncertainty management in expert systems using fuzzy Petri nets," *IEEE Trans. on Knowledge and Data Engineering*, vol. 8, no. 1, pp. 96-105, Feb. 1996.

[9] Looney, C. G., "Fuzzy Petri nets for rule-based decision making," *IEEE Trans. on Systems, Man and Cybernetics*, vol. 18, no. 1, pp. 178-183, Jan/Feb. 1988.

[10] Murata, T., "Petri nets: properties, analysis and applications," *Proceedings of the IEEE*, vol. 77, no. 4, pp. 541-580, April 1989.

[11] Nilsson, N., *Principles of Artificial Intelligence*, Morgan Kaufmann, San Francisco, 1980.

[12] Pal, S. and Konar, A., "Modeling cognition using fuzzy neural nets," In *Fuzzy Theory Systems: Techniques & Applications*, Leondes, C. T. (Ed.), Academic Press, 1999.

[13] Pearl, J., "Distributed revision of composite beliefs," *Artificial Intelligence*, vol. 33, 1987.

[14] Pedrycz, W. and Gomide, F., "A generalized fuzzy Petri net model," *IEEE Trans. on Fuzzy Systems*, vol. 2, no. 4, pp. 295-301, Nov. 1994.

[15] Quillian, R., "Semantic memory," In *Semantic Information Processing*, Minsky, M. (Ed.), MIT Press, Cambridge, MA, 1968.

[16] Russel, S. and Norvig, P., *Artificial Intelligence*, Prentice-Hall, Englewood-Cliffs, NJ, 1995.

[17] Scarpelli, H., Gomide, F. and Yager, R.R., "A reasoning algorithm for high-level fuzzy Petri nets," *IEEE Trans. on Fuzzy Systems*, vol. 4, no. 3, pp. 282-294, Aug. 1996.

[18] Wang, H., Jiang, C. and Liao, S., "Concurrent reasoning of fuzzy logical Petri nets based on multi-task schedule," *IEEE Trans. on Fuzzy Systems*, vol. 9, no. 3, June 2001.

[19] Yu, S. K., "Knowledge representation and reasoning using fuzzy Pr/T net-systems," *Fuzzy Sets and Systems*, vol. 75, pp. 33-45, 1995.

Chapter 4

Belief Propagation and Belief Revision Models in Fuzzy Petri Nets

In the last chapter, we introduced several forward-reasoning models of fuzzy Petri nets. This chapter addresses the issues of forward reasoning for both acyclic and cyclic networks. In an acyclic network, beliefs of the axioms are propagated downstream to compute the beliefs of a desired goal proposition. This has been referred to as belief propagation. Reasoning in a cyclic network is performed through local computation of beliefs in the places of the networks. The local computation involves recurrent updating of beliefs, called belief revision. Sometimes the belief revision process results in sustained periodic oscillation (limit cycles) in the beliefs of the propositions lying on a cycle of the network. No decisions can be arrived at from a reasoning system with such behavior. To circumvent this problem the condition for the existence of limit cycles has been derived, and an algorithm for the elimination of limit cycles has been proposed. The principles of inconsistency management, called non-monotonic reasoning, have also been outlined for the belief revision model.

4.1 Introduction

An expert system (ES) generally embodies a knowledge base (KB), a database (DB) of facts, and an inference engine (IE) for interpreting the DB with the help of the KB. DB of real-world problems often contain imprecise data; KB also include pieces of knowledge, the certainty of which is not always guaranteed. Further, due to nonavailability of data from authentic sources and incompleteness of data or knowledge, inconsistent data elements often creep into the reasoning space. This chapter aims at developing a unified approach for reasoning in ES in presence of imprecise and inconsistent DB and uncertain KB.

In order to increase the reasoning efficiency [18] of the IE for such an ES, all the above forms for inexactness of DB and KB become mandatory to be mapped onto a structured parallel distributed architecture. A Petri net (PN) [9], which is a directed bipartite graph with a high degree of structural parallelism and pipelining, is an ideal choice for the proposed IE. The input and the output places of each transition in a PN can be used to represent the antecedent-consequent pairs of the pieces of knowledge, represented by production rules (PR). The degree of precision of the propositions can be mapped as markings of the corresponding places representing the propositions. The certainty factor (CF) of the PR is assigned to the appropriate transitions. A threshold, which is also assigned to each transition, represents a lower bound on the degree of propositions that the input places of the transition should possess in order to fire the transition. After a transition fires, the new markings at its output places are evaluated. It may, however, be added that the markings from the input places of the transitions are not removed after transition-firing, as is done in classical PN. The well-known problem of *structural conflict* [13], therefore, does not arise in the present context.

It is to be noted that in the proposed PN model, when one (or more) consequent proposition of an ith PR and antecedent propositions of a jth PR is (are) common, the ith transition always precedes the jth transition. The pipelining in the execution of the rules based on their dependence relationship is thus established in the PN models. In addition, the highest possible degree of parallelism underlying the IE is fully supported by the PN models, since the transition in the PN, where the enabling conditions are satisfied simultaneously, can fire concurrently. The throughput [18] of the IE could thus be improved significantly by implementing it on a PN.

Fuzzy logic [6], [21] has been employed here for its simplicity coupled with its adequate power of humanlike reasoning. The degree of truth of propositions, CF, and thresholds are, therefore, all mapped into the fuzzy truth scale [0, 1]. Further, the operations to be carried out on the proposed PN are simple fuzzy AND and OR operations. The PN thus used for reasoning is called fuzzy PN (FPN), following Looney, who pioneered the concept of imprecision management in decision-making problems using acyclic FPN [11]. Chen et al., too, attempted to develop an algorithm for reasoning [5] in a more generic class of FPN, which, however, is plagued by a number of shortcomings. For example, the cycles were virtually opened up in the algorithm itself. Secondly, the algorithm cannot deal with FPN having intermediate places (defined in Section 4.2) with more than one input arc. Finally, the dependence of the degree of precision of the concluding place on the length of the reasoning paths is undesirable. There exist an extensive works on FPN [1-14], each addressing one or more issues related to this new discipline. In this chapter, we, however, discuss aspects of forward reasoning in an FPN following the first author's previous works [8] on the subject.

Two distinct models of FPN constructed for automated reasoning in ES are presented in this chapter. The first model deals with the computation of the degree of precision, hereafter called fuzzy belief of a given concluding proposition, based on the preassigned fuzzy beliefs of the independent starting propositions in an acyclic FPN. The algorithm proposed for reasoning in ES using this model is free from the limitations found in [5].

The second model is concerned with the computation of steady state fuzzy beliefs at the places in the network from their preassigned (initial) fuzzy beliefs. An algorithm for reasoning in ES based on the model has been developed by extending Looney's algorithm to encompass a generic class of FPN that may include cycles. Belief updating process in the proposed algorithm is carried out for all places synchronously. It is possible that after a finite number of belief revision in the FPN, fuzzy beliefs at each place on the cycle may be repeated, exhibiting periodic temporal oscillation of belief at the places on the cycle. No inference can be derived from an FPN with sustained periodic oscillation (PO), hereafter called limit cycles (LC). A condition for the existence of LC has been derived and the principle used for its elimination has also been discussed. Finally, the principle of inconsistency management has been presented with reference to the proposed model.

The chapter has been subdivided into four sections. The algorithms for belief propagation in acyclic FPN have been discussed in Section 4.2. A proof of deadlock freedom of the algorithms is also presented in this section. The algorithms for belief revision in cyclic nets have been presented in Section 4.3. The algorithms address the issues of limit cycle elimination and non-monotonic reasoning. Concluding remarks are listed in Section 4.4.

4.2 Imprecision Management in an Acyclic FPN

Before describing the technique for imprecision management in acyclic FPN, we present a few definitions.

4.2.1 Formal Definitions and the Proposed Model

Definition 4.1: *AN FPN can be defined as a 9-tuple:*

$$FPN=\{P, Tr, D, I, O, cf, th, n, b\}$$

where

- $P = \{p_1, p_2, ..., p_m\}$ *is a finite set of places,*
- $Tr = \{tr_1, tr_2, ..., tr_n\}$ *is a finite set of transitions,*

- $D = \{d_1, d_2, \ldots\ldots, d_m\}$ *is a finite set of propositions,*
- $P \cap T \cap D = \phi$, $|P| = |D|$,
- *I: Tr* $\rightarrow P^{\infty}$ *is the input function, representing a mapping from transitions to bags of (their input) places,*
- *O: Tr* $\rightarrow P^{\infty}$ *is the output function, representing a mapping from transitions to bags of (their output) places,*
- *cf, th: Tr* $\rightarrow$ *[0, 1] are association functions, representing a mapping from transitions to real values between 0 and 1,*
- *n: P* $\rightarrow$ *[0, 1] is an association functions, representing a mapping from places to real values between 0 and 1,*
- *b: P* $\rightarrow D$ *is an association function, representing a bijective mapping from places to propositions.*

In realistic terminology, n_i represents the fuzzy beliefs of place p_i i.e., $n_i = n(p_i)$; $cf_j = cf(tr_j)$ and $th_j = th(tr_j)$ represent the CF and threshold of transition tr_j respectively. Further $d_i = b(p_i)$.

Definition 4.2: A *transition tr_i is enabled if AND $\{n_i : p_i \in I(tr_j)\} > th_j$ where $n_i = n(p_i)$ and $th_j = th(tr_j)$. An enabled transition is fired. On firing it results in fuzzy truth tokens (FTTs) at all its output arcs. The value of the FTTs is a function of the CF of the transition and fuzzy beliefs of its input places.*

The technique for computing the fuzzy beliefs at a place can be conveniently represented by a model. Based on Definitions 4.1 and 4.2 we propose a model, called the Belief Propagation Model.

4.2.2 Proposed Model for Belief Propagation

Consider a place p_i which is one of the common output places of transitions tr_j, where $1 \leq j \leq m$. Now, for computing fuzzy beliefs n_i at place of p_i, first the condition for transition firing is checked for all tr_j. If a transition fires then the fuzzy beliefs of its input places are ANDed and then the result, called FTT is saved. If the transition does not fire, then the FTT corresponding to this transition is set to zero. The fuzzy belief of place p_i can now be computed by ORing the FTT associated with the tr_j for $1 \leq j \leq m$.

For the sake of brevity, the AND (minimum of inputs) and OR (maximum of inputs) operators are represented by $\wedge$ and $\vee$, respectively.

Example 4.1: The parameters defined above are illustrated with reference to Figure 4.1, which describes Production Rule (PR) PR1, where PR1: IF((d_1) AND (d_2)) THEN ((d_3) OR (d_4)). Here $d_1 =$ it is hot, $d_2 =$ the sky is cloudy, $d_3 =$ it will rain, $d_4 =$ humidity is high, P = \{p_1, p_2, p_3, p_4\}, T=\{tr_1\}, D=\{d_1, d_2, d_3, d_4\},

$I(tr_1) = \{p_1, p_2\}$, $O(tr_1) = \{p_3, p_4\}$. Let $cf_1 = 0.9$, $th_1 = 0.1$, $n_1 = 0.9$, $n_2 = 0.5$, and n_3 and n_4 are to be determined. Now as $n_1 \wedge n_2 = 0.5 > th_1$, therefore the transition fires and $n_3 = n_4 = [(n_1 \wedge n_2) \wedge cf_1] = 0.5$.

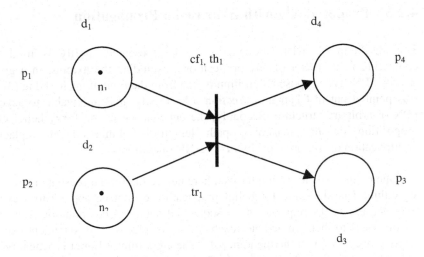

$P = \{p_1, p_2, p_3, p_4\}$, $T = \{tr_1\}$, $D = \{d_1, d_2, d_3, d_4\}$, $cf_1 = 0.9$, $th_1 = 0.1$, $I(tr_1) = \{p_1, p_2\}$, $O(tr_1) = \{p_3, p_4\}$, $n_1 = 0.9$, $n_2 = 0.5$ n_3 and n_4 are to be determined.

Fig. 4.1: An FPN representing PR_1.

Definition 4.3: *A place with no input arcs is called an* **axiom**.

Definition 4.4: *A path, connected by a set of directed arcs between any two distinct places, in an FPN is called a* **reasoning path**. *The* **length of an acyclic single** *reasoning path between two given places is estimated by counting the number of transitions between the said places on the path. Further, for estimating the length of the reasoning path with cycles, the number of transitions on each cycle is counted only once. If there exists more than one reasoning paths between two places, then the length of the largest path is the measure of the reasoning path length between the said places.*

Definition 4.5: *Any place excluding the starting and terminating place on the reasoning path is called an* **intermediate place**.

Definition 4.6: *If place $p_i \in I(tr_a)$ and $p_k \in O(tr_a)$, then p_k is called* **immediately reachable** *from p_i. The set of places which is immediately reachable from p_i is called* **immediate reachability set** *of p_i and is denoted by IRS (p_i) [3].*

Definition 4.7: *If p_k is immediately reachable from p_i and p_j is immediately reachable from p_k then p_j is* **reachable** *from p_i. The reachability relationship is* *the reflexive and transitive closure of the immediately reachability relationship. The set of places, which is reachable from p_i is called **the reachability set** of p_i and is denoted by RS (p_i) [3].*

4.2.3   Proposed Algorithm for Belief Propagation

The algorithm of belief propagation to be presented shortly is used for computing fuzzy belief of any propositions, excluding the axioms, in a given acyclic FPN. The proposed algorithm is free from the limitations found in [3], as was pointed out in Chapter 3 (Section 3.5). Firstly, it is applicable to acyclic FPN of arbitrary structure. Secondly, the dependence of the fuzzy belief of a proposition on the reasoning path length is eliminated by replacing multiplication operation in [3] by fuzzy AND operation.

One point needs to be emphasized here before the formal presentation of the algorithm. The algorithm for belief propagation cannot be applied to a cyclic FPN because of the requirement of sequential nature of computation from one set of places to their immediate reachable set of places. This statement can be easily proved based on the theorem [6]: "The algorithm of Belief propagation, if applied on a cyclic FPN, does not lead to successful completion because of the non-availability of fuzzy belief at least one parent of every place selected for belief computation." However, the problem of belief propagation for a cyclic FPN can be easily tackled by the belief revision algorithm (vide Section 4.3) with initial zero beliefs at all places excluding the axioms.

The proposed algorithm for belief propagation is comprised of two main procedures, namely "reducenet" and "evaluatenet." The procedure "reducenet" is required for reducing an FPN of a given structure to a smaller dimension by isolating the reasoning paths, originating at the axioms and terminating at the given conclusion (Goal) place. The procedure "evaluatenet" is used for sequentially computing fuzzy beliefs of all the propagation lying on the reasoning paths of the reduced network.

The procedure "reducenet" works on the following principle. The FPN along with the marked axioms and the concluding proposition (Goal) is submitted to the computer as the input of the procedure "reducenet." The procedure continues backtracking, starting from the Goal place until the axioms are reached. The paths thus traced by the procedure are reproduced as the resulting network.

Procedure reducenet (FPN, Axioms, Goal, Parents)
Begin
　　Nonaxioms:= Goal ;
　　Repeat
　　　Find-parents (Nonaxioms); / / Find parents of Nonaxioms. / /

Mark the generated parent places, hereafter called Parents and the transitions connected between parents and Nonaxioms;
Nonaxioms: = Parents – Axioms; // Nonaxiom-Parents detection //
Until Parents ⊆ Axioms ;
Trace the marked places and transitions;
End.

In the procedure: Evaluatenet presented below, "Noncomputed" represents the set of places where fuzzy beliefs have not been computed yet; "Computed" represents the set of places, where fuzzy beliefs have been computed so far; "Unsuccessful-Parents" set represents the set of parents places, whose all children do not have fuzzy beliefs yet; "Successful-Children" denotes the set of places, whose all parents do have fuzzy beliefs. The other variables are self-explanatory in the present context.

Procedure Evaluatenet (Axioms, Noncomputed)
Begin
Unsuccessful-Parents: = φ ; Computed: = φ ;
For all q ∈ Noncomputed Flag (q): = false ;
B: = Axioms;
While Noncomputed ≠φ **do Begin**
Successful-children: = φ;
For all p ∈ B **do Begin**
Found-Succ.-Children: = φ;
Found-Unsucc.-Parents: = φ;
For all q ∈ IRS (p) **do Begin**
If (Flag (q) ≠ True)
Then
If (all parents of q possess fuzzy beliefs)
Then
Compute FTT at the transition between p and q for all p
where IRS (p) = q and compute fuzzy belief at q , following
Belief propagation model;
Found-Succ.-Children: = Found-Succ.-Children ∪ {q};
Flag (q):= True;
End If
Else keep record of p in Found-Unsucc.-Parent set;
Successful-Children: = Found-Succ.-Children ∪ {q};
Flag (q): = True;
End If
Else keep record of p in Found-Unsucc.-Parent set;
Successful-Children:= Successful-Children ∪ Found-Succ.-Children;
Computed: = Computed ∪ Successful-Children;
Noncomputed: = Noncomputed – Successful-Children;
Unsuccesful-Parents: = Unsuccessful-Parents ∪

Found-Unsucc.-Parents;
If (IRS (p) $\subseteq$ Computed) **Then** Unsuccessful-Parents: =
Unsuccessful-Parents - {p};
End If;
End If;
End For;
End For;
B: = Successful-Children $\cup$ Unsuccessful-Parents;
End While;
End.

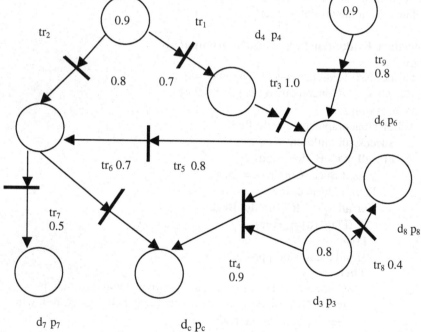

Fig. 4.2: An FPN used for utilizing evaluatenet procedure, after elimination of places p_7 and p_8 by using reducenet procedure.

Procedure Evaluatenet works on the following principle. For a given parent place $p \in B$ and a child place $q \in$ IRS(p), we need to check whether all the parents of q possess tokens. If yes, fuzzy belief is computed at q, and the place q is declared as a successful child. Similarly, if at least one child of p is not found to be successful, then p is called a found-unsuccessful-parent. After generating unsuccessful children from all $p \in B$, we redefine B as the union of Unsuccessful-Parents and Successful-Children. The redefinition of B ensures completeness of the algorithm. The procedure is terminated when fuzzy beliefs at all possible places in the FPN have been computed.

Table 4.1: Results of Execution of Procedure Evaluatenet for FPN of Figure 4.2

beginning of execution

Unsuccesful-Parents:=Φ, Computed:= Φ, B=Axioms=$\{p_1, p_2, p_3\}$,
Start of while noncomputed$\neq \Phi$ loop, Successful-Children:= Φ.

Start of for all p ε loop

p	Found-Successful Children (FSC)	Successful-children	Computed	Unsuccesful-Parents	Noncomputed	Fuzzy belief of FSC
p_3	$\{\Phi\}$	$\{\Phi\}$	$\{\Phi\}$	$\{p_3\}$	$\{p_4, p_5, p_8, p_c\}$	nil
p_2	$\{p_4\}$	$\{p_4\}$	$\{p_4\}$	$\{p_2, p_3\}$	$\{p_5, p_6, p_c\}$	$0.7 \wedge 0.9 = 0.7$
p_1	$\{p_6\}$	$\{p_4, p_6\}$	$\{p_4, p_6\}$	$\{p_2, p_3\}$	$\{p_5, p_c\}$	$(0.7 \wedge 1.0) \vee (0.9 \wedge 0.8) = 0.8$

end of for all p ε B loop, B= $\{p_2, p_3, p_4, p_6\}$

Noncomputed$\neq \Phi$, so While Noncomputed$\neq \Phi$ loop is continued
Successful-Children := Φ. Start of for all p ε loop

p	Found-Successful-Children (FSC)	Successful-Children	Computed	Unsuccessful-Parents	Noncomputed	Fuzzy belief of FSC
p_2	$\{p_5\}$	$\{p_5\}$	$\{p_4, p_5, p_6\}$	$\{p_3\}$	$\{p_c\}$	$(0.9 \wedge 0.8) \vee (0.8 \wedge 0.8) = 0.8$
p_3	$\{p_c\}$	$\{p_5, p_c\}$	$\{p_4, p_5, p_6, p_c\}$	$\{\Phi\}$	$\{\Phi\}$	$(0.8 \wedge 0.7) \vee \{(0.9) \wedge (0.8 \wedge 0.8)\} = 0.8$
p_4	← Nil since flag (q) =1 for all q ε IRS(p_4) →					
p_6	← Nil since flag (q) =1 for all q ε IRS(p_6) →					

end of for all p ε B loop, B= $\{p_5, p_c\}$

Since Noncomputed = $\varnothing$, While loop is terminated.

End of execution

Example 4.2: The procedure evaluatenet has been executed for the FPN given in Figure 4.2, where the fuzzy beliefs of the axioms and CF of transitions are shown in the figure, and thresholds of all transitions are assumed to be 0.2. The results obtained at different steps of the procedure are given in Table 4.1.

Time-Complexity: Time-complexity of the reducenet algorithm in the worst case is $a \times M$, where a and M, respectively, represent the number of axioms and places in the FPN before reduction.   The evaluatenet algorithm can be implemented on a uniprocessor as well as multiprocessor architecture with (M + N) number of processors corresponding to M places and N transitions. The worst time-complexity of the uniprocessor and multiprocessor-based evaluatenet algorithm are $3M^2$ and $(2M^2/pr)$, respectively, where pr represents the minimum number of places or transition where computation can take place in parallel and $M \geq N$.

4.2.4   Properties of FPN and Belief Propagation Scheme

In this section, we present some interesting properties of FPN and the proposed belief propagation scheme.

Theorem 4.1: *In an FPN, any place that itself is not an axiom, is always reachable from the set of axioms.*

Proof: Let the contradiction of the statement be true. Thus, if p be a place such that $p \notin$ set of axioms A, then $p \notin RS(\alpha)$ where $\alpha \in A$. Now, since $p \notin RS(\alpha)$ then none of the parents of p belong to $RS(\alpha)$ i.e., $q \notin RS(\alpha)$ where IRS (q) = p. Analogously, $r \notin RS(\alpha)$ where IRS (r) = q. Thus, backtracking in the FPN, starting from p without traversing the same place more than once (for avoiding traversal on the cyclic path more   than once), we ultimately find some place z such that

$$p = IRS\ (IRS\ (IRS....(\ IRS\ (Z\ )\ )\ )\ )$$

where Z has no input arcs.

Therefore, by definition, Z is an axiom. So, the initial assumption that p $\notin RS(Z)$ where $Z \in A$ is violated. Hence, the theorem is proved.

Corollary 4.1: *Any place p with more than one input arc is reachable from one or more axioms, through all the reasoning paths terminating at p.*

Theorem 4.2: *If there is no cycle in an FPN, then there should exist at least one place immediately reachable from any of the axioms such that all its parents belong to the set of axioms.*

Proof: The theorem will be proved by the method of contradiction. So, let us assume that there does not exist even a single place, immediately reachable from any of the axioms, such that all its parents belong to the set of axioms in an FPN without cycles.

Let $S = \{u, v, w, x\}$ be the set of places immediately reachable from the set of axioms, where each element of S has more than one parent (since with one parent the proof is obvious). Now, according to the initial assumptions, place u cannot have all its parents in the set of axioms. Moreover, since place u is reachable from axioms through all the reasoning paths terminating at place u (vide corollary 4.1) and S is the exhaustive set of places immediately reachable from the axioms, therefore, u is reachable from at least one place in S. In an analogous manner, it can be proved that any place $p \in$ S is reachable from some other place $p' \in$ S. This proves the existence of at least one cycle in the network (passing through elements of S), which, however, is against the initial assumption. Therefore, the initial assumption is wrong and the statement of the theorem is true.

Theorem 4.3: *Any place for which fuzzy belief has not been computed yet is reachable from place $b \in B$.*

Proof: For B = the set of axioms A, the statement of the theorem follows directly from theorem 4.1. However, for $B \neq A$, let the converse of the statement be true, i.e., for any place p for which fuzzy belief has not been computed, let $p \notin$ RS (b) for $b \in B$. Now, since p is reachable from places belonging to A, vide theorem 4.1, therefore, p is reachable from set X where the fuzzy belief of each place $x \in$ X has already been computed. Now, following, the procedure "evaluatenet" it is clear that $B \subset X$. So, p which is not reachable from the places in B is reachable from places in $X - B$. Now, since p is reachable only from places in X–B (and not from places in B), therefore, there should exist at least one place p' such that $p \in$ RS (p') and all the parents of p' belong to $X - B$. Further, since all the parents of p' possess fuzzy beliefs, therefore fuzzy belief at p' should be computed currently. However, from the algorithm it is clear that set B always keeps track of the parents corresponding to the children where fuzzy beliefs have to be computed currently. So, there cannot be any place p' whose fuzzy belief is yet to be computed with all its parents in the set $(X - B)$. Consequently, there cannot be any place p, which satisfies the initial assumption. Therefore, the initial assumption is wrong and the statement of the theorem is true.

Definition 4.8: *The algorithm for belief propagation is said to reach **a state of deadlock** if for a given set B, no q is readily identified for belief computation prior to termination of the algorithm (i.e., Noncomputed $\neq \phi$).*

Theorem 4.4: *The computation process in the algorithm for belief propagation never reaches a state of deadlock.*

Proof: When B = A = the axioms, we always get a place q ready for belief computation (vide Theorem 4.2). When B = A, it is clear from theorem 4.3 that the places having no fuzzy beliefs in the FPN are reachable from places belonging to B. So, set B can be treated like set of axioms A, from the point of view of reachability of the places having no fuzzy beliefs. Therefore, the statement of the theorem follows from the statement of Theorem 4.2. The theorem can be formally proved by the method of induction.

4.3 Imprecision and Inconsistency Management in a Cyclic FPN

This section presents a new technique for imprecision and inconsistency management in a cyclic FPN using a model for belief-revision, based on the central concept of fuzzy reasoning presented in [11]. The basis of reasoning of this model is humanlike and is different from the existing probabilistic models [12], [17] for belief-revision.

4.3.1 Proposed Model for Belief-Revision

Let $n_i(t)$ be the fuzzy beliefs of a place p_i and $t_q(t)$ be the fuzzy truth token (FTT) associated with the transition tr_q at time t. Let p_i be one of the common output places of transition tr_q for $1 \leq k \leq u$ (Figure 4.3) and p_j, $1 \leq j \leq v$ are the input places of transition tr_k (Figure 4.4). The updating of fuzzy values at transition tr_q and place p_i can then be represented by (4.1) and (4.2), respectively.

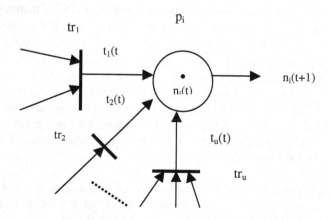

Fig. 4.3: An FPN exhibiting connectivity from transitions to a place.

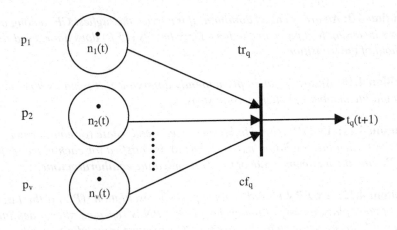

Fig. 4.4: An FPN exhibiting connectivity from places to transitions.

Let $x= \wedge \{n_j(t)\}$. Now if $x>$ threshold marked at tr_q
$\quad\quad 1 \leq j \leq v$

then the transition fires with the resulting FTT as given by

$$t_q(t+1)= x \wedge cf_q \quad\quad (4.1)$$

where cf_q is the time invariant CF of the transition, else $t_q(t+1) = 0$.

If p_i is an axiom then

$$n_i(t+1) = n_i(t), \quad\quad (4.2)$$

else $\quad n_i(t+1)= \vee \; t_k(t+1) \quad\quad (4.3)$
$\quad\quad\quad\quad 1 \leq k \leq u$

Initial fuzzy beliefs are assigned at all places of a given FPN. Therefore, by definition of enabling of transition, all the transition in the FPN may be enabled simultaneously. The conditions of transition firing are, therefore, checked and the temporal FTT are computed concurrently at all transition using (4.1). Then fuzzy beliefs at all the places in the FPN can be revised in parallel by using (4.2) and (4.3) as is appropriate. This is regarded as one complete belief revision step. For computing fuzzy beliefs at all places up to the *m*th step, one should repeat the belief-revision process m times recursively.

4.3.2 Stability Analysis of the Belief Revision Model

The definitions of a few more terms, which will be referred to frequently in the rest of the chapter, are in order.

Definition 4.9: *An arc is called* **dominant** *if it carries the highest CF among all the arcs incoming to a place or the least fuzzy beliefs (LFB) does not exceed the threshold of the transition.*

Definition 4.10: *An arc is called* **permanently dominant** *if it remains dominant for an infinite number of belief revision steps.*

Definition 4.11: *AN FPN is said to have reached* **steady-state** *(or equilibrium condition) when the condition: $n_i(t^* + 1) = n_i(t^*)$ is satisfied for each place p_i in the FPN and the minimum value of $t = t^*$ is called the* **equilibrium time.**

Definition 4.12: *AN FPN is said to have periodic oscillation (PO) if the fuzzy belief of each place on any of the cycles in the FPN is repeated after a definite number of belief-revision steps. In case the PO sustains for an infinite number of belief-revision steps, we say* **limit cycles (LC)** *exist at each place on the cycle.*

The properties of the belief-revision model, which have been obtained analytically are listed in Theorems 4.5 through 4.7.

Theorem 4.5: *If all the arcs on a cycle with n transition remain dominant from r_1 th to r_2 th belief-revision steps, then the fuzzy beliefs at each place on the cycle would exhibit i) at least 'a' number of periodic oscillations where $a = \lfloor (r_2 - r_1)/n \rfloor$ and ii) limit cycles if r_2 approaches infinity.*

Proof: If all the arcs on a cycle become dominant at the r_1th belief-revision step, then the fuzzy beliefs at each place on the cycle will be transferred to its immediate reachable place on the cycle. Now, if all the arcs remain dominant until the r_2th belief-revision step, then transfer of belief from one place to its immediately reachable place on the cycle would continue for $(r_2 - r_1)$ times. Let us now assume without any loss of generality, that

$$r_2 - r_1 = a.n + b$$

where n is the number of transition on the cycle under consideration, $b \leq n - 1$, and a and b are nonnegative integers such that both of these cannot be zero simultaneously. It is clear from (4.3) that the fuzzy beliefs at each place on the cycle would retrieve the same value after every n number of belief-revision steps. So, the fuzzy beliefs at each place on the cycle would exhibit at least a number of periodic oscillations, where $a = \lfloor (r_2 - r_1)/n \rfloor$ as $b \leq n - 1$. This proves the first part of the theorem.

Now, to prove the second part, we substitute the limiting value of r_2 to infinity in the expression for a, which yields the value of a to be infinity, proving the existence of limit cycles at each place on the cycle.

The following two definitions are useful to explain Theorem 4.6

Definition 4.13: *The **fuzzy multiplication (max-min composition)** of an (m ×* *n) matrix by an (n × 1) column vector is carried out in the same way as in conventional matrix algebra with substitution of multiplication and summation operators by fuzzy AND (Min) and OR (Max) operators, respectively.*

Definition 4.14: *The **complementation of a fuzzy vector** is obtained by replacing elements of the vector by their one's complements (i.e., one minus the value of the element).*

Definition 4.15: *The **fuzzy AND operations** of two vectors is carried out in the same way as in case of addition of two vectors in conventional Matrix Algebra with the substitution of addition by fuzzy AND operation.*

Definition 4.16: *A matrix P is called binary **transition to place connectivity** (TPC) matrix if its element p_{ij} is 1, when $p_i \in O$ (tr$_j$) and 0, otherwise. A matrix Q is called binary **place to transition connectivity** (PTC) matrix if its element $q_{ij} = 1$ $p_j = I$ (tr$_i$) and 0 otherwise.*

Definition 4.17: *A **belief vector** N(t) is a column vector, the ith component of which represents the fuzzy beliefs n_i (t) of place p_i at time t. The belief vector at t = t*, the equilibrium time is called the steady-state belief vector (SSBV).*

Definition 4.18: *A **CF vector** represents a time-invariant column vector, the ith component of which represents the CF of the transition tr$_i$ in an FPN .*

Theorem 4.6: *If the transitions remain enabled at each belief revision step, then the SSBV N* satisfies the following expression, where P and Q are TPC and PTC matrices respectively, CF is the certainty factor vector, c over a vector represents its fuzzy complementation operation, and the o denotes the fuzzy AND-OR composition operator.*

$$N^* = P \, o \, [(Q \, o \, (N^*)^c \,)^c \wedge CF]$$

Proof: Assume that there exist m places and n transition in the FPN. Now, since transitions are always enabled, therefore, by (4.1), FTT of transition tr$_i$, at time (t + 1) is given by

$$t_i(t + 1) = \{ \underset{1 \le \exists j \le m}{\wedge} (n_j(t))\} \wedge cf_i$$

$$= \{ \underset{1 \le \exists j \le m}{\vee} (n_j^c(t)\}^c \wedge cf_i$$

$$= \{ \underset{1 \le \forall j \le m}{V} (q_{ij} \wedge n_j^c(t)\}^c \wedge cf_i, \qquad (4.4)$$

Thus for $1 \leq i \leq n$, we get a column vector $T(t + 1)$, the ith component of which is $t_i(t + 1)$. Here,

$$T(t + 1) = (Q \circ N^c(t))^c \wedge CF. \tag{4.5}$$

Now, in order to represent belief-revision at both axioms and non-axioms by a single expression like (4.3), we construct, without any loss of generality, a self-loop around each axiom through a transition with threshold $= 0$ and $CF = 1$. Thus, one can easily use expression (4.3) for computing fuzzy beliefs at any place at time $(t + 1)$. It can now be easily proved that

$$N(t + 1) = P \circ T(t + 1). \tag{4.6}$$

Now, combining (4.5) and (4.6) and substituting $t = t^*$, the equilibrium time, in the derived expression, the theorem can be proved.

Example 4.3: Assuming the thresholds and CF associated with all the transition to be zero and one, respectively, the computation of belief vector $N(1)$ and $N(2)$ are illustrated in this example with reference to Figure 4.5. First, virtual transitions are included in the self-loops, represented by the dotted line-segments in Figure 4.5, constructed around each axiom. Then the matrices P, Q, and vector $N(0)$ are obtained from Figure 4.5 as given below.

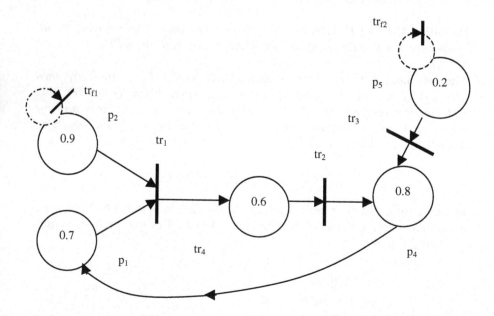

Fig. 4.5: An FPN having virtual transitions, constructed around the axioms, for illustrating the formation of P and Q matrices.

From Transitions

To Places	tr_1	tr_2	tr_3	tr_4	tr_{f1}	tr_{f2}
p_1	0	0	0	1	0	0
p_2	0	0	0	0	1	0
p_3	1	0	0	0	0	0
p_4	0	1	1	0	0	0
p_5	0	0	0	0	0	1

$P=$ (matrix above)

From Places

To Trans.	p_1	p_2	p_3	p_4	p_5
tr_1	1	1	0	0	0
tr_2	0	0	1	0	0
tr_3	0	0	0	0	1
tr_4	0	0	0	1	0
tr_{f1}	0	1	0	0	0
tr_{f2}	0	0	0	0	1

$Q=$ (matrix above)

and $N(0) = [0.7\ 0.9\ 0.6\ 0.8\ 0.2]^T$, where T denotes the transposition operator over the row vector.

Combining (4.5) and (4.6) and substituting $t = 0$ and $t = 1$ in the resulting expression, we find

$$N(1) = P \circ (Q \circ N^c(0))\}^c = [\,0.8\ 0.9\ 0.7\ 0.6\ 0.2]^T,$$

$$N(2) = P \circ (Q \circ N^c(1))^c = [\,0.6\ 0.9\ 0.8\ 0.7\ 0.2]^T.$$

In case steady state is reached in an FPN, the number of belief-revision steps required to reach steady state is computed below:

Let

   A = set of axioms,

   $Cycle_i$ = set of places on cycle i,

   N = set of cycles,

   $(l_{ja})_i$ = minimum reasoning path length from a given axiom a to a given place j on cycle i,

   $(l_1)_i = Max \{Max (l_{ja})_i\}$,

        $a \in A$ $j \in Cycle_i$

   $l_{max1} = Max(l_1)_i$,

        $i \in N$

   n = number of transitions on the largest cycle,

   $(l_{jq})_i$ = minimum reasoning path length from a given place j on cycle i to a given terminal place q,

   $(l_2)_i = Max\{Max (l_{jq})_i\}$, where Y = set of terminals,

        $q \in Y$ $j \in Cycle_i$

   $l_{max2} = Max(l_2)_i$

        $i \in N$

   i = any cycle,

   l_{aq} = minimum reasoning path length from an axiom a to a given terminal place q without touching any cycle,

   $l_3 = Max \{ Max (l_{aq})\}$

        $a \in A$ $q \in Y$

Theorem 4.7: In case steady-state condition can be reached in an FPN, then the number of belief revisions required to reach steady-state in the worst case (T_{SS}) is given by

$$T_{SS} = Max\{l_3, (l_{max1} + l_{max2} + n - 1)\}.$$

Proof: It can be easily shown [8] that T_{SS} is equal to the belief revision steps needed for the transfer of fuzzy beliefs from the set of axioms to all the terminal places of the FPN. So, T_{SS} is equal to the number of belief revision required to traverse the largest reasoning path length between the axioms and the terminal places. In case the largest reasoning path does not pass through any cycle, then $T_{SS} = l_3$. On the other hand, if it passes through cycles, then we need to compute three terms:

   1)   Belief revision steps required for the fuzzy beliefs of the axioms to reach all possible places on all the cycles,

2) Belief revision steps required for the transfer of fuzzy beliefs of axioms from a few places on the largest cycle to all places on the same cycle, and

3) The maximum number of belief revision required for the propagation of beliefs from the places on the cycles to all the terminal places.

1) and 3) have already been estimated. Now, to compute 2, we use one property, which follows from Theorem 4.5, that for reaching steady-state in an FPN, at least one arc on each cycle should be permanently nondominant. Thus for the largest cycle with n transitions, belief revision required in 2) in the worst case = $(n - 1)$. So adding 1), 2) and 3), we get T_{ss} for the largest reasoning path that passes through cycle. Keeping option for the largest reasoning path either to pass or bypass cycles, we get the result presented in the theorem.

Corollary 4.2: *The number of belief revisions required to reach steady-state is always $\leq$ number of transitions in the FPN.*

4.3.3 Detection and Elimination of Limit Cycles

It follows from Corollary 4.2 that, in case limit cycles sustain in an FPN, then the steady-state condition will not be reached even after z number of belief revisions, where z denotes the number transition in the FPN. Procedure Belief revision has been developed based on this concept

1) either to detect existence of limit cycles or

2) to determine the steady state values of fuzzy beliefs in an FPN in absence of LC .

Procedure Belief-revision (FPN, Initial beliefs, thresholds, total-no-of-trans)
Begin
    Belief-revision-step-count : = 0 ;
    Repeat
        1) Update FTT of all transitions in parallel by using (4.1) ;
        2) Update fuzzy beliefs of all places in parallel by using (4.2) or (4.3), whichever is appropriate;
        3) Increment Belief-revision-step-count by one;
    Until
        For each place in the FPN
        (Current-value-of-fuzzy-belief = last-value-of-fuzzy-belief) OR
        (belief-revision-step-count = total-no-of-trans + 1)
        End For;
    If (belief-revision-step-count < total-no-of-trans)
    Then print (current value of fuzzy beliefs of all places)

End If
      Else print ("limit cycle exists.")
End.

It may be mentioned here than an alternative belief-revision algorithm in vector-matrix form can be easily developed using (4.5) and (4.6).

Belief-revision algorithm too can be implemented on a uniprocessor as well as multiprocessor architecture with (M + N) number of processing elements located at M places and N transition, respectively. Computational time needed for sequential and parallel belief-revision algorithm in the worst case are of the order of $2M^3$ and $2M^2$, respectively.

Now, before describing the method of elimination of LC, we define fuzzy gain.

Definition 4.19: *Fuzzy gain of an acyclic single reasoning path between two places is computed by taking the minimum of the fuzzy belief of the places and the CF of the transition on that path. In case there exists parallel reasoning paths between two places, then the overall fuzzy gain of the path is defined as the maximum of the individual path gains. Fuzzy gain of a reasoning path, starting from axioms up to a given transition is defined as the maximum of the fuzzy gains from each axiom up to each of the input places of the transition.*

Once the existence of LC is detected in a given FPN, its elimination becomes imperative for arriving at any conclusion. Elimination of LC, however, implies the elimination of permanent dominance of at least one arc of the cycle, containing the places, which exhibit LC behavior. This can be achieved by permanently preventing one of the transitions on the cycle from firing by setting an appropriate threshold at the transition. One possible choice of threshold is the largest fuzzy beliefs on the reasoning path from the axioms to the selected transition.

Questions may, however, be raised: can a user adjust the thresholds? The answer to this is affirmative if the user can ensure that the inferences derived by the system are least affected by threshold adjustment. Again, it is clear from the discussions in [7], [8] that the reasoning path with the least fuzzy gain is ignored in most cases, if there exists multiple reasoning paths leading to the concluding place. It is, therefore, apparent that in order to eliminate LC, one should select a transition that lies on the path with the least fuzzy gain.

However, adjustment of threshold at a transition on the cycle with least fuzzy gain may sometimes induce LC into a neighborhood cycle. This is due to the fact that a transition selected for threshold adjustment may cause permanent dominance of the output arc of another transition, both of which have a common output place, shared by two neighborhood cycles. For example, selecting transition tr_1 in cycle $p_1 p_2 p_1$ (Fig. 4.6) causes permanent dominance of the arc d_1 on cycle $p_2p_3p_4p_5p_2$ where p_2 is the common output place of tr_1 and tr_3. In case all the arcs excluding d_1 on the cycle $p_2 p_3 p_4 p_5 p_2$ were permanently dominant, then permanent dominance of d_1 would cause LC to be introduced in the cycle.

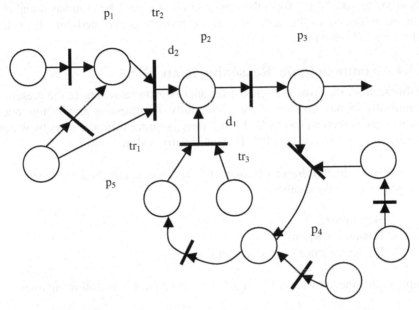

Fig. 4.6: An FPN for illustrating the possibility of induction of LC into neighborhood cycle.

To select the right transition on the cycle, one has to satisfy two criteria jointly:

1) The transition to be selected should lie on the reasoning path with the lowest possible fuzzy gain, as far as practicable and

2) The selected transition should not induce LC into neighborhood cycles.

     To satisfy these two criteria jointly, we compute the threshold of all the transition on the cycle, based on the guideline presented above and construct a set of ordered pairs (of transition and fuzzy gain of the path containing the transition), such that the elements of the set are sorted in ascending order using fuzzy gain as the key field. Thus, the former the position of the transition–fuzzy gain pair, smaller is the fuzzy gain of the path containing the transition. The transition-fuzzy gain pair, which occupies the first position in the set, is picked up, the threshold is assigned at the transition following the method discussed earlier, and the existence of LC is checked in the neighboring cycle. If LC does not exist in the neighborhood cycle, then the LC-elimination process for the proposed cycle is terminated. In case LC is induced in the neighborhood cycle, the transition from the next ordered pair is picked up and the possibility of induction of LC in the neighborhood cycle is examined again. The process is thus continued for each transition in the set according to the position of its corresponding ordered pair in the set, until a suitable one, which does not cause induction of LC in the neighborhood cycle is detected. If no such transition is found in the set, the LC from the proposed cycle cannot be eliminated and the same is informed to the user. The above process is repeated for all cycles exhibiting LC behavior.

4.3.4   Nonmonotonic Reasoning in an FPN

In this section, we present a scheme for approximate reasoning in the presence of mutually inconsistent set of facts. Naturally, the question arises: how does inconsistency creep into an FPN? In fact, there are three different ways by which inconsistency enters into an FPN. They are briefly outlined below.

Inconsistency in database: Consider, for instance, a murder history with the following facts in the database.

    Has-alibi(ram)
    Has-no-alibi(ram)
    Has-precedence-of-murder(ram)

Suppose that the knowledge base for the system includes the following rules.

    Rule 1: Has-precedence-of-murder (X),
            Has-alibi(X) →Not-suspect(X).

    Rule 2: Has-precedence-of-murder (X),
            Has-no-alibi(X) →Suspect(X).

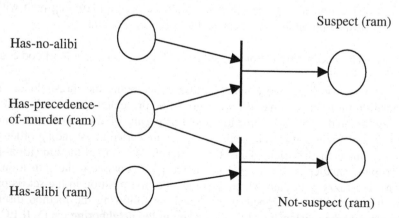

Fig. 4.7: Entry of inconsistency into the FPN due to inconsistent facts of the database.

Since both the rules satisfy the pre-condition of firing, they are encoded into FPN terminology (Fig. 4.7). As a consequence, inconsistent facts like has-no-alibi(ram) & has-alibi(ram), and Suspect(ram) & Not-suspect(ram) appear in the FPN.

Lack of specificity in production rules: The production rules used in most decision-support/ expert systems are not specific, i.e., they do not specify all the necessary preconditions in the antecedent part of the rules. Because of the lack of specificity, many rules that include mutually inconsistent consequent parts may fire, resulting in inconsistency in the reasoning space. The reasoning space in the present context being an FPN, contradictory information appears in the network. As an example, consider the following facts and rules.

Facts: Loved (r, s), Loves (s, m), proposed-to-marry(r, s).

Rule 1: Proposed-to-marry(X,Y), Loved(Y,Z) →Hates(X,Y)

Rule 2: Loved(X, Y) → Proposed-to-marry(X, Y).

Since both the rules are firable in presence of the given facts, they are transformed into Petri net terminology as given in Figure 4.8.

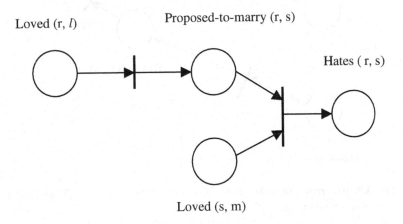

Loved (r, *l*)   Proposed-to-marry (r, s)   Hates (r, s)

Loved (s, m)

Fig. 4.8: Entry of inconsistency into an FPN because of the presence of nonspecific rules in the knowledge base.

It is indeed important to note that Loved (r, s) is contradictory to Hates(r, s) in Fig. 4.8.

Lack of precision of facts in the database: In case the facts available in the database have a very poor degree of precision, we may presume their contradiction to be true. Thus, two rules, one with the given fact as the premise and the other with its contradiction as the premise, may fire resulting in a conflict. As an example, consider the following facts and rules.

Facts: Loved (r, s), Loved (m, s), Loses-social-recognition(s), Husband(r), Another-man (m).

Rule 1: Loved (Husband, Wife), Loved(Another-man, Wife)→ Loses-social-recognition(Wife)
Rule 2: Loses-social-recognition (Wife)→Commits-suicide (Wife)
Rule 3: Tortured (Husband, Wife) → Commits-suicide (Wife)
Rule 4: Hated (Husband, Wife) →Tortured (Husband, Wife).

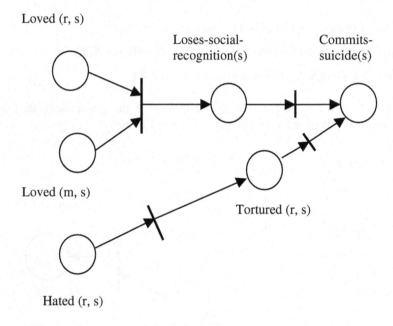

Loved (r, s)

Loses-social-recognition(s)

Commits-suicide(s)

Loved (m, s)

Tortured (r, s)

Hated (r, s)

Fig. 4.9: Illustrating the entry of inconsistency of facts into an FPN because of the lack of precision of data.

The FPN, corresponding to the above rules and facts, is presented in Figure 4.9. Suppose the fact: Loved(r, s) is found to have a poor degree of precision, say 0.001. In that case, we may consider its contradiction: Hated(r, s) to be true. However, we cannot ignore Loved(r, s) since it is present in the database. Thus because of the lack of precision of Loved (r, s), the same fact and its contradiction appear in the FPN.

So far we discussed different ways by which inconsistency creeps into an FPN, but we did not discuss any algorithm for reasoning in presence of the contradictions in the reasoning space. This section introduces an algorithm for nonmonotonic reasoning that judiciously elects one fact from each set of mutually inconsistent facts by arranging a voting scheme. The fact with the highest fuzzy vote is regarded as the winner, and other facts in the same set are discarded from the reasoning space. The voting scheme in the present context is realized with belief revision algorithm. Details of the nonmonotonic reasoning algorithm is available elsewhere [7]. We just outline the main steps of the algorithm for the sake of completeness of our discussion.

Algorithm for nonmonotonic reasoning

1. Open the output arcs of each fact in the set of mutually inconsistent information, so that they themselves cannot participate in the voting process of their election.

2. Call belief revision algorithm. If limit cycles are detected, eliminate limit cycles and repeat belief revision algorithm, and display the steady-state beliefs of the facts in the mutually inconsistent set of information.

3. The Information with the highest steady-state belief is elected, and thus all other information in the set are discarded by setting their beliefs permanently to zero. The output arcs opened in step 1 should be restored to allow the winner to participate in the subsequent steps of reasoning.

4. Repeat steps 1, 2, and 3 for each set of mutually inconsistent information.

Step 1 of the above algorithm is required to prevent the contradictory set of information from participating in their election process. Since the votes are weighted, this is essential for a fair election. The second step is the main step to identify the winning fact among the contradictory competitors. The third step provides a method of elimination of contradictory set of information by keeping only one of them in the FPN, and disregarding others. The step 4 is needed to repeat the first three steps for each set of contradictory information one by one. After the fourth step is over, the FPN thus constructed for automated reasoning becomes free from all types of inconsistency.

4.4 Conclusions

This chapter introduced the models of belief propagation and belief revision for approximate reasoning in acyclic and cyclic networks respectively. The analysis of deadlock freedom of the belief propagation algorithm ensures that the algorithm does not terminate pre-maturely. The belief revision algorithm on the other hand may undergo limit cyclic behaviors. The principles of the limit cycle elimination technique may then be undertaken to keep the reasoning free from getting stuck into limit cycles. Possible sources of entry of nonmonotonicity into the reasoning space and their elimination by a special voting arrangement has also been undertaken.

Unlike the existing methodology of reasoning on FPN, the most important aspect of the chapter lies in the representation of the belief updating policy by a single vector-matrix equation. Thus, if the belief revision does not get trapped into limit cycles and nonmonotonicity, the steady state solution of the reasoning system can be obtained by recurrently updating the belief vector until the vector converges to a stable state.

Exercises

1. Label the places and transitions in the FPN shown in Figure 4.10. Mark the axioms and use *reducenet* algorithm to back-trace the necessary part of the FPN for the computation of belief at place p_c.

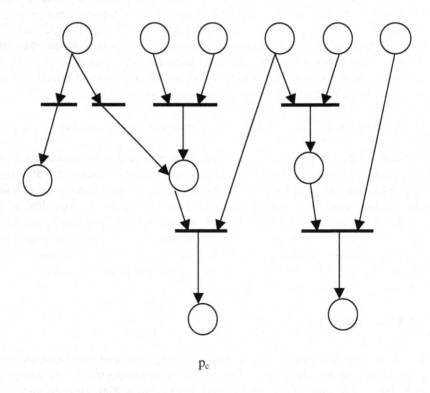

p_c

Fig. 4.10: An FPN used to illustrate problem 1.

2. Assign arbitrary beliefs at the axioms, and determine the belief at the concluding place p_c by invoking procedure *evaluatenet*. Assume thresholds and CF of all transitions to be 0 and 1, respectively.

3. In the FPN shown in Fig. 4.11, the initial beliefs are: $n_1(0) = 0.9$, $n_2(0) = 0.8$ and $n_3(0) = 0.4$, thresholds: $th_1 = th_2 = 0.2$, and certainty factors: $cf_1 = cf_2 = 0.9$.

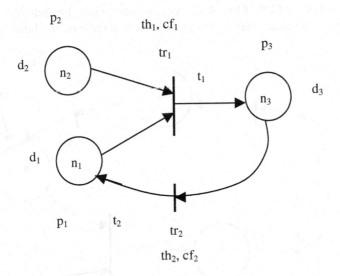

Fig. 4.11: A hypothetical FPN for the rules: d_1, $d_2 \rightarrow d_3$ and $d_3 \rightarrow d_1$, used to illustrate exercise 3.

Verify the results given below (Table 4.2) by executing belief revision algorithm on the FPN shown in Fig. 4.11 and hence show that limit cycles exist in the belief revision process of the given net.

Table 4.2: Results of belief revision on FPN of Fig. 4.11

Time t	$t_1(t+1)$	$t_2(t+1)$	$n_1(t+1)$	$n_2(t+1)$	$n_3(t+1)$
0	0.8	0.4	0.4	0.8	0.8
1	0.4	0.8	0.8	0.8	0.4
2	0.8	0.4	0.4	0.8	0.8

4. Suppose $n_2(0)$ in Fig. 4.11 has been changed from 0.8 to 0.3, and all other parameters remain unchanged. Will there be any change in the dynamic behavior of the reasoning system?

[**Hints:** Here, since $n_2(0) < n_1(0)$ and $n_2(0) < n_3(0)$, the network after two belief revision cycles will reach steady-state with belief values:

$(n_1)_{\text{steady-state}} = (n_2)_{\text{steady-state}} = (n_3)_{\text{steady-state}} = 0.3.]$

5. Adding a self-loop around the axiom p_2 in the FPN of Fig. 4.11 through a virtual transition of threshold = 0, represent the belief updating process in the FPN by a state equation. Use the initial beliefs as supplied in problem 3 and set thresholds and CF of all transitions to be 0 and 1 respectively.

6. Consider the FPN of Fig. 4.12. Assign suitable initial beliefs such that FTT value t_5 is greater than t_2 at all iterations. Will there be any limit cycles?

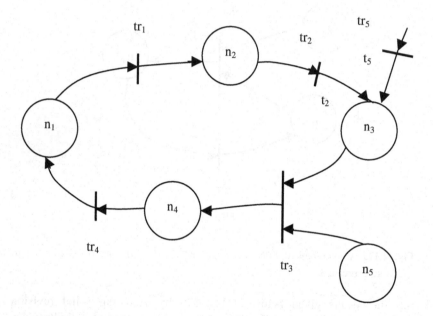

Fig. 4.12: An FPN used to illustrate exercise 6.

How can we adjust thresholds of transitions to ensure that $t_5 > t_2$ at all iterations? Suppose, we set threshold th_2 of transition tr_2 very high (close to 1), so that t_2 is zero, then $n_3 = t_5$. Suggest a suitable value of th_2 that can ensure elimination of limit cycles in Fig. 4.12.

7. Construct an FPN with two sets of contradictory pair of information, and use nonmonotonic reasoning algorithm to eliminate one from each pair of contradictions.

References

[1]   Bernard, P., *An Introduction to Default Logic.* Springer Verlag, pp. 13-30, 1989.

[2]   Buchanan, B. G. and Shortliffe, E. H., *Rule-Based Expert System: The MYCIN Experiments of the Stanford University Heuristic Programming Projects,* Addison-Wesley, Reading, MA, 1984.

[3]   Bugarin A. J., "Fuzzy reasoning supported by Petri nets," *IEEE Trans. on Fuzzy Systems,* vol. 2, no. 2, pp. 135-150, 1990.

[4]   Cao, T. and Sanderson, A. C., "Task sequence planning using fuzzy Petri nets," *IEEE Trans. on Systems, Man and Cybernetics,* vol. 25, no. 5, pp. 755-769, 1995.

[5]   Chen, S., Ke, J. and Chang, J., "Knowledge representation using fuzzy Petri nets," *IEEE Trans. Knowledge and Data Engineering,* vol. 2, no. 3, pp. 311-390, Sept.1990.

[6]   Eshera, M. A. and Brash, S. C., "Parallel rule based fuzzy inference engine on mess connected systolic array," *IEEE Expert,* Winter 1989.

[7]   Konar, A. and Mandal, A. K., "Non-monotonic reasoning in expert systems using fuzzy Petri nets," *Advances in modeling and Analysis,* vol. 23, no. 1, pp. 51-63, 1992.

[8]   Konar, A., *Uncertainty Management in Expert Systems Using Fuzzy Petri Nets,* Ph. D. Thesis, Jadavpur University, India, 1994.

[9]   Konar, A. and Mandal, A. K., "Uncertainty management in expert systems using fuzzy Petri nets," *IEEE Trans. on Knowledge and Data Engineering,* vol. 8, no. 1, pp. 96-105, 1996.

[10]   Lipp, H. P. and Gunther, G., "A fuzzy Petri net concept for complex decision making process in production control," *Proc. of First European Congress on Fuzzy and Intelligent Technology (EUFIT'93),* Aachen, Germany, vol. I, pp. 290-294, 1993.

[11]   Looney, G. C., "Fuzzy Petri nets for rule based decision making," *IEEE Trans. Systems, Man, and Cybernetics,* vol. 18, no. 1, Jan.-Feb. 1988.

[12]   Pearl, J., "Distributed revision of composite beliefs," *Artificial Intelligence*, vol. 33, pp. 173-213, 1987.

[13]   Peterson, J. L., *Petri Net Theory and Modeling Systems*, Prentice-Hall, Englewood Cliffs, NJ, pp. 7-30, 1981.

[14]   Pedrycz, W. and Gomide, F., "A generalized fuzzy Petri net model," *IEEE Trans. on Fuzzy Systems*, vol. 2, no. 4, pp. 295-301, 1994.

[15]   Reichenbach, H., *Elements of Symbolic Logic*, Macmillan, New York, pp. 218, 1976.

[16]   Scarpelli, H., Gomide, F. and Yager, R., "A reasoning algorithm for high level fuzzy Petri nets," *IEEE Trans. on Fuzzy Systems*, vol. 4, no. 3, pp. 282-295, 1996.

[17]   Shafer, G. and Logan, R., "Implementing Dempster's rule for hierarchical evidence," *Artificial Intelligence*, vol. 33, 1987.

[18]   Toagi, M. and Watanabe, H., "Expert system on a chip: an engine for real time approximate reasoning," *IEEE Expert*, Fall 1986.

[19]   Wang, H., Jiang, C. and Liao, S., "Concurrent reasoning of fuzzy logical Petri nets based on multi-task schedule," *IEEE Trans. on Fuzzy Systems*, vol. 9, no. 3, 2001.

[20]   Yu, S. K., Knowledge representation and reasoning using fuzzy Pr/T net-systems," *Fuzzy Sets and Systems*, vol. 75, pp. 33-45, 1995.

[21]   Zadeh, L. A., "Knowledge representation in fuzzy logic," *IEEE Trans. on Knowledge and Data Engineering*, vol. 1, no. 1, Mar. 1988.

Chapter 5

Building Expert Systems Using Fuzzy Petri Nets

The belief revision model we introduced in Chapter 4 has been employed in this chapter to design an expert system for criminal investigation. Special emphasis is given to the architectural issues of expert system design so as to reason in presence of imprecision and inconsistency of data and uncertainty of knowledge. Circularity in reasoning space, which acts as a bottleneck in most of the well-known reasoning systems, has been taken care of by the belief revision model. A case study on the simulated murder/suicide history of a housewife is included to give readers a clear view about the functional behavior of the modules in the architecture. A performance analysis of the proposed expert system reveals that reasoning efficiency of the system greatly depends on the topology of the network and the count of mutually inconsistent sets of propositions that jointly participate in the reasoning process. It has further been shown that more than ninety percent of the total reasoning time is consumed by the nonmonotonic reasoning and the limit cycle elimination modules.

5.1 Introduction

A number of well-known methods for expert system design [2], [12], [17], [18] is readily available in the current literature of artificial intelligence [7]. Unfortunately, these methods are inadequate to handle the complexity of many real-world problems, suffering from incompleteness of data and knowledge. Self-reference in the knowledge base sometimes forces the expert system to repeatedly fire a set of rules in sequence by a fixed set of data elements. This is informally known as circular reasoning [15]. Criminal investigation is one of such problems, which suffer from incompleteness of data and knowledge and circular reasoning. In this chapter, we take up criminal investigation as an illustrative problem-domain to substantiate the design issues of the proposed expert system.

Fingerprint, facial image, and voice matching are usual methods for identifying suspects in a criminology problem. Occasionally, most of this information is either not available or is found to be severely contaminated with measurement noise, and is thus inadequate for the investigation process. Under such circumstances, criminology experts have to depend on the incidental descriptions. Reasoning about the suspects from the incidental description is a complex problem as the information sources often add their own views with the facts witnessed or experienced by them. Further, the information received from multiple sources is often found to be imprecise and contradictory. Automatic identification of criminals from stray incidental description remained an unsolved problem until today. This chapter provides a novel scheme for the automatic detection of the criminals by handling the imprecision and the inconsistency of facts (data) and the uncertainty of knowledge by a unified approach.

The methodology of criminal investigation presented in this chapter is centered on the belief revision model of fuzzy Petri nets. Because of self-references in the rules, the Petri net constructed by instantiating the rules with suitable data/facts may include cycles. The belief revision process on a cyclic net may further result in limit cyclic behavior, when the singleton belief of the propositions mapped on the places of a cycle continue rolling on the cycle, without being affected by the propositions lying outside the cycle. A solution to the limit cyclic behavior is given in the last chapter. Besides limit cycles, contradictory evidences obtained from multiple sources may also paralyze the reasoning process in a fuzzy Petri net. The chapter eliminates contradiction from the reasoning space by invoking the nonmonotonic reasoning algorithm discussed in the last chapter.

The objective of this chapter is to design an expert system (ES) for criminal investigation through realization of the inference engine (IE) by a suitable Petri net model [3-5], [11], [13-14], [16], [19-20]. To take care of the diversity of the authenticity level of the sources, the degree of precision of the facts received from the sources are first normalized. Further, the database of such system being quite large, an organization of the database in the form of a hierarchical (tree) structure is needed for efficient access of the database. The pre-processing of the database will be covered in Section 5.2. The principles of default reasoning for preliminary identification of the suspects is covered in Section 5.3. The design aspects of the inference engine includes FPN-construction, belief revision, limit cycle elimination, nonmonotonic reasoning, and explanation tracing, some of which have been covered in the last chapter. An introduction to the inference engine is given in Section 5.4. A case study highlighting the functional operation of the modules for the proposed expert system is presented in Section 5.5 with reference to the murder/suicide history of a housewife. Performance analysis of the expert system is given in Section 5.6. Conclusions are listed in Section 5.7.

5.2 The Database

Fuzzy memberships of the information, collected from sources with various degrees of authenticity levels are first normalized using a set of intuitively constructed **grade of membership functions**, shown in Figure 5.1. The information with their normalized values are then recorded in a database in the form of a data-tree, to be described shortly.

As an illustration of the normalization process, it is observed from Figure 5.1 that an information with a fuzzy membership of 0.12, collected from a source with very poor authenticity level (AL), gives rise to a normalized membership of 0.85 (shown by dotted lines).

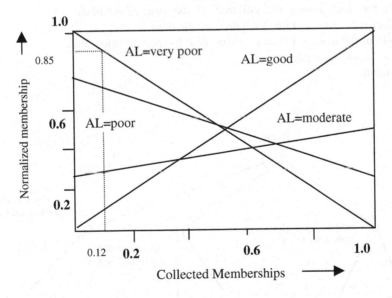

Fig. 5.1: Normalization of fuzzy memberships using membership functions.

5.2.1 The Data-tree

The database in the proposed ES has been organized in the form of a data-tree (Fig. 5.2) having a depth of three levels [8-10]. The root simply holds the starting pointer, while the second level consists of all the predicates, and the third level contains the relevant facts corresponding to each predicate of the second level. The normalized memberships corresponding to each fact are also recorded along with them at the third level. Such organization of the data-tree helps in efficient searching in the database. To illustrate the efficiency of searching, let us consider that there exist P number of distinct predicates and at most L number of facts under one predicate. Then to search a particular clause, say has-alibi (jadu) in the data-tree, we require P + L number of comparisons in the worst case, instead of P × L number of comparisons in a linear sequential search. Now, we present the algorithm for the creation of the data-tree.

Procedure create-tree (facts, fuzzy-membership)
Begin
Create root of the data-tree;
Open the datafile, i.e., the file containing the database;
Repeat
  i) Read a fact from the datafile; //after reading, the file pointer increases//
  ii) **If** the corresponding predicate is found at the second level of
      the data-tree, **Then** mark the node **Else** create a new node at the
      second level to represent the predicate and mark it;
  iii) Search the fact among the children of the marked predicate;
  iv) **If** the fact is not found, **Then** include it at the third level
      as a child of the marked predicate along with its membership value;
Until end-of-datafile is reached;
 Close the datafile;
End.

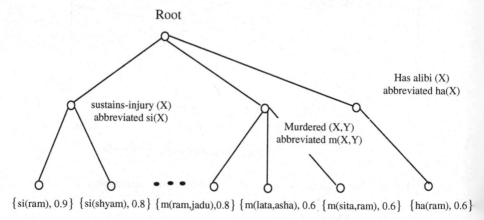

Fig. 5.2: The data-tree representation of the database.

Example 5.1: Given six facts with representative membership values, how should we organize them in the form of a data-tree? We first identify the corresponding predicates and define them as the offspring of the root node and label them. Next we attach the facts along with their memberships as the children of the respective predicates (Fig. 5.2).

5.3 The Knowledge Base

The KB of the proposed ES comprises of a set of default rules (DR) [1], a set of production rules (PR) and two working memory (WM) modules. DR are used to guess about the suspects. The typical structure [15] of the DR is as follows:

$$DR1: \frac{SR\,(Y,\,X)\;OR\;(M\,(X)\wedge M\,(Y)\wedge L\,(X,\,w\,(Y)))}{S\,(Y,\,murderer\text{-}of(X))}$$

which means that if (Y had strained relations with X) or ((X and Y are males) and (X loves wife of Y)) then unless proved otherwise, (suspect Y as the murderer of X).

The variables X, Y, ... of a predicate are instantiated by users through a query language. A working memory WM2 (Fig. 5.3) in the proposed system [8] is used to hold all the instantiated antecedents of a default rule. If all the bound variables in the antecedent part of a rule are consistent, then the rule is fired and the values bound to Y for the predicate suspect (Y, murderer-of (X)) are recorded. These values of Y together with a particular value of X (=x say) are used subsequently as the space of instantiation for each variable in each PR.

The list of suspects, guessed by the DR, is also used at a subsequent stage for identifying the suspect-nodes (places) of the FPN. The process of collecting data from users through query language and representation of those data in the form of a tree, called default-data-tree, can be best described by the following procedure.

Procedure default-data-tree formation (no-of-predicates)
    //no-of-predicates denotes number of predicates
        in the entire set of default rules//
Begin
  Create root-node;
  i:= 1;
  While i ≤ no-of-predicates **do**
  Begin
    write "Does predicate p_i (X,Y) exist ?" ;

read (reply);
If reply = no **Then** create a predicate p_i (X, Y) at the
second level under the root;
Repeat
    write "predicate p_i (X,Y), X = ? ,Y = ?" ;
    read (X, Y);
    If (X =a) and (Y =b) **Then** check whether p_i(a, b) exist
    at the third level below predicate p_i;
    If no **Then** create a data node p_i(a, b) under the predicate p_i;
    Until (X= null) and (Y= null);
    //X=Y= null indicates no more p_i data clauses//
    i= i+ 1;
End While;
End.

Procedure default-data-tree is similar to the create-tree procedure with a small difference in the input/output. In case of create-tree, data are read from a file, while data for the default-data-tree is directly received from the users online. Secondly, the create-tree procedure stores the facts along with their memberships at the leaves of the tree. In case of default-data-tree, facts are binary, so question of storing memberships does not arise.

Once a default-data-tree is formed, the following procedure for suspect-identification may be used for guessing the suspects with the help of the file of default rules (DRs) and the default-data-tree.

Procedure suspect-identification (DRs, default-data-tree)
Begin
i : = 1;
Open the file of DRs;
  Repeat
    If (all the predicates connected by AND operators OR at least
      one of the predicates connected by or operator in
      DR_i are available as predicates in the default-data-tree)
    Then do
    Begin
      Search the facts (corresponding to the predicates) of
      the DR_i under the selected predicates in the tree so
      that   the possible variable bindings satisfy the rule;
      Fire the rule and record the name of suspects in a list;
    End;
    i = i + 1;
  Until end-of-file is reached;
  Close the file of DRs;
End.

The structure of a production rule (PR$_1$) used in our ES is illustrated through the following example.

PR$_1$: murdered (X, Y):-
    suspect (Y, murderer-of(X)),
    has-no-alibi (Y),
    found-with-knife ((Y), on-the-spot-of-murder-of (X)).

The above rule states that *"if Y is a suspect in the murder history of X, and Y has no alibi, and Y has been found with knife on the spot of murder of X then declare Y as the murderer of X."* Variables of each predicate under a PR, in general, have to be instantiated as many times as the number of suspects plus one. The last one is due to the person murdered. However, some of the variable bindings yield absurd clauses like has-alibi (ram) where 'ram' denotes the name of the person murdered. Precautions have been taken to protect generation of such absurd clauses. The procedure for instantiating variables of PRs with the name-list of suspects and the person murdered is presented below. To keep the procedure simple, we intentionally did not take any attempt to prevent the generation of absurd clauses.

Procedure variable-instantiation-of-PRs (PRs, suspect, person- murdered);
Begin
  For i := 1 to no-of-PRs **do**
  Begin
  Pick up the PR $(X_1, X_2, ..., X_m)$;
  Form new PRs by instantiating $X_1, X_2, ...,X_m$ by elements from the set of
  suspects and the person murdered such that $X_1 \neq X_2 \neq ... \neq X_m$;
  Record these resulting rules in a list;
  End For;
End.

Example 5.2: Consider the rule PR$_1$ given above and the following instances of $(X, Y) \in \{(r, s), (m, s)\}$. After instantiation with $(X, Y) = (r, s)$ and (m, s), we get two instances of the rule as given below:

    murdered(r, s):- suspect(s, murderer-of(r)), has-no-alibi (s), found-with-knife ((s), on-the-spot-of-murder-of (r)), and

    murdered(m, s):- suspect(s, murderer-of(m)), has-no-alibi(s), found-with-knife ((s), on-the-spot-of-murder-of (m)).

Antecedent atomic clauses of these instantiated rules need to be searched in the data-tree to test the firing ability of a rule.

Once the instantiation of the variables in the production rules based on the procedure presented above is over, the facts of the resulting rules are searched in the static database, represented in the form of a data-tree. A working memory WM1 keeps track of the instantiated clauses of production rules, after these are detected in the data-tree. Moreover, if all the AND clauses of a PR are found in the data-tree, the rule is fired and the consequence of the PR is also recorded in the WM1. The information stored in WM1 is subsequently used to encode the place-transition pairs of FPN, corresponding to the PR under consideration. The WM1 is then cleared for future storage.

5.4 The Inference Engine

The inference engine (IE) of the said ES comprises of five modules, namely, i) the module for searching the antecedents of PRs on the data-tree, ii) the FPN formation module, iii) membership-revision and limit-cycle detection and elimination module, iv) nonmonotonic reasoning module, and v) decision making [10] and explanation tracing module.

5.4.1 Searching Antecedent Parts of PR in the Data-tree

The following procedure describes the technique for searching the instantiated clauses of a PR in a data-tree. The algorithm first searches the predicates corresponding to the antecedent clauses in the data-tree. On finding the predicate, the algorithm searches the clauses under the found predicate. The rest of the algorithm is self-explanatory.

Procedure search-on-data-tree (data-tree, key-data)
//searches the antecedents of PRs on the data-tree//
Begin
  Search the predicate corresponding to the key-clause in the
  second level of the data-tree;
  If search is successful **Then** do
  Begin
    Search the key-clause among the children of the
    predicate under consideration;
    If search is successful **Then** print "key found"
    Else print "key-clause not found";
  End,
  Else print "key-predicate not found";
End.

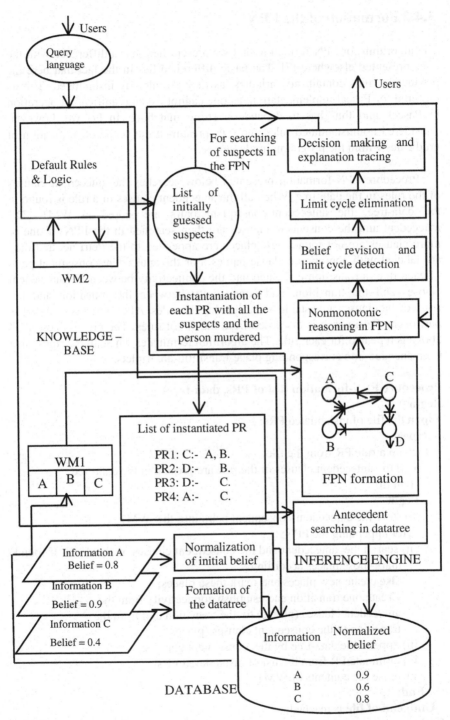

Fig. 5.3: The schematic architecture of the overall system.

5.4.2 Formation of the FPN

The algorithm for FPN-formation that we present here has a difference with the ones presented elsewhere [7]. The basic difference lies in the fact that here the production rules contain no variables, as they are already instantiated. This is required for Pascal implementation, as one cannot execute unification operation in Pascal, and thus has to depend on string matching. In [7] we, however, considered noninstantiated rules and thus presumed the scope of unification of predicates within the programming language.

Procedure FPN-formation presented below searches the antecedent clauses of the rules in the data-tree. When all the antecedent clauses of a rule is found in the data-tree, the antecedent-consequent pairs are stored in WM1. The antecedent and the consequent clauses are then searched in the FPN. If one or more clauses are not found, new places are appended to the Petri net and they are labeled with appropriate clause names. For the antecedent-consequent pairs of each rule, a transition is created and the connectivity between the antecedent clauses and the transition, and connectivity between the transition and the consequent clauses are established following the rules. After a rule is encoded as part of the Petri net, it is discarded from the set of rules. The process described above is repeated for each rule. The algorithm terminates when all the rules have been encoded into corresponding place-transition-arc triplets.

Procedure FPN-formation (list of PRs, data-tree)
Begin
Open the file of instantiated PRs;
  Repeat
   Pick up a rule PR from the file;
    If all the antecedent clauses of the rule are present in the data-tree
    Then do
    Begin
      i) store the antecedent-consequent pairs into the WM1
       for appending the FPN;
      ii) search the antecedent and the consequent clauses on the FPN; **If** found,
       mark those places accordingly,
       Else create new places and mark these places;
       Create one transition and establish connectivity from the
       antecedent clauses (places) to the transition and from the
       transition to the consequent clauses (places);
      iii) append the data-tree by the consequent clauses;
      iv) eliminate PR from the list of instantiated PRs;
      v) erase the contents of WM1;
    End;
  Until end of file is reached;
End.

After the construction of the FPN is over, we initialize the beliefs of the propositions/predicates mapped at the appropriate places. Usually, the beliefs are normalized to keep them free from subjective bias of the information sources prior to their assignment at the places. An expert of the subject domain then assigns the thresholds and certainty factors to the respective transitions. In case the Petri net includes contradictory evidences, nonmonotonic-reasoning procedure is executed to eliminate contradictory/mutually inconsistent information from the reasoning space. The belief revision algorithm is then invoked. If limit cycles are detected, the limit cycle elimination algorithm is executed to eliminate limit cycles. Details of these steps have been covered in the last chapter.

5.4.3 Decision Making and Explanation Tracing

After an equilibrium (steady state) condition of fuzzy beliefs is attained at all places of the FPN, the decision-making and explanation-tracing module is invoked for identifying the suspect-predicate with the highest steady-state fuzzy belief among a number of competitive suspects. The suspect thus identified is declared as the culprit. The module then starts backtracking in the network, starting from the selected culprit-place and terminating at the axioms and thus the reasoning paths are identified. Fuzzy gains are then computed for all the reasoning paths, terminating at the culprit-place. The path with the highest fuzzy gain, defined in the last chapter, is then identified and information on this path is then used for providing an explanation to the user.

5.5 A Case Study

The ES presented above is illustrated in this section with reference to a simulated criminal investigation problem. The problem is to detect a criminal from a set of suspects and to give a suitable explanation for declaring the suspect as the criminal.

A set of default rules is first used to find the set of suspects. Let the set of suspects be S = {sita, ram, lata} and the name of the person murdered is 'sita'. The database for the system along with the abbreviated forms is given below.

Database	**Abbreviations**
Husband (ram, sita)	H(r, s)
Wife (sita, ram)	W(s, r)
Girl-friend (lata, ram)	GF (*l*, r)
Boy-friend (madhu, sita)	BF (m, s)
Loved (sita, ram)	L (s, r)
Loved (sita, madhu)	L (s, m)
Loved (ram, lata)	L (r, *l*)

Loved (lata, ram)	L (l, r)
Hated (sita, ram)	Ht (s, r)
Lost-social -recognition (sita)	LSR (s)
Commits-suicide (sita)	S (s)
Tortured (ram,sita)	T (r, s)
Had-strained-relations-with (ram, sita)	SR (r, s)
Murdered (ram, sita)	M (r, s)
Proposed-to-marry (ram, lata)	PTM (r, l)
Accepted-marriage-offer-from (lata, ram)	AMO (l, r)
Has-precedence-of-murder (lata)	HPM (l)
Murdered (lata, sita)	M (l, s)

The generic production rules for the system with the abbreviated predicates and variables, as defined in the database, are presented below.

Production rules

1. LSR (W):- L(W, H), AMO (GF, H), L (W, BF).
2. S (W):- LSR (W).
3. S (W):-T (H, W)..
4. PTM (H, GF):- L (H, GF).
5. Ht (W, H):- PTM (H, GF).
6. .AMO (GF, H):- L (GF, H), PTM (H, GF).
7. M(GF,W):- AMO(GF, H), HPM(GF).
8. T(H, W):- Ht(W, H).
9. T(H, W):- SR(H, W).
10. SR(H, W):-T(H, W).
11. M(H,W):- SR(H, W), PTM(H, GF).

The variables used in the above PRs are first instantiated with the elements of set S. The resulting set of rules, thus obtained, is then stored into a file. A data-tree corresponding to the proposed database is then formed. The data-tree together with the file of instantiated PRs is then used for the formation of the FPN as shown in Figure 5.4. The initial normalized fuzzy memberships corresponding to each predicate are shown in Figure 5.4. It is assumed that the threshold associated with all transitions except tr_6 is zero. The threshold of tr_6 is set to 0.95 arbitrarily.

Through consultation with a dictionary, the system identifies a pair of contradictory evidences namely L(s, r) and Ht(s, r). The nonmonotonic reasoning procedure, now, opens the output arcs of these two predicates and then detects the existence of limit cycles in the FPN. The limit cycles are then eliminated by setting the threshold of tr_4 to 0.9, by using the guidelines of limit cycle elimination. The steady-state fuzzy memberships of the propositions L (s, r) and Ht(s, r) are then found to be 0.4 and 0.6, respectively. Therefore, according to the nonmonotonic reasoning procedure, belief of L (s, r) is permanently set to zero, the opened arcs are reconnected and the initial fuzzy memberships at all places except at L (s, r) are returned. The initial thresholds are also returned to appropriate transitions.

The limit cycle detection procedure, which is then invoked, detects limit cycles. The limit cycle elimination module again eliminates limit cycles by setting the value of threshold for transition tr_4 to 0.9 [8].

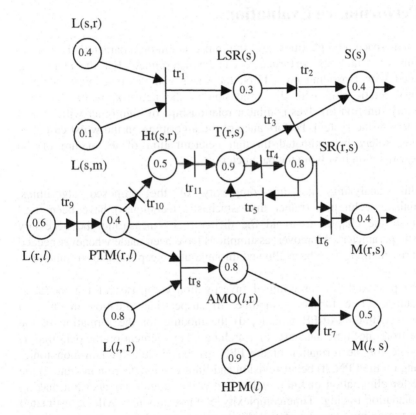

Fig. 5.4: A FPN representing a murder history of a housewife s, where the husband r and the girlfriend *l* of r are the suspects.

The steady-state fuzzy memberships in the network are then computed. Now, out of the three suspect nodes S (s), M (r, s) and M (*l*, s) the steady-state value of M (*l*, s) is found to be the highest (0.6). So, 'lata' is considered to be the culprit. The decision-making and explanation-tracing module then continues backtracking in the network, starting from the place labeled with M (l, s) until the axioms are reached. The reasoning paths L(r, l)-PTM(r, l)-AMO(l, r)-M(l, s), HPM (l)-M(l, s) and L(l, r)-AMO(l, r)-M(l, s) are then identified. Fuzzy gain of all these three paths is then found to be 0.6. Thus the following conclusion and explanation is reported to the users.

Conclusion: Lata is the criminal.

Explanation: *Ram loved Lata. Ram proposed to marry Lata. Lata loved Ram. Lata accepted marriage offer from Ram. Lata has precedence of murder. Therefore, Lata murdered Sita.*

5.6 Performance Evaluation

Time and space complexities are the most common characteristics of an algorithm by which its performance can be determined. For the algorithms associated with reasoning on a FPN, the space complexity exhibits a linear functional relationship with the dimension of the network, while the time-complexity function involves nonlinear relationships of high order with several parameters of the system. Performance of the algorithms in the worst case can, therefore, solely be estimated through determination of the nature of the nonlinear function involving the parameters.

While analyzing the time-complexity of the proposed algorithms, deterministic techniques (rather than stochastic techniques) have been used throughout the chapter to avoid the difficulty of measuring the necessary stochastic parameters. Moreover, assumptions have been made whenever needed and approximations have been allowed throughout to keep the analysis simple.

The proposed ES for criminal investigation comprises of ten modules: a) default-data-tree formation module, b) suspect-identification module, c) variable-instantiation-of-PR module, d) the module for the formation of the datatree from the static database, e) searching a key element on the data-tree, f) the module for the formation of the FPN, g) the module for non-monotonic-reasoning on the FPN, h) belief-revision and limit-cycle-detection module, i) the module for elimination of limit cycles, and j) the module for decision-making and explanation-tracing. Time-complexity of these modules will be estimated now for computing the order of complexity of the entire ES.

5.6.1 Time-Complexity for the Default-Data-Tree-Formation Procedure

For computing the time-complexity of the default-data-tree formation procedure, let us assume that there exists altogether N_P predicates and C_i number of clauses corresponding to the predicate P_i. It is further assumed that the creation of one node or one comparison in the tree requires unit time.

Now, the number of nodes in the tree and hence the time required for the formation of the tree without considering the time spent for searching in the procedure is given by T-node, where

$$\text{T-node} = 1 + N_P + \sum_{i=1}^{N_p} C_i \tag{5.1}$$

Moreover, the condition $X = Y = 0$ is checked $\sum_{i=1}^{N_p} (C_i + 1)$ times.

Further, the total time spent for checking the existence of predicates $= N_P$. So, the overall time-complexity of the procedure is given by TDDT where

$$\text{TDDT} = (1 + N_P + \sum_{i=1}^{N_p} C_i) + \sum_{i=1}^{N_p} (C_i + 1) + N_P$$

$$= 1 + 3N_P + 2 \sum_{i=1}^{N_p} C_i \tag{5.2}$$

If C represents maximum number of clauses/predicates, then the worst case time-complexity of this procedure is given by

$$\text{TDDT (worst case)} < 1 + N_P (3 + 2C) \tag{5.3}$$

$$= O (N_P . C) \tag{5.4}$$

i.e., the order of the time-complexity of the default-data-tree formation procedure in the worst case is given by the product of the maximum number of clauses per predicate and the number of predicates in the tree.

5.6.2 Time-Complexity for the Procedure Suspect-Identification

Before computing the time-complexity, let us consider a general default rule as stated below.

$$\text{DR: } \frac{\overset{(1)}{(A1 \wedge A2 \wedge \ldots \wedge An)} V \overset{(2)}{(B1 \wedge B2 \wedge \ldots \wedge Bm) \ldots . V \overset{(Z)}{(W1 \wedge W2 \wedge \ldots \wedge Wp)}: Q}}{Q}$$

From the above DR, it is clear that a general structure of DR may have say Z sets of AND predicates (i.e., predicates connected by AND operators) and obviously $(Z-1)$ number of OR operators. Further, let us assume that there exist a_i, b_j and w_k sets of clauses corresponding to A_i, B_j, and W_k predicates, respectively. So, number of comparisons, required for checking the consistency of clauses in a DR in the worst case is computed as follows:

number of searches/rule (worst case)

$$= \prod_{i=1}^{N} a_i + \prod_{j=1}^{m} bj + \ldots\ldots + \prod_{k=1}^{p} w_k \qquad (5.5)$$

If C is the maximum number of clauses/predicate and there exists at most r number of AND predicates (within one of the Z-sets of AND predicates) then equation (5.5) reduces to

number of searches/ rule (worst case) $< Z \cdot C^r.$ $\qquad (5.6)$

If the set of DRs contain N_d rules, the worst case time-complexity of the procedure suspect-identification is given by TSI, where

$$TSI = N_d \, Z \, C^r. \qquad (5.7)$$

It may be noted that the time required for checking the existence of AND and OR predicates in the tree has been neglected in the above analysis.

5.6.3 Time-Complexity for the Procedure Variable-Instantiation-of-PRs

Let us assume that in one PR there exist v number of variables and in the list of suspects there exist s number of suspects. So, the variables of PRs have to be instantiated by $(s + 1)$ number of data. Generally $v < (s + 1)$.

Now, the first variable in one PR can be instantiated in $(s + 1)$ ways. Once the first variable is fixed, the second variable can be instantiated in s ways, while the third and fourth variables can be instantiated in $(s - 1)$ and $(s - 2)$ ways respectively. Proceeding in this way, all the v variables in the worst case can be instantiated in W ways, say, where

$$W = (s + 1) \cdot (s) \cdot (s - 1) \cdot (s - 2) \ldots\ldots \{(s + 1 - v) + 1\} \qquad (5.8a)$$

$$= \text{factorial } (s + 1) \, / \, \text{factorial } (s + 1 - v) \qquad (5.8b)$$

Further, from expression (5.8a) it may be noted that

$$W \ll (s + 1)(s + 1)(s + 1) \ldots (s + 1) \text{ v-times} = (s + 1)^v \qquad (5.9)$$

If the list of (non-instantiated) PRs contains Npr number of rules, then the number of instantiation in the worst case is given by TVI, where

$$\text{TVI} = \text{Npr} \frac{\text{factorial } (s + 1)}{\text{factorial } (s + 1 - v)} \ll \text{Npr} \ (s + 1)^v. \qquad (5.10)$$

5.6.4   Time-Complexity for the Procedure Create-tree

For computing the time-complexity of the procedure, let us assume that

p = number of distinct predicates in database,
C_i = number of clauses corresponding to predicate Pi,
$t = \sum\limits_{i=1}^{p} C_i$ = total number of clauses in the database.

Analysis of time-complexity of this procedure will be considered in two steps. First, the number of comparisons required for searching the predicates on the data-tree will be evaluated. Then the number of comparisons required for searching the clauses in the tree will be determined.

Since Predicate$_i$ (Pre$_i$) in the static database occurs C_i times (in the form of clauses) and each time a clause has to be positioned in the tree, the predicate is searched once, therefore, Pre$_i$ is searched C_i times in the tree.

For the first occurrence of Pre$_i$ (corresponding to the first clause of Pre$_i$), the search calls for $(i - 1)$ comparisons, while subsequent searching for Pre$_i$ requires i comparisons each. So, total number of comparison of Pre$_i$ in the second level (corresponding to all C_i clauses)

$$= (i - 1) + (i + i + \ldots + i \text{ up to } (C_i - 1) \text{ times})$$
$$= i \ C_i - 1. \qquad (5.11)$$

So, total number of comparisons for all the predicates

$$t = \sum\limits_{i=1}^{p} (i \ C_i - 1) = N1, \text{ say.} \qquad (5.12)$$

Now, let us consider the number of searches for clauses at the third level of the data-tree.

Let the clauses in the database corresponding to Pre_i are in the sequence Cl_{1i}, Cl_{2i}, ... , Cl_{Ci} from the standpoint of their occurrences in the data0tree. So, no comparisons are required for Cl_{1i} while Cl_{2i} and Cl_{Ci} requires one and two comparisons respectively. So, the total number of comparisons of C_i number of clauses corresponding to Pre_i

$$= 0 + 1 + 2 + ... + (C_i - 1)$$
$$= (1/2) C_i (C_i - 1).$$

Now, total number of comparisons in the third level corresponding to the clauses of all the predicates

$$(1/2) \sum_{i=1}^{p} C_i (C_i - 1) = N2, \text{ say.} \tag{5.13}$$

So, total number of comparisons for the formation of the entire data-tree is given by TCT where

$$TCT = N1 + N2 = (1/2) \sum_{i=1}^{p} [C_i^2 + (2i - 1) C_i - 2]. \tag{5.14}$$

Now, since $\sum_{i=1}^{p} C_i = t,$ \hfill (5.15)

$$\sum_{i=1}^{p} C_i^2 = C_1^2 + C2^2 + ... + Cp^2 < (C_1 + C_2 + ... + Cp)^2 = t^2. \tag{5.16}$$

Further,

$$\sum_{i=1}^{p} i \, C_i = C_1 + 2 C_2 + 3 C_3 + ... + p \, Cp$$
$$< p (C_1 + C_2 + C_3 + ... + Cp) = p \, t. \tag{5.17}$$

Substituting expressions (5.16) and (5.17) in (5.14) we obtain:

$$TCT < t^2/2 + p \, t - t / 2 - p \tag{5.18}$$

$$< t (p + t/2). \tag{5.19}$$

5.6.5 Time-Complexity for the Procedure Search-on-Data-Tree

Searching a clause on the data-tree requires searching of the corresponding predicate at the second level and searching for the clause under that predicate at the third level.

For analyzing the time-complexity, let us use the same notations such as p, t and C_i as used in the complexity analysis for the procedure create-tree. Further, we assume C to be the maximum number of clauses/ predicate.

Now, number of searches at the second level of the data-tree in the worst case = p.

Moreover, number of searches at the third level of the data-tree in the worst case = Max $(C_1, C_2, C_3, \ldots, Cp)$ = C.

So, the time-complexity of the procedure search-on-data-tree in the worst case is given by TSDT where

$$TSDT = p + C. \qquad (5.20)$$

In words, the time-complexity of the procedure search-on-data-tree in the worst case can be computed by summing the number of predicates and the maximum number of clauses per predicate.

5.6.6 Time-Complexity for the Procedure FPN-Formation

Most of the time required for the procedure FPN-foundation is spent in searching the antecedent clauses of the instantiated PRs on i) the data-tree and ii) the FPN. Searching on the FPN is needed to avoid duplication of the nodes on the network.

The following assumptions, which can happen in the worst case will be considered in our analysis.

i) Each time the instantiations of the antecedent clauses of a rule is available, one rule becomes fireable and the data-tree is expanded with one consequence predicate. This is called one pass. After each pass, the fireable rule is discarded from the list of rules.

ii)'n' such passes are required for the complete execution of the procedure FPN-formation.

We further assume the following list of notations:

P_o = initial number of predicates in the tree,
t_0 = initial number of clauses in the tree,
C = maximum number of clauses/predicate in the datatree,
Npr = number of instantiated PRs,
Pp = maximum number of predicates/PR,

l =   a variable, representing the number of PRs for which searching have to be continued in the FOR loop,

v =   number of antecedent clauses/ PR in the worst case.

Since number of searches required in the data-tree/ predicate of the PR in the worst case = number of predicates in the tree + maximum number of clauses/ predicate (see section 5.6.5), therefore, the number of searches required for the first pass of the FOR loop is given by l Pp $(P_0 + C)$. Further, since l = Npr in the first pass, therefore, the number of comparisons required in the first pass becomes Npr. Pp. $(P_0 + C)$.

Once the first pass is over, the consequence clause (and hence the predicate) of the first PR are included in the data-tree and the fireable rule is discarded from the list; so l becomes equal to (Npr − 1). Therefore, the number of comparisons required in the second pass is given by $(Npr − 1)$ Pp $[(P_0 + 1) + C]$. So, the number of comparisons at the i-th pass on the data-tree is given by $\{Npr − (i − 1)\}$ Pp $[\{ P_0 + (i − 1)\} + C]$.

Therefore, after Npr passes, the total number of comparisons required on the data-tree is given by SDT where

$$\text{SDT} = \sum_{i=1}^{p} \{Npr − (i − 1)\} \, Pp \, [\, \{ P_0 + i − 1) \} + C \,] \qquad (5.21)$$

$$= Pp \, (Npr/2) \, [(Npr + 1) \, P_0 + C \, (Npr − 1) + (Npr^2 − 1)/3]. \qquad (5.22)$$

Since Npr >> 1, we assume $(Npr − 1) \rightarrow Npr$ and $(Npr + 1) \rightarrow Npr$. So, the expression (5.22) reduces to

$$\text{SDT} = (Npr^2/2) \, Pp \, (P_0 + C + Npr/3). \qquad (5.23)$$

Now, we compute the number of searches required on the FPN during formation of the network.

For the first fireable PR no search on the FPN is needed, since there exists no places on the network. For the next fireable PR, having v antecedents in the worst case, $(v + 1) \, v$ comparisons are required, since each antecedent clause of the second rule requires $(v + 1)$ comparisons. For the third fireable PR, number of comparisons required = $\{2 \, (v + 1)\}v$.

So, for the i-th fireable rule the number of comparisons required on the FPN is given by $\{(i − 1) \, (v + 1)\}v$.

Therefore, total number of comparisons required on the FPN is given by

$$\text{SFPN} = \sum_{i=1}^{p} (i-1) \ v \ (v+1)$$
$$= (\text{Npr}/2) \ (\text{Npr} - 1) \ (v+1) \ v \quad\quad\quad\quad (5.24)$$
$$= (\text{Npr}^2/2) \ v \ (v+1). \quad\quad\quad\quad\quad\quad (5.25)$$

The total number of comparisons required in the worst case for the FPN-formation procedure is thus given by

$$\text{TFFPN} = \text{SDT} + \text{SFPN}$$
$$= (\text{Npr}^2/2) \ [\text{Pp} \ (\text{P}_0 + \text{C} + \text{Npr}/3) + v \ (v+1)]. \quad\quad (5.26)$$

Since $(\text{P}_0 + \text{C}) \ll (\text{Npr}/3)$ and v is negligible in comparison to v^2 expression (5.26) further reduces to

$$\text{TFFPN} = (\text{Npr}^2/2) \ [\text{Pp} \ \text{Npr}/3 + v^2]. \quad\quad\quad (5.27)$$

5.6.7 Time-Complexity for the Belief-Revision and Limit-Cycle-Detection Procedure

The following list of notations has been used for analyzing the time-complexity of the belief-revision procedure.
Let
    M = number of places in a FPN,
    N = number of transitions in a FPN,
    a = number of axioms in a FPN,
    e = number of input arcs of a place in a FPN,
    f = number of input arcs of a transition in a FPN, and
    l = largest reasoning path-length in a FPN without cycle.

A look at the belief-revision algorithm (see chapter 4) reveals that number of operations at a place = $(e - 1)$ OR. Number of operations at a transition = $(f-1)$ AND + 1 AND corresponding to the CF of the transition + 1 comparison for thresholding = $f + 1$.

So, the time-complexity of the algorithm, is given by T_{BU} where

$$T_{BU} = [\text{Me} + \text{N} \ (f+1) \ ] \ l. \quad\quad\quad\quad (5.28)$$

Now, in the worst case, $e \rightarrow N$ and $f \rightarrow M$. Therefore, the worst case value for T_{LU} is given by

$$T_{BU} = (2M + 1) Nl \qquad (5.29a)$$
$$\approx 2M \, Nl \qquad (5.29b)$$

5.6.8 Time-Complexity Analysis for the Procedure Limit-Cycle-Elimination

First, let us compute the time-complexity, considering a single cycle on the FPN which exhibits limit cyclic behavior. However, before analyzing the time-complexity, a set of notations, which will be used throughout the analysis, are presented below.

Let

n = number of transitions on the cycle,

M = total number of places on the FPN,

m_i = number of places on the reasoning paths, starting from the axioms and terminating at the i-th transition lying on the cycle.

It can be shown that the number of comparisons required to obtain the reasoning paths, starting from the axioms and terminating at the i-th transition on the cycle, lying on the cycle is given by $m_i M + M - m_i$ [6-8]. For n-transitions on the cycle, the overall number of searches is given by

$$\sum_{i=1}^{p} (m_i M + M - m_i).$$

Now, the number of comparisons required to find the maximum fuzzy belief on a seasoning path, terminating at the i-th transition on the cycle = $m_i - 1$. The time required for counting the reasoning path length corresponding to this part = m_i.

So, the overall time-complexity for counting the path-length and finding the maximum belief on the paths =

$$\sum_{i=1}^{p} m_i + \sum_{i=1}^{p} (m_i - 1) = \sum_{i=1}^{p} (2 m_i - 1). \qquad (5.30)$$

Now, the formation of the set TAT requires n multiplication and sorting of the elements of TAT requires n^2 comparisons.

So, the overall time-complexity for the limit-cycle-elimination with one cycle only is Coverall where

$$\text{Coverall} = n + n^2 + \sum_{i=1}^{n} (2m_i - 1) + \sum_{i=1}^{n} (m_i M + M - m_i). \qquad (5.31)$$

In the worst case m_i approaches M. So, expression (5.31) reduce to

$$\text{Coverall} = n^2 + 2\,n\,M + n\,M^2. \tag{5.32}$$

Since n is a fraction of M, let n = f. M, where f ≤ 1. Substituting n = f. M in expression (5.32) we obtain:

$$\text{Coverall} = f\,M^3 + (f^2 + 2\,f)\,M^2. \tag{5.33}$$

Since maximum value of f = 1, therefore, (5.33) further reduces to

$$\text{Coverall (worst case)} = M^2\,(M + 3) = O(M^3). \tag{5.34}$$

If two cycles have a shared place, then elimination of limit cycle from one cycle may cause limit cycles to grow in another cycle. Under such situation, the algorithm for detection of limit cycle is undertaken, after eliminating limit cycle from the first cycle. Since the worst case time needed for the detection of limit cycles = (2 M+1) M, therefore, the overall time-complexity is given by

$$\{M^2\,(M + 3)\,\} + \{\,(2M + 1)\,M\,.$$

If an adaptation of threshold of a transition causes introduction of limit cycles into the neighborhood cycles, then the time-complexity in the worst case is given by TLCE, where

TLCE =   worst-case time for elimination of limit-cycle + (n − 1) times the worst case time for detection of limit-cycles.

$$= \quad M^3 + (2\,n + 1)\,M^2 + (n - 1)\,M. \tag{5.35}$$

Since $M^3 + 2\,n\,M^2 \gg M^2 + (n - 1)\,M$, expression (5.35) further reduced to

$$\text{TLCE} = M^2\,(M + 2\,n). \tag{5.36}$$

5.6.9 Time-Complexity for the Procedure Nonmonotonic Reasoning

The following notations are used in the analysis.

y = number of contradictory pair of evidences,
M = number of places in the entire FPN,
n = number of transitions on the cycle.

Now, the time-complexity due to opening and closing of 2y arcs = 4y.

In case of existence of limit cycles in the network, the worst case time-complexity for limit-cycle elimination and belief revision are given by $M^2 (M + 2 n)$ and $M (2 M + 1)$, respectively. Further, the number of comparisons required to identify a predicate from its competitive contradictory counterparts = y.

So, the overall time-complexity of this procedure is given by TNR where

$$
\begin{aligned}
\text{TNR} &= 4y + M^2 (M + 2 n) + M (2 M + 1) + y \\
&= M^3 + (2 n + 2) M^2 + M + 5 y.
\end{aligned} \tag{5.37}
$$

Approximating $(2n + 2)$ by $2n$ and neglecting $(M + 5 y)$ from the last expression, the approximate time-complexity of the procedure is given by TNR (approximate) where

$$
\text{TNR (approximate)} = M^2 (M + 2 n). \tag{5.38}
$$

It may be noted that the approximate time-complexity of non-monotonic–reasoning procedure is equal to that of the limit-cycle elimination procedure.

5.6.10 Time-Complexity for the Procedure Decision-Making and Explanation-Tracing

The following notations are used in the analysis.

s = number of suspects in the murder-history,
mmax = number of places that lie on the reasoning path leading to the conclusion,
x = number of parallel reasoning paths leading to the conclusion,
M = total number of places in the network.

Now, number of comparisons required for identifying the suspect-node with the highest fuzzy belief in the worst case = s – 1.

Moreover, the time-complexity for identifying the backward reasoning paths, starting from the culprit-node and leading towards the axioms = time-complexity for identifying the nodes on the reasoning path through pointers + time-complexity for checking whether these nodes are axioms

$$
\begin{aligned}
&= O(\text{mmax}) + O(\text{mmax} - 1) \\
&< 2\, O(M) - 1.
\end{aligned}
$$

Again, the time-complexity for finding the fuzzy gains = $O(\text{mmax})$ whose worst case value can never exceed $O(M)$.

Lastly, the number of comparisons required for determining the path with the highest fuzzy gain out of x-parallel paths = $x - 1$.

So, the overall time-complexity of the procedure is given by TDM where

$$TDM = (s - 1) + [2\ O(M) - 1] + O(M) + (x - 1)$$
$$= O(3M) + (s + x - 3). \tag{5.39}$$

Since $(s + x - 3) << 3\ M$, expression (5.39) reduces to

$$TDM = 3\ M. \tag{5.40}$$

5.6.11 Time-Complexity of the Overall Expert System

The worst-case time-complexity of the proposed ES can be computed by adding the time-complexities of its components. It is given by TES where

$$TES = 2\ Np\ C + N_d\ Z\ C^r + Npr\ s^v +\ t\ (p + t/2) +$$
$$(Npr^2/2)\ (Pp\ Npr/3 + v^2) + 2\ M^2\ (M + 2\ n) + 3\ M \tag{5.41}$$

$$= 2\ M^3 + 4\ n\ M^2 + (1/6)\ Pp\ Npr^3 + (1/2)\ Npr^2\ v^2 + (1/2)\ t^2 + t\ p + 2\ Np\ C$$
$$+ Nd\ Z\ C^r + Npr\ s^v + 3M. \tag{5.42}$$

It is observed that the first two terms in expression (5.42), which are due to limit cycle elimination and nonmonotonic, have a major contribution in the overall time-complexity. In order to measure their degree of contribution in the overall time-complexity, we consider one typical set of data, appropriate for a simulated criminal investigation problem as follows:

Let $M = 500$, $n = 10$, $Pp = 6$, $Npr = 100$, $v = 6$, $t = 200$, $p = 40$, $Np = 10$, $C = 5$, $N_d = 10$, $Z = 4$, $r = 3$ and $s = 5$.

Substituting the above set of data in expression (5.42), we find

$$TES = 262.7771 * 10^6 \text{ time units,}$$

where as $2\ M^3 + 4\ n\ m^2 = 260 * 10^6$ time units.

The above results show that more than 90% of the overall time-complexity is consumed together by the non-monotonic reasoning and limit-cycle-elimination modules.

5.7 Conclusions

The FPN-based expert system has the merits of representing self-referential rules like A →B, B→C, and C→A, D. When the instantiations of the rules are available, they form cycles in the reasoning space. Complexity of the system grows further when it includes more number of touching cycles. FPN with cycles may exhibit limit cyclic behavior. Elimination of limit cycles is possible for simple cases. Elimination of limit cycles from a FPN with complex multiple touching cycles is an open-ended research problem.

Presence of inconsistent data in cyclic nets makes the reasoning problem much more complex. The analysis of time-complexity reveals that maximum time for reasoning in the proposed ES is consumed by nonmonotonic reasoning and limit cycle elimination procedures.

The proposed ES was realized in Pascal (a procedural language). Pascal does not have the provision of unification like PROLOG. To avoid unification of predicates, we generated all possible instantiations of the variables in the rules by the list of guessed suspects and the person murdered. A Pascal/PROLOG combination at the executable file level can overcome this problem.

Exercises

1. Construct a program for FPN-formation from a set of facts and rules. Assign beliefs and FTTs, CFs and thresholds and write a function for belief revision in the network.

2. Construct a program to check the existence of limit cycles in a FPN and a function to eliminate limit cycles. Test your program by hand calculation for a small cyclic net.

3. Write a function for unification of two predicates in a procedural language that supports string matching.

4. Implement create-tree procedure by your favorite language.

5. Design a program for nonmonotonic reasoning. Your program should call limit cycle detection and elimination modules when needed.

6. Design a knowledge base for weather forecasting and use a Petri net approach to predict the climate at a given geographical location on the earth.

7. Design a general strategy for elimination of limit cycles from a FPN with multiple touching cycles. Construct an algorithm using your strategy. [*open-ended problem*]

8. An alternative approach to non-monotonic reasoning is to submit additional facts and rules. Design an alternative algorithm to eliminate inconsistency from the reasoning space by additional facts/ rules. Test your algorithm by hand computation. [*open-ended problem*]

References

[1] Bernard, P., *An Introduction to Default Logic*. Springer-Verlag, New York, pp. 13-30, 1989.

[2] Buchanan, B. G. and Shortliffe,E. H., *Rule-Based Expert System: The MYCIN Experiments of the Stanford University Heuristic Programming Projects,* Addison-Wesley, Reading, MA, 1984.

[3] Bugarin A. J., "Fuzzy reasoning supported by Petri nets," *IEEE Trans. on Fuzzy Systems*, vol. 2, no. 2, pp. 135-150, 1990.

[4] Cao, T. and Sanderson, A. C., "Task sequence planning using fuzzy Petri nets," *IEEE Trans. on Systems, Man and Cybernetics*, vol. 25, no. 5, pp. 755-769, 1995.

[5] Chen, S., Ke, J. and Chang, J., "Knowledge representation using fuzzy Petri nets," *IEEE Trans. on Knowledge and Data Engineering*, vol. 2, no. 3, pp. 311-390, Sept.1990.

[6] Eshera, M. A. and Brash, S. C., "Parallel rule based fuzzy inference engine on mess connected systolic array," *IEEE Expert*, Winter 1989.

[7] Konar, A., *Artificial Intelligence: Behavioral and Cognitive Modeling of the human Brain*, CRC Press, Boca Raton, FL, 1999.

[8] Konar, A., *Uncertainty Management in Expert Systems Using Fuzzy Petri Nets*, Ph. D. Thesis, Jadavpur University, India, 1994.

[9] Konar, A. and Mandal, A. K., "Non-monotonic reasoning in expert systems using fuzzy Petri nets," *Advances in Modeling and Analysis,* vol. 23, no.1, pp. 51-63, 1992.

[10]   Konar, A. and Mandal, A. K., "Uncertainty management in expert systems using fuzzy Petri nets," *IEEE Trans. on Knowledge and Data Engineering*, vol. 8, no. 1, pp. 96-105, 1996.

[11]   Looney, G. C., "Fuzzy Petri nets for rule based decision making," *IEEE Trans. on Systems, Man, and Cybernetics*, vol. 18, no. 1, Jan.-Feb. 1988.

[12]   Pearl, J., "Distributed revision of composite beliefs," *Artificial Intelligence*, vol. 33, pp. 173-213, 1987.

[13]   Peterson, J. L., *Petri Net Theory and Modeling Systems*, Prentice-Hall, Englewood Cliffs, NJ, pp. 7-30, 1981.

[14]   Pedrycz, W. and Gomide, F., "A generalized fuzzy Petri net model," *IEEE Trans. on Fuzzy Systems*, vol. 2, no. 4, pp. 295-301, 1994.

[15]   Reichenbach, H., *Elements of Symbolic Logic*, Macmillan, New York, pp. 218, 1976.

[16]   Scarpelli, H., Gomide, F. and Yager, R., "A reasoning algorithm for high level fuzzy Petri nets," *IEEE Trans. on Fuzzy Systems*, vol. 4, no. 3, pp. 282-295, 1996.

[17]   Shafer, G. and Logan, R., "Implementing Dempster's rule for hierarchical evidence," *Artificial Intelligence*, vol. 33, 1987.

[18]   Toagi, M. and Watanabe, H., "Expert system on a chip: an engine for real time approximate reasoning," *IEEE Expert*, Fall 1986.

[19]   Wang, H., Jiang, C. and Liao, S., "Concurrent reasoning of fuzzy logical Petri nets based on multi-task schedule," *IEEE Trans. on Fuzzy Systems*, vol. 9, no. 3, 2001.

[20]   Yu, S. K., Knowledge representation and reasoning using fuzzy Pr/T net-systems," *Fuzzy Sets and Systems*, vol. 75, pp. 33-45, 1995.

Chapter 6

Distributed Learning Using Fuzzy Cognitive Maps

Mammals perform spatial reasoning by a specialized structure called cognitive maps located in the hippocampus region of their forebrain. In the treaties of machine intelligence, the phrase cognitive maps, however, have a wider meaning. They include encoding of knowledge about causal events and their automated recall. Modeling of cognitive maps by fuzzy logic is apparent because of the inherent fuzziness in most of the real-world data and knowledge bases. This chapter provides an overview of various models of cognitive maps and their learning behavior. The dynamics of the learning models have been analyzed to determine the condition for their stability. The chapter concludes with a discussion of the scope of application of the proposed models in practical engineering systems.

6.1 Introduction

The word "cognition" generally refers to a faculty of mental activities dealing with abstraction of information from a real-world scenario, their representation and storage in memory and automatic recall [1-15], [31], [34]. It also includes construction of high-level percept [33] from primitive/low-level information/knowledge, hereafter referred to as perception. There is plenty of literature dealing with psychological models of cognition [16-22]. The models include representation of human thoughts on memory [24-28], [36] iconic and echoic representation [21-29], [31-32], [34-35] of knowledge, part and whole relationship of objects in human memory, understanding complex problems [24] and many others.

In the treaties of brain sciences, the phrase *cognitive map* has a traditional meaning, referring to a specialized structure of the mammalian brain. Computational scientists, however, use the phrase cognitive map in a lucid sense to represent cause-effect relationships of events in a knowledge base. For instance, *a unit rise in temperature that will cause a severe degree of malfunctioning in the system* can easily be modeled with a cognitive map [30], which by classical models of knowledge representation cannot be described for their limitation in representing *causal* relations.

The *fuzzy cognitive map* has very recently been introduced in the fields of machine intelligence. Coined by Bart Kosko, the phrase *fuzzy cognitive map* refers to a graph theoretic structure capable of encoding knowledge by employing the logic of fuzzy sets. The chapter outlines different models of fuzzy cognitive maps and their learning paradigm.

The chapter has been divided into eight sections. Section 6.2 outlines Axelrod's model of cognitive maps. The extension of Axelrod's model by Bart Koko is presented in Section 6.3. Further extensions of Kosko's classical model are presented in Sections 6.4 and 6.5. Zhang, Chen, and Bezdek's model is presented in Section 6.6. Pal and Konar's model is outlined in Section 6.7. Conclusions are listed in Section 6.8.

6.2 Axelrod's Cognitive Maps

Axelrod (see [19]) introduced cognitive maps in the 1970s for representing social scientific knowledge. These maps are signed directed graphs where "nodes" denote concepts (like "social instability") and the directed edges denote causal connections. A positive or negative sign is attached with the edges to denote causal increase or decrease, respectively. For instance, a positive (negative) edge from node A to node B implies that A causally increases (decreases) B.

Axelrod's Model

Axelrod employed adjacency matrix to represent cognitive maps. Let e_{ij} be an edge describing the causal relation from concepts c_i to c_j. Then

$$e_{ij} = 1, \text{ if } c_i \text{ causally increases } c_j,$$
$$= -1, \text{ if } c_i \text{ causally decreases } c_j,$$
$$= 0, \text{ if } c_i \text{ imparts no causality to } c_j.$$

Thus, for a given map, shown in Figure 6.1, we get the adjacency matrix E.

To

From		c_1	c_2	c_3	c_4	c_5	c_6
E=	c_1	0	-1	1	0	0	0
	c_2	0	0	0	1	0	0
	c_3	0	0	0	0	1	0
	c_4	0	0	0	0	0	-1
	c_5	0	0	0	-1	0	-1
	c_6	0	0	0	0	0	0

$c_1, c_2, \ldots, c_6$ in the above adjacency matrix correspond to the concepts in Figure 6.1.

A question then naturally arises: what is the utility of the adjacency matrix E? Suppose we want to see the causal effect of c_1 and c_4 jointly; in that case, we can multiply the row vector C given below by the matrix E.

$$C = [\; 1 \quad 0 \quad 0 \quad 1 \quad 0 \quad 0 \;]$$
$$ c_1 \quad c_2 \quad c_3 \quad c_4 \quad c_5 \quad c_6$$

Then

$$C E = [\; 0 \quad -1 \quad 1 \quad 0 \quad 0 \quad -1 \;]$$
$$ c_1 \quad c_2 \quad c_3 \quad c_4 \quad c_5 \quad c_6$$

which indicates the effect of c_1 on both c_2 and c_3 and the effect of c_4 on c_6.

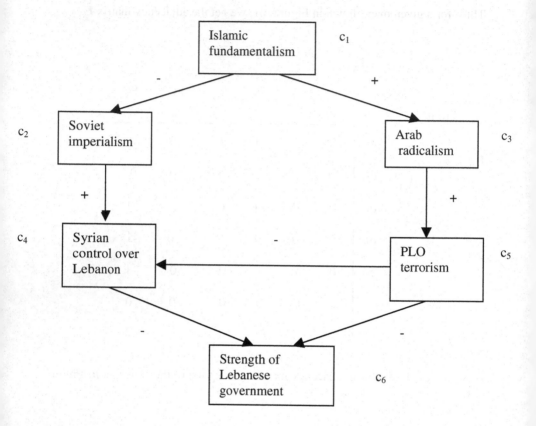

Fig. 6.1: A cognitive map describing the political relations for Middle East peace.

6.3 Kosko's Model

Kosko [19] formalized the definition of causality in a fuzzy cognitive map. According to him,

Let

    c_i = a concept, and

    Q_i = a fuzzy linguistic set (much c_i, more or less c_i, etc.) of c_i.

Then, for any two concepts c_i and $c_{j,}$, c_i causes c_j iff

    i) $Q_i \subset Q_j$ and $\neg\, Q_i \subset \neg\, Q_j$

    ii) $Q_i \subset \neg Q_j$ and $\neg\, Q_i \subset Q_j$,

"⊂" stands for logical implication. In rule (i) c_i causally increases c_j, whereas in rule (ii) c_i causally decreases c_j.

Example 6.1: Let c_1 = Islamic fundamentalism, c_2 = Soviet imperialism, c_3 = Arab radicalism. Then, vide Figure 6.1, we have

$$\overset{+}{c_1 \rightarrow c_3} \quad \text{and} \quad \overset{-}{c_1 \rightarrow c_2}$$

where labels + or − represent positive or negative causality.

Suppose we want to express the knowledge:

1) Extensive Islamic fundamentalism increases massive Arab radicalism.
2) Extensive Islamic fundamentalism causes a severe fall in Soviet imperialism.

Let Q_1= Extensive, Q_2 = Severe, and Q_3= Massive. Then, the above two rules by the definition of causality should satisfy:

1. $Q_1(c_1) \subset Q_3(c_3)$ and
   $\neg Q_1(c_1) \subset \neg Q_3(c_3)$;

2. $Q_1(c_1) \subset \neg Q_2(c_2)$ and
   $\neg Q_1(c_1) \subset Q_2(c_2)$.

In extending FCMs, Kosko considered fuzzy labels of the edges in the map. According to him, if the labels satisfy a *partial ordered relation* with respect to ≤ operator, then we can evaluate the causal effect of one node on the desired node. For instance, let

$$P = \{none \leq some \leq much \leq a\ lot\}$$

be the *partially ordered set* of attached label of the edges. Then, for the given map (vide Fig. 6.2) we find the causal effect of c_1 over c_5 through three paths:

1. $c_1 - c_2 - c_4 - c_5$
2. $c_1 - c_3 - c_5$
3. $c_1 - c_3 - c_4 - c_5$.

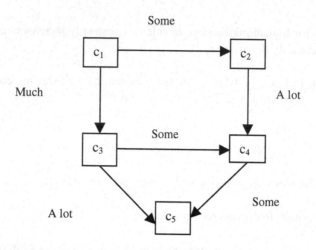

Fig. 6.2: A cognitive map with fuzzy labels at the edges.

Here, the causal effect of c_1 over c_5 is determined by taking the minimum of the attached labels of the individual paths. Let I_1, I_2, and I_3 denote the effect of c_1 on c_5 through the paths 1 to 3, respectively, and e_{ij} be the label attached with edge from node i to node j. Then

$$I_1(c_1, c_5) = \text{Min}\{e_{12}, e_{24}, e_{45}\}$$
$$= \text{Min}\{\text{Some, A lot, Some}\}$$
$$= \text{Some,}$$

$$I_2(c_1, c_5) = \text{Min}\{e_{13}, e_{35}\}$$
$$= \text{Min}\{\text{Much, A lot}\}$$
$$= \text{Much,}$$

$$I_3(c_1, c_5) = \text{Min}\{e_{13}, e_{34}, e_{45}\}$$
$$= \text{Min}\{\text{Much, Some, Some}\}$$
$$= \text{Some.}$$

To determine the total effect of c_1 on c_5, we take the maximum of I_1 through I_3. Thus, total effect of c_1 on c_5, denoted by $T(c_1, c_5)$ is computed below:

$$T(c_1, c_5) = \text{Max}\{ I_1(c_1, c_5), I_2(c_1, c_5), I_3(c_1, c_5)\}$$
$$= \text{Max}\{\text{Some, Much, Some}\}$$
$$= \text{Much.}$$

In other words, c_1 imparts *much* causality to c_5.

6.4 Kosko's Extended Model

Kosko extended the elementary model of Axelrod by including a nonlinear function in the model [20]. To be specific, let $E_{(n \times n)}$ be the incidence matrix of an FCM, and $C = [c_1 \ c_2 \ ......c_n]$ be the given state vector for the FCM. Here c_i, the i-th component of vector C, denotes the strength of concept c_i. Then the next state vector can be evaluated by

$$C(t + 1) = S [C(t) . E] \tag{6.1}$$

where

      S is a nonlinear function applied over the individual components of the vector-matrix product;

and t denotes the time.

The incorporation of nonlinearity sometimes forces the cognitive map to recycle through states.

Example 6.2: Let E be the incidence matrix for a given FCM of five concepts: c_1, c_2, c_3, c_4, c_5. Let

	To	c_1	c_2	c_3	c_4	c_5
From						
E =	c_1	0	1	0	-1	0
	c_2	0	0	1	0	-1
	c_3	0	-1	0	1	-1
	c_4	1	0	-1	0	1
	c_5	-1	1	0	-1	0

Let S be a binary hard-type nonlinearity given by

$$S(a) = +1 \quad \text{for } a > 0,$$
$$= 0 \quad \text{for } a \leq 0.$$

Then for a given initial state:

$$C_1 = [0\ 0\ 0\ 1\ 0].$$

We find limit cyclic behavior of the system through the following states:

$$c_1 - c_2 - c_3 - c_4 - c_1.$$

6.5 Adaptive FCMs

Kosko further extended his FCM by incorporating adaptive learning behavior of their edges [20]. Let e_{ij} be the fuzzy strength of an edge from node i to node j. Let x_i and x_j be the signal strength of concept c_i and c_j, respectively. Let S be a nonlinear function. Then the Hebbian-type learning law used to adapt the edge strength e_{ij} is given below:

$$\frac{de_{ij}}{dt} = -\beta\, e_{ij} + S\,(x_i)\,.\,S\,(x_j),\ \beta > 0 \tag{6.2}$$

where

$$S(x_k) = 1/\,[1 + \exp(-\alpha\, x_k)],\ \alpha > 0,\ k \in \{i,j\} \tag{6.3}$$

is a Sigmoid function.

The $-\beta e_{ij}$ term in expression (6.2) denotes a natural decay in the fuzzy edge strength, and the second term in the right side ensures strengthening of edge connectivity with the nodes.

A discrete version of equation (6.2) is given by

$$e_{ij}(t+1) = (1 - \beta)\, e_{ij}(t) + S\,(x_i)\,.\,S\,(x_j). \tag{6.4}$$

For a given $e_{ij}(0)$, we can iterate the above expression until e_{ij} attains a steady value e_{ij}^*. Once $e_{ij} = e_{ij}^*$ for all i. j, the encoding process is terminated. For recall in an FCM, Kosko used equation (6.1), as discussed earlier. The recall model, however, gets trapped within limit cycles.

Among the other adaptive learning equations, the differential Hebbian learning law presented below needs mention:

$$\frac{de_{ij}}{dt} = -\beta\, e_{ij} + \frac{dS(x_i)}{dt}\ \frac{dS(x_j)}{dt}\qquad \text{for } \beta > 0 \tag{6.5}$$

The advantage of the differential Hebbian model over the Hebbian model lies in learning higher-order causal relations in case former model correlates multiple cause changes with effect changes.

The discrete version of the model has successfully been used for encoding edge strengths in "virtual reality" systems. This encoding model too is used in conjunction with the recall model (6.1), which, because of its limit cyclic behavior, helps cycling the events in virtual reality systems.

6.6 Zhang, Chen, and Bezdek's Model

Zhang et al. [40] in late 1980s presented a novel scheme for cognitive reasoning using FCM. They defined **N**egative **P**ositive **N**eutral (NPN) logic for both crisp and fuzzy variables. In the case of a crisp variable, NPN logic allows a valuation space of $\{-1, 0, 1\}$, whereas for a fuzzy variable the valuation space is $[-1, 1]$. The following operations are essentially needed to use this logic.

1. $x * y = \text{Sign}(x)\,\text{Sign}(y)\,(|x| * |y|)$ (6.6)

    where $*$ is any T-norm in $[-1, 1]$. Typically, $*$ is the fuzzy AND (min) operator.

2. $(x, y) * (u, v) = [\,\min(x * u, x * v, y * u, y * v),$
    $\qquad\qquad\quad \max(x * u, x * v, y * u, y * v)]$ (6.7)

3. $(x, y)\ \text{OR}\ (u, v) = [\min(x, u), \max(y, v)]$.   (6.8)

In NPN logic-based FCM, the edges are labeled with a doublet (u, v), where u and v are the minimum and the maximum fuzzy strength of the edge. For formalization, let us consider a fuzzy relation R in $X \times Y$ where $X = \{x_i\}$ and $Y = \{y_j\}$ are finite sets. Then $\mu_R(x_i, y_j)$ is the membership of the edge connecting nodes x_i and y_j. Now, suppose for a given edge e_{ij} a set of experts assigned different values of $\mu_R(x_i, y_j)$. Let there be k number of experts, m of which assigned negative values $u_1, u_2, \ldots, u_m$, say, and $(k - m)$ number of experts assigned positive values $v_1, v_2, \ldots, v_{k-m}$, say. Then we define

$$a = \Sigma u_i / m \qquad\qquad\qquad (6.9)$$

and $b = \Sigma v_i / (k - m)$ (6.10)

and attach (a, b) as the label of the edge e_{ij}.

One important characteristic of NPN relation is the transitivity. Let $X = \{x_1, x_2, \ldots, x_n\}$ be a finite set. An NPN relation R is **transitive** iff for all i, j, k, such that $0 < i, j, k \leq n$,

$$\mu_R(x_i, x_k) \geq \max_{x_j} [\mu_R(x_i, x_j) * \mu_R(x_j, x_k)] \qquad (6.11)$$

The (max-*) composition of two NPN relations $R \subseteq (X \times Y)$ and $Q \subseteq (Y \times Z)$, denoted by R o Q is defined below:

$$\mu_{R \circ Q} = \max [\mu_R(x, y) * \mu_Q(y, z)] \text{ for } x \in X, y \in Y \text{ and } z \in Z \qquad (6.12)$$

In the last expression, max is equivalent to OR. The n-fold composition of R is denoted by R^n, where

$$R^n = R \circ R \circ \ldots \circ R \text{ (n-times)}.$$

The transitive closure R^{TC} of an NPN relation R in X is the smallest (max-*) transitive NPN relation containing R. It can be shown that for $R_{(n \times n)}$,

$$R^{TC} = R^1 + R^2 + \ldots + R^{2n} \qquad (6.13)$$

Zhang et al. categorized their work into three components: i) cognitive map building, ii) cognitive map understanding, and iii) decision making. The first component is concerned with the fusion of multiple experts' opinion to determine the doublet of the fuzzy edges. The second component deals with determination of the heuristic transitive closure (HTC) of an NPN relation by a HTC algorithm, and finds the two most effective paths between any two elements of the FCM by invoking a heuristic path (HP) finding algorithm. Given $\mu_R(i, k) = (x, y)$ and $\mu_R(k, j) = (u, v)$, say, the HTC algorithm determines

$$\mu_R (i, j) = (x, y) * (u, v) \qquad (6.14)$$

for all i, j, and k = 1 to n. Once the computation is over, the algorithm evaluates

$$R^{TC} = R \circ ([I] + R) \qquad (6.15)$$

where $R = [\mu_R (i, j)]_{(n \times n)}$.

The HP algorithm, on the other hand, explores depth first search on the graph to determine the most effective paths that cause a positive maximum and a negative minimum effect.

The third component is needed to answer the users' query from the resulting outcome of the second phase. Suppose, for instance, the second phase returns a label (−0.3, 0). Then the answer to the query: "whether there is any positive causal effect of x_i over x_j?" should be answered in the negative.

6.7 Pal and Konar's FCM Model

Pal and Konar presented an alternative model [29] of fuzzy cognitive map with a slightly different nomenclature. According to them, a cognitive map is an associative structure consisting of nodes and directed arcs where the nodes carry fuzzy beliefs and the arcs or edges carry connectivity strength.

Let

$n_i(t)$ and $n_j(t)$ be the fuzzy beliefs of node N_i and N_j, respectively, and

δw_{ij} (t) be the change in connectivity strength (CS) w_{ij} of an edge E_{ij} connected from node N_i to N_j.

Then, following the principles of Hebbian learning and considering self-mortality of w_{ij}, we can write

$$\delta w_{ij}(t) = -\alpha w_{ij}(t) + S(n_i(t)) \cdot S(n_j(t)) \qquad (6.16)$$

where
$$S(n_k) = 1 / [1 + \exp(-n_k)] \quad \text{for } k = \{i, j\} \qquad (6.17)$$

and α represents the forgetfulness (mortality) rate.

It may be noted that that the encoding model (6.16) is similar to Kosko's model (6.2). But there is a significant difference between the recall models of Pal and Konar and that of Kosko. The recall model of Pal and Konar in point-wise notation is given below.

$$n_i(t + 1) = \text{Max}[n_i(t), \text{Max}_k \{(n_k \text{ Min } w_{ki})\}] \qquad (6.18)$$

The above equation states that the cognitive system will attempt to restore the last value of N_i through a memory refresh cycle. The fuzzy belief of node N_i thus will assume either i) the last value or ii) the supremum of (n_k, w_{ki}), whichever is larger.

To study the stability of the cognitive map, Pal and Konar [30] derived some interesting results, presented below.

Property 1: *For all t, $n_i(t +1)$ in expression (6.18) is bounded between 0 and 1.*

Proof: Let us consider three possible ranges of w_{ki}: (i) $w_{ki} < 0$, (ii) $0 < w_{ki} \leq 1$, and (iii) $w_{ki} > 1$.

Case I: When $w_{ki}(t) < 0$, for $\exists(\forall)k$, let us assume that $\max_k \{\min (n_k(t), w_{ki}(t))\} < 0$. However, since $0 < n_i(0) < 1$, recursive use of expression (6.18) reveals that $\forall t, 0 < n_i(t+1) < 1$.

Case II: When $0 < w_{ki}(t) < 1$, $\exists(\forall)k$, the proof of property 1 is obvious.

Case III: When $w_{ki}(t) > 1$, $\exists(\forall)k$, since $0 < n_k(0) < 1$, $0 < \min (n_k(0), w_{ki}(0)) < 1$. So $\max_k \{(\min(n_k(0), w_{ki}(0))\} < 1$. Again, since $0 < n_i(0) < 1$ by expression (6.18), $0 < n_i(1) < 1$. Property 1 thus can easily be proved by the method of induction.

Property 1, therefore, holds for all the above cases.

Theorem 6.1: *The recall model given by expression (6.18) is unconditionally stable.*

Proof: It is clear from expression (6.18) that $n_i(t + 1) \geq n_i(t)$ for all t. Since an oscillation requires both an increase and decrease in value and $n_i(t + 1)$ is never less than $n_i(t)$, for any t, therefore, $n_i(t + 1)$ cannot exhibit oscillatory time-response. Further, since $0 < n_i(t + 1) < 1$ for all t, vide property 1, and $n_i(t)$ does not exhibit oscillation, so $n_i(t)$ must reach steady state with value between 0 and 1.

Since the model represented by expression (6.18) need not require to satisfaction of any condition for its stability, the statement of the theorem follows.

Corollary 1 follows from the above theorem. In this corollary, we consider a node vector **N**, the i-th component of which represents the fuzzy belief of node N_i.

Corollary 6.1: *Given a node vector* N *such that its i-th component corresponds to the belief of node* N_i *for all i. Also, given an edge connectivity matrix* W *whose (i, j)-th component* w_{ij} *denotes connectivity strength from node* N_j *to node* N_j. *The condition for steady-state for the node vector* N *then satisfies the following inequality:*

$$N^* \geq W^{*T} \text{ o } N^* \qquad\qquad (6.19)$$

where N^* and W^* represent the steady-state values of N and W, respectively.

Proof: The vector-matrix form corresponding to expression (6.18) is given by

$$N(t + 1) = N(t) \vee (W^T o N(t)) \tag{6.20}$$

where $\vee$ and o denote fuzzy or (max) and max-min composition operators respectively. The $\vee$ operator between two vectors computes the component-wise maximum of the two vectors like addition of two column vectors.

Now to satisfy $N(t +1) = N(t) = N^*$ at time $t = t^*$ we are required to satisfy

$$N^* \geq W^{*T} o N^*.$$

Thus, the result follows.

Theorem 6.2: *The condition for stability, limit cycles, and instability of the dynamics (6.16) depends on the choice of α as prescribed below:*

Stable: when $0 < \alpha < 2$,
Limit cyclic: when $\alpha = 2$, and
Unstable, when $\alpha > 2$.

Proof: Replacing δ by $(E - 1)$ in expression (6.16) we obtain:

$$(E - 1 + \alpha) w_{ij} = X \tag{6.21}$$

where

$$X = [1/ \{1 + \exp (-n_i)\}] [1/ \{1 + \exp (-n_j)\}]. \tag{6.22}$$

The auxiliary equation for expression (6.21) is given by

$$(E - 1 + \alpha) w_{ij} = 0 \tag{6.23}$$

or, $E = 1 - \alpha.$ \hfill (6.24)

Therefore, the complementary function is given by

$$w_{ij}(t) = A (1 - \alpha)^t \tag{6.25}$$

where A is a constant to be determined from boundary conditions.

Let steady-state value of w_{ij} and X be denoted by w_{ij}^* and X^*, respectively. Now, for computing w_{ij}^* we first need to compute X^*. The steady-state value of w_{ij} can now be obtained from the particular integral

$$w_{ij}(t) = X^*/ (E - 1 + \alpha). \tag{6.26}$$

Since X^* is constant, therefore, we substitute $E = 1$ in expression (6.26) to determine w_{ij}^*. Thus,

$$w_{ij}^* = X^*/\alpha. \qquad (6.27)$$

The complete solution for (6.16) is given by

$$w_{ij}(t) = A (1 - \alpha)^t + X^*/\alpha. \qquad (6.28)$$

Substituting $t = 0$ in (6.28) we obtain:

$$A = w_{ij}(0) - X^*/\alpha. \qquad (6.29)$$

Combining (6.28) and (6.29), we obtain the complete solution for $w_{ij}(t)$ given below:

$$w_{ij}(t) = [w_{ij}(0) - X^*/\alpha] (1 - \alpha)^t + X^*/\alpha. \qquad (6.30)$$

The condition for stability, limit cycles, and instability as stated in the theorem follows directly from (6.30).

Theorem 6.3: *The steady-state value of X always satisfies the following inequality:*

$$0.25 \leq X^* \leq 0.48. \qquad (6.31)$$

Proof: From (6.22) we have

$$X^* = [1/\{1 + \exp(-n_i^*)\}] [1/\{1 + \exp(-n_j^*)\}], \qquad (6.32)$$

where n_i^* and n_j^* denote the steady-state values of n_i and n_j, respectively.

Since the right-hand side of (6.32) is monotonic for $0 \leq n_i^*, n_j^* \leq 1$, we determine the range of X^* by substituting $n_i^* = n_j^* = 0$ and $n_i^* = n_j^* = 1$ respectively in (6.32). Hence, the result follows.

Corollary 6.2: *The steady-state value of w_{ij} always satisfies the following inequality: $0.25/\alpha \leq w_{ij}^* \leq 0.48/\alpha$.*

Proof: Proof of the corollary directly follows from (6.27) and (6.31).

Corollary 6.3: *$w_{ij}^* > 0$ for all weights $w_{ij}(0) \neq 0$ follows directly from Theorem 6.3.*

The significance of the above results is that the cognitive learning system never destroys a weight of a link permanently. In other words, the cognitive learning model maintains stability of the structure for the cognitive map.

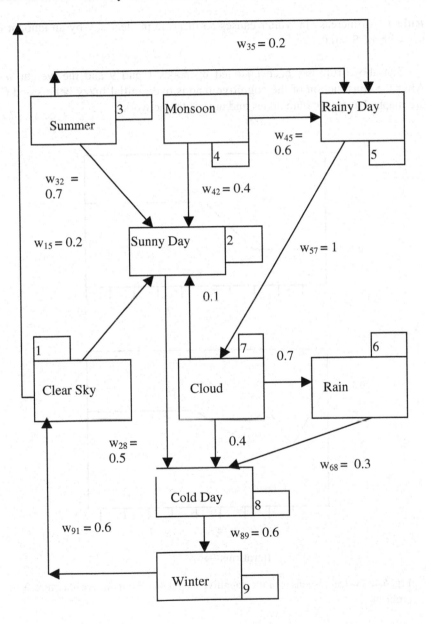

Fig. 6.3: Cognitive map of a typical weather forecasting system.

Computer Simulation: Pal and Konar [30] illustrated the behavior of their cognitive model with reference to a simulated weather forecasting system. To describe the construction of their cognitive map, let us consider one typical rule embedded in the map using nodes and arcs.

Rule 1: An increase in winter causes an increase in clear sky by an amount of 0.8 with a CS = 0.6.

The above rule has been encoded by nodes 1 and 9 and the weight w_{91}. After the construction of the cognitive map is over, initial fuzzy beliefs and CS are mapped at appropriate nodes and arcs of the network.

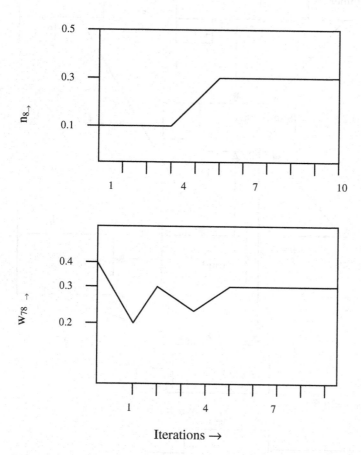

Fig. 6.4: Typical responses of the cognitive map representing the weather forecasting problem.

The CS of the arcs then updated in parallel and with the new CS the fuzzy beliefs at all nodes are updated in parallel. This may be termed as one complete "encode-recall" cycle. A number of cycles are repeated until the steady-state condition is reached. The steady-state beliefs thus obtained represent a permanent nonvolatile image of cognitive system's memory.

The cognitive map presented in Figure 6.3 has been simulated on a computer and the temporal variations of beliefs and CS of nodes and arcs are obtained for a selected value of α Plots of belief for one typical node (say node 9) and one weight (say w_{78}) are given in Figure 6.4 for a value of $\alpha(=1.6)$ in the stable range.

The cognitive model presented by Pal and Konar [30] thus is found to be stable for $0 < \alpha < 2$. Further, the steady-state values of weights are never zero indicating the stability of the structure.

6.8 Conclusions

This chapter presented various models of fuzzy cognitive maps and their learning behavior. Most of these models employ unsupervised learning. Kosko's model and its extension by Pal and Konar include Hebbian-type encoding. The model by Zhang et al. is of completely different type and is used for reasoning using a specialized algebra, well known as NPN algebra. All the models introduced in the chapter provide scope for fusion of knowledge of multiple experts. The principle of knowledge fusion has been illustrated in exercise 2.

Exercises

1. Given a cognitive map (Fig. 6.5) with signed edges. Construct an edge matrix E, whose elements $e_{ij} \in \{-1, 0, 1\}$. What is the causal effect of C_1 and C_4 jointly on others?

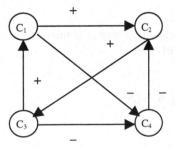

Fig. 6.5: An illustrative cognitive map.

2. Given four concepts C_1, C_2, C_3, and C_4 and three cognitive maps generated by three experts incorporating these concepts (Fig. 6.6), how can you fuse the concepts of the experts?

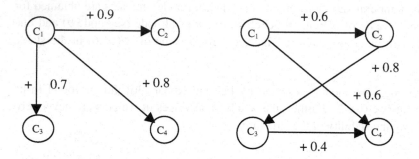

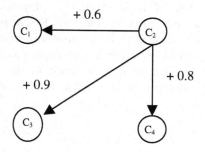

Fig. 6.6: Three cognitive maps with shared concepts.

[**Hints:** Convert the cognitive maps (a) through (c) into matrix representations and take their average]

3. Given an FCM (Fig.6.7) [20]:

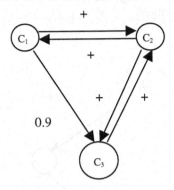

Fig. 6.7: An FCM.

a)   Find the edge matrix.
b)   With a threshold of ½, check what state transition take place from the
     starting state [1   1   0].
c)   Repeat (b) for initial state [0 1 0].

4. Given a fuzzy cognitive map with fuzzy labels at the edges (Fig. 6.8).
   Determine the total effect of C_1 on C_6. Further given: {none ≤ some ≤ much ≤
   a lot}

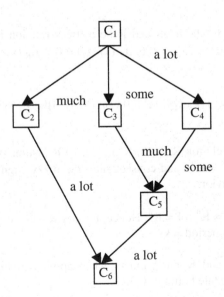

Fig. 6.8: A cognitive map with fuzzy labels at the edges.

5. Given an FCM with two sets of fuzzy strengths of the edges, find the composite fuzzy strength of the edges.

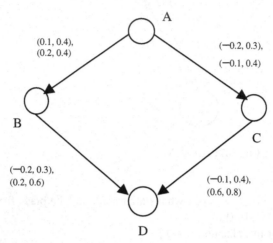

Fig. 6.9: An FCM with 2 sets of fuzzy strength of weights.

[**Hints:** Find average of the first labels and the second labels against each arc.]

6. Let $X = \{x_1, x_2, x_3, x_4, x_5\}$ be a set and R be a fuzzy relation in $X \times X$. Given $\mu_R(x_1, x_2) = 0.2$, $\mu_R(x_2, x_4) = 0.6$, $\mu_R(x_1, x_5) = 0.8$, $\mu_R(x_5, x_4) = 0.9$. Find $\mu_R(x_1, x_4)$.

[**Hints:** $\mu_R(x_1, x_4) = \text{Max} \; [\{\mu_R(x_1, x_2) * \mu_R(x_2, x_4)\}, \{\mu_R(x_1, x_5) * \mu_R(x_5, x_4)\}]$]

7. Show that for any fuzzy relation R, if $R \circ R \circ \ldots \ldots \circ R$ (k-times) $= R^k = I$ for integer k, then $R^{n+k} = R^n$, and consequently the fuzzy cognitive map will have limit cyclic behavior.

[**Hints:** $R^{n+k} = R^n \circ R^k = R^n \circ I = R^n$. Hence, the cognitive map will have sustained oscillation with period = k]

8. Verify the result of Pal and Konar's model when applied to Figure 6.3. Check that n_8 has a steady state belief of 0.3.

References

[1] Atkinson, R. C. and Shiffrin, R. M., "Human memory: A proposed system and its control process," In *The Psychology of Learning and Motivation: Advances in Research and Theory*, Spence, K. W. and Spence, J. T. (Eds.), vol. 2, Academic Press, New York, 1968.

[2] Baddeley, A. D., "The fractionation of human memory," *Psychological Medicine*, vol.14, pp. 259-264, 1984.

[3] Bharick, H. P., "Semantic memory content in permastore: Fifty years of memory for Spanish learned in school," *Journal of Experimental Psychology: General*, vol. 120, pp.20-33, 1984.

[4] Biswas, B. and Konar, A., *A Neruro-Fuzzy Approach to Criminal Investigation*, Physica-Verlag, 2005 (in press).

[5] Biswas, B., Konar, A. and Mukherjee, A. K., "Image matching with fuzzy moment descriptors," *Engineering Applications of Artificial Intelligence*, Elsevier, vol. 14, pp. 43-49, 2001.

[6] Chang, T. M., "Semantic memory: facts and models," *Psychological Bulletin*, vol. 99, pp. 199-220, 1986.

[7] Downs, R. M. and Davis, S., *Cognitive Maps and Spatial Behavior: Process and Products*, Aldine Publishing Co., 1973.

[8] Duncan, E. M. and Bourg, T., "An examination of the effects of encoding and decision processes on the rate of mental rotation," *Journal of Mental Imagery*, vol. 7, pp. 33-56, 1983.

[9] Goldenberg, G., Podreka, I., Steiner, M., Suess, E., Deecke, L. and Willmes, K., "Pattern of regional cerebral blood flow related to visual and motor imagery, Results of Emission Computerized Tomography," In *Cognitive and Neuropsychological Approaches to Mental Imagery*, Denis, M, Engelkamp, J. and Richardson, J. T. E. (Eds.), Martinus Nijhoff Publishers, Dordrecht, The Netherlands, pp. 363-373, 1988.

[10] Greeno, J. G., "Process of understanding in problem solving," In *Cognitive Theory*, Castellan, N. J., Pisoni, Jr. D. B. and Potts, G. R. (Eds.), Hillsdale, Erlbaum, NJ, vol. 2, pp. 43-84,1976.

[11] Gross, C. G. and Zeigler, H. P., *Readings in Physiological Psychology: Learning and Memory*, Harper & Row, New York, 1969.

[12] Farah, M. J., "Is visual imagery really visual? Overlooked evidence from neuropsychology," *Psychological Review*, vol. 95, pp. 307-317, 1988.

[13] Jolicoeur, P., "The time to name disoriented natural objects," *Memory and Cognition*, vol. 13, pp. 289-303, 1985.

[14] Kintsch, W., "The role of knowledge in discourse comprehension: A construction-integration model," *Psychological Review*, vol. 95, pp. 163-182, 1988.

[15] Kintsch, W. and Buschke, H., "Homophones and synonyms in short term memory," *Journal of Experimental Psychology*, vol. 80, pp. 403-407, 1985.

[16] Konar, A. and Pal, S., "Modeling cognition with fuzzy neural nets," In *Fuzzy Logic Theory: Techniques and Applications*, Leondes, C. T. (Ed.), Academic Press, New York, 1999.

[17] Konar, A., *Artificial Intelligence and Soft Computing: Behavioral and Cognitive Modeling of the Human Brain*, CRC Press, Boca Raton, FL, 1999.

[18] Konar, A, *Computational Intelligence: Principles, Techniques and Applications,* Springer-Verlag, Heidelberg, 2005 (in press).

[19] Kosko, B., "Fuzzy cognitive maps," *Int. J. of Man-Machine Studies*, vol. 24, pp. 65-75, 1986.

[20] Kosko, B., *Fuzzy Engineering*, Prentice-Hall, NJ, ch. 15, pp. 499-528, 1996.

[21] Kosslyn, S. M., Mental imagery, In *Visual Cognition and Action: An Invitation to Cognitive Science*, Osherson, D. N. and Hollerback, J. M. (Eds.), vol. 2, pp. 73-97, 1990.

[22] Kosslyn, S. M., "Aspects of cognitive neuroscience of mental imagery," *Science*, vol. 240, pp. 1621-1626, 1988.

[23] Marr, D., *Vision*, W. H. Freeman., San Francisco, 1982.

[24] Matlin, M. W., *Cognition*, Harcourt Brace Pub., Reprinted by Prism Books Pvt. Ltd., Bangalore, India, 1994.

[25] Milner, B., "Amnesia following operation on the temporal lobe," In *Amnesia Following Operation on the Temporal Lobes*, Whitty, C. W. M. and Zangwill O. L. (Eds.), Butterworth, London, pp. 109-133, 1966.

[26] McNamara, T. P., Ratcliff, R. and McKoon, G., "The mental representation of knowledge acquired from maps," *Journal of Experimental Psychology: Learning, Memory and Cognition*, pp. 723-732, 1984.

[27] Moar, I. and Bower, G. H., "Inconsistency in spatial knowledge," *Memory and Cognition*, vol. 11, pp.107-113, 1983.

[28] Moyer, R. S. and Dumais, S. T., "Mental comparison," In *The Psychology of Learning and Motivation*, Bower, G. H. (Ed.), vol. 12, pp. 117-156, 1978.

[29] Paivio, A., "On exploring visual knowledge," In *Visual Learning, Thinking and Communication*, Randhawa, B. S. and Coffman, W. E. (Eds.), Academic Press, New York, pp. 113-132, 1978.

[30] Pal, S. and Konar, A., "Cognitive reasoning with fuzzy neural nets," *IEEE Trans. on Systems, Man and Cybernetics*, Part - B, August, 1996.

[31] Peterson, M. A., Kihlstrom, J. F., Rose, P. M. and Glisky, M. L., "Mental images can be ambiguous: reconstruals and reference-frame reversals," *Memory and Cognition*, vol. 20, pp. 107-123, 1992.

[32] Pylyshyn, Z. W., "Imagery and artificial intelligence," In Minnesota Studies in the Philosophy of Science, *Perception and Cognition Issues in the Foundations of Psychology*, vol. 9, University of Minnesota Press, Minneapolis, pp. 19-56, 1978.

[33] Rumelhart, D. E., McClelland, J. L. and the PDP research group, *Parallel Distributed Processing: Explorations in the Microstructure of Cognition*, vol. 1 and 2, MIT Press, Cambridge, MA, 1968.

[34] Shepard, R. N. and Cooper, L.A. (Eds.), *Mental Images and Their Transformations*, MIT Press, Cambridge, MA, 1982.

[35] Shepherd, R. N. and Metzler, Z., "Mental rotation of three dimensional objects," *Science*, vol. 171, pp. 701-703, 1971.

[36] Tulving, E., "Multiple memory systems and consciousness," *Human Neurobiology*, vol. 6, pp. 67-80, 1986.

[37] Zhang, W. R., Chen, S. S. and Bezdek, J. C., "Pool2: a generic system for cognitive map development and decision analysis," *IEEE Trans. on Systems, Man and Cybernetics*, vol. 19, no. 1, pp. 31-39, 1989.

[38] Zhang, W. R. "Pool–a semantic model for approximate reasoning and its application in decision support," *Management Information System*, vol. 3, no. 41, pp. 65-78, 1986.

[39] Zhang, W. R., "NPN fuzzy sets and NPN qualitative algebra: a computational network for bipolar cognitive modeling and multi-agent decision analysis," *IEEE Trans. Systems, Man and Cybernetics*, vol. 26, no. 4, 1996.

[40] Zhang, W. R., Chen, S. S., Wang, W. and King, R., "A cognitive map based approach to the co-ordination of distributed co-operative agents," *IEEE Trans. on Systems, Man and Cybernetics*, vol. 22, no. 1, pp. 103-114, 1992.

Chapter 7

Unsupervised Learning by Fuzzy Petri Nets

This chapter presents two distinct models of unsupervised learning on a special type of cognitive maps, realized with fuzzy Petri nets. The learning models in the present context adapt the weights of the directed arcs from the transition to the places in a Petri net. The first model employs a simple Hebbian-type adaptation with a natural decay in weights. The model is proved to be conditionally stable for a selected range of the mortality rate. In the recall (reasoning) phase, a pretrained Petri net with stable weights is capable of determining the beliefs of the desired propositions from the supplied beliefs of the axioms. An application of the proposed learning model to automated car driving in an accident-prone environment has been studied. The second model employs a minor variation of Hebbian learning with no passive decay in weights, and is found to be stable in both the learning and the recall phases. The chapter concludes with a discussion of the relative merits of the proposed models in knowledge acquisition/refinement problems.

7.1 Introduction

Kosko's pioneering work on fuzzy cognitive maps [15-16] stands as a milestone in the field of machine intelligence. However, his model has two limitations. First, the model cannot describe many-to-one (or -many) relations. Secondly, the recall model gets trapped within limit cycles and is therefore not applicable in real-time decision-making problems. This chapter provides two alternative extensions of Kosko's model to enable it to represent many-to-one (or -many) casual relations by using Petri nets. The proposed recall models are free from limit cyclic behaviors.

In the proposed models, many-to-one (or -many) causal relations are described as a joint occurrence of a (conjunctive) set of antecedent clauses and one or more disjunctive consequent clauses. Further, instead of representing positive and negative causal relations separately, we can describe them by a uniform notion without attaching any "sign" label to the arcs in the fuzzy cognitive map. Readers at this stage may wonder: how is it possible? To answer this, suppose there is a negative causal relation:

$$P \xrightarrow{-} Q$$

between concepts P and Q, which means P causally decreases Q. We can represent the same causal relation as:

$$P \rightarrow \neg Q$$

which means P causally increases the dis-concept Q. The $\neg$ symbol is a logical negation operator. Thus, we can represent all negative causal relation as equivalent positive causal relations and attach no sign to the arcs to label the positive/negative causality.

Another point that needs to be addressed before formally introducing the scheme is why do we select fuzzy Petri nets (FPN) [1-14], [17], [22-26] to model the fuzzy cognitive maps? The answer to this is that FPN supports the necessary many-to-one (or -many) causal relations and it has already proved itself successful as an important reasoning [9-14] and learning tool [22]. The principle of reasoning and learning that we shall introduce in the chapter, however, is different from the existing works.

The chapter is classified as follows. The encoding and the recall for the first model of the cognitive map are introduced in Section 7.2. The state-space representation of the model is given in Section 7.3. An analysis of stability of the proposed model is performed in Section 7.4. A computer simulation of an automated collision-free car driving system based on the proposed model is presented in Section 7.5. Implication of the results of computer simulation is given in Section 7.6. An alternative learning model and its stability analysis is presented in Section 7.7. A case study to illustrate the concept of knowledge refinement is also included in the same section. Concluding remarks and comparison of the two models are given in Section 7.8.

7.2 The Proposed Model for Cognitive Learning

The model for cognitive learning has two main components: (i) the encoding and (ii) the recall modules. The encoding is needed to store stable fuzzy weights at the arcs, connected between the transition and the places [17] of a FPN, while the recall is needed for the purpose of deriving stable fuzzy inferences. Before presenting the model, we formally define a FPN [18].

Definition 7.1: *A FPN is a 9-tuple, given by,*

$$FPN = <P, Tr, T, D, I, O, Th, n, W > \qquad (7.1)$$

where

P = $(p_1, p_2, ..., p_n)$ *is a finite set of places;*

Tr = $(tr_1, tr_2, ..., tr_m)$ *is a finite set of transitions;*

T = $(t_1, t_2, .., t_m)$ *is a set of tokens in the interval [0,1] associated with the transitions $(tr_1, tr_2, .., tr_m)$, respectively;*

D = $(d_1, d_2, ..., d_n)$ *is a finite set of propositions, each proposition d_k corresponds to place p_k;*

$P \cap Tr \cap D = \varnothing$; *cardinality of (P) = Cardinality of (D);*

I: $Tr \rightarrow P^\infty$ *is the input function, representing a mapping from transitions to bags of (their input) places;*

O: $Tr \rightarrow P^\infty$ *is the output function, representing a mapping from transitions to bags of (their output) places;*

Th = $(th_1, th_2, ..., th_m)$ *represents a set of threshold values in the interval [0,1] associated with transitions $(tr_1, tr_2, ..., tr_m)$, respectively;*

n: $P \rightarrow [0,1]$ *is an association function hereafter called fuzzy belief, representing a mapping from places to real values between 0 and 1; $n(p_i)$ = n_i, say; and*

W = $\{w_{ij}\}$ *is the set of weights from j-th transition to i-th place, where i and j are integers.*

We now present an example to illustrate the above parameters in a typical FPN structure.

Example 7.1: The FPN in FigURE 7.1 represents the following causal rule (CR):

$$CR: (it\text{-}is\text{-}hot), (the\text{-}sky\text{-}is\text{-}cloudy) \rightarrow (it\text{-}will\text{-}rain).$$

The above rule means that the events (concepts), it-is-hot and the-sky-is-cloudy, causally influence the event, it-will-rain. For all purposive action, this rule is identical with typical If-Then (implication) relation.

Here, $P = \{p_1, p_2, p_3\}$, $Tr = \{tr_1\}$, $T = \{t_1\}$, $D = \{d_1, d_2, d_3\}$ where d_1 = it-is-hot, d_2 = the-sky-is-cloudy, and d_3 = it-will-rain, $I(tr_1) = \{p_1, p_2\}$, $O(tr_1) = \{p_3\}$, $Th = \{th_1\}$, $n = \{n_1, n_2, n_3\}$, and $w = \{w_{31}\}$.

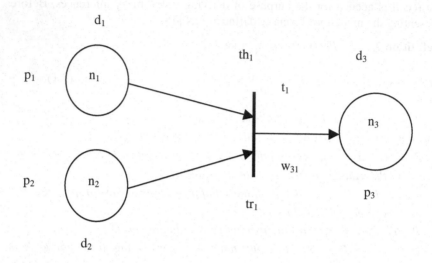

Fig. 7.1: A fuzzy Petri net representing the given causal rule.

7.2.1 Encoding of Weights

The encoding model, represented by equation (7.2) and presented below has been designed based on the Hebbian learning concept [27], which states that *the strength of the weight of an arc should increase with the increasing signal levels at the adjacent nodes (here the places and the transitions).*

$$\frac{d\,w_{ij}(t)}{dt} = -\alpha\,w_{ij}(t) + t_j(t)\,n_i(t) \tag{7.2}$$

where w_{ij} denotes the weight of the arc connected between transition tr_j and place p_i, α denotes the forgetfulness (mortality) rate, t_j represents the token associated with transition tr_j (also called truth token), and n_i represents the fuzzy belief associated with proposition d_i. The $-\alpha\,w_{ij}(t)$ term indicates a natural decay of the weight $w_{ij}(t)$ in the cognitive map. This has relevance with the natural tendency of forgetfulness of human beings. The discrete version of expression (7.2) takes the form of expression (7.3)

$$w_{ij}(t+1) = (1-\alpha)\,w_{ij}(t) + t_j(t).\,n_i(t) \tag{7.3}$$

7.2.2 The Recall Model

The recall process aims at determining fuzzy beliefs at the places. The recall model in the present context has two applications. First, it is part of the encoding cycle and, secondly, it is an independent recall procedure. The natural question that automatically arises is why does the recall model appear as part of the encoding cycle? The answer to this is really interesting. As the weights are updated once by the recurrent equation (7.3), some weights are strengthened and some are weakened. Spontaneously then, some concepts should be strengthened and some should be weakened. A little thinking reveals that a similar *metamorphosis* of concepts always takes place in the psychological mind of the human beings. This justifies the inclusion of the recall model in the encoding phase. On the other hand, after the weights are stabilized, the recall/reasoning model is needed in its own course of action to derive new fuzzy inferences from the known beliefs of a set of axioms (places with no input arcs).

Before describing the recall of tokens at places, we first define the firing condition of the transition. *A transition fires if the ANDed value of fuzzy beliefs of its input places exceeds the threshold associated with it.* On firing of a transition, the ANDed (Min) value of belief of its input places is transmitted to its output [18]. This is described below by expression (7.4) (Fig. 7.2):

$$t_q(t+1) = \left(\bigwedge_{k=1}^{n} n_k(t) \right) \wedge u\left[\left(\bigwedge_{\exists k=1}^{n} n_k(t) \right) - th_q \right] \tag{7.4}$$

where $n_k(t)$ represents the fuzzy belief of place p_k, which is an input place of transition tr_q and u is a unit step function. "$\wedge$" denotes fuzzy AND (min) operator. In the present context, the first $\wedge$ operator takes minimum of the beliefs n_k of the corresponding input places p_k in the FPN. The u operator checks the firing condition of the transition. If the transition fires, then u returns a one value, else it is zero. The second $\wedge$ operator combines the first part of token computation with the third part of firing condition checking. The third $\wedge$ operator is self-explanatory. The notation $\exists k$ below the third $\wedge$ operator represents that the transition tr_q possesses at most n-number of input places.

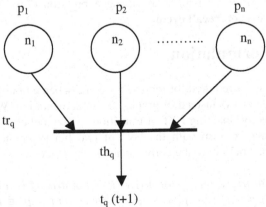

Fig. 7.2: A FPN representing connectivity from n places to transition tr_q.

Belief updating at place p_i is explained below by expression (7.5) (Fig. 7.3):

$$n_i (t + 1) = n_i(t) \; V \; (\; \overset{n}{\underset{j=1}{V}} \; (w_{ij}(t)) \wedge t_j(t + 1))) \qquad (7.5)$$

where $w_{ij}(t)$ denotes the weight of the arc connected between transition tr_j and place p_i. Here, $n_i(t + 1)$ is computed by taking the maximum (V) of the current belief $n_i(t)$ and the influence from the other transitions that can directly affect the belief of place p_i.

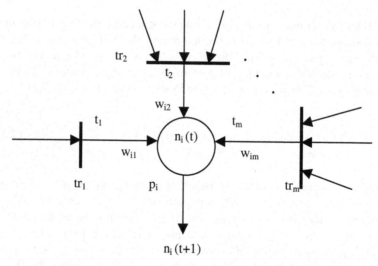

Fig. 7.3: A FPN representing connectivity from m transitions to a place p_i .

It may be added here that during the recall cycle, fuzzy truth tokens (FTT) at all transitions are updated concurrently, and the results, thus obtained, are used for computing beliefs at all places in parallel by using expression (7.5). This is referred to as the **belief revision/recall cycle.**

7.3 State-Space Formulation

The state-space formulation is required for representing the scheme of cognitive reasoning by using simple arrays instead of graphs for implementation. We will develop three expressions for updating FTT of transitions, fuzzy belief at places and weights in vector- matrix form with the help of two binary connectivity matrices P and Q, redefined below for the convenience of the readers.

Definition 7.2: *A **place to transition connectivity (PTC) matrix** Q is a binary matrix whose elements $q_{jk} = 1$, if for places p_k and transition tr_i, $p_k \in I \; (tr_j)$;*

otherwise $q_{jk} = 0$. If the FPN has n places and m transitions, then the Q matrix is of (m $\times$ n) dimension [10].

Definition 7.3: *A **transition to place connectivity (TPC) matrix** P is a binary matrix whose element $p_{ij} = 1$, if for places p_i and transition tr_j, $pi \in O\ (tr_j)$; otherwise $p_{ij} = 0$. With n places and m transitions in the FPN, the P matrix is of (n $\times$ m) dimension [11].*

7.3.1 State-Space Model for Belief Updating

The equation for belief updating at places is already presented in expression (5). For n places in the FPN, there will be n number of expressions analogous to (5). Further, if we have m number of transitions in the FPN, all of them may not be connected to place p_i like the configuration shown in Figure 7.3. So, to represent connections from a few transitions to a place $p_{i,}$ we use connectivity matrix W, and thus j is replaced by $\exists j$ (some j) in the range 1 to m. Thus, we find the following expression.

$$n_i(t + 1) = n_i(t) \text{ V} \left[\underset{\exists j\ =\ 1}{\overset{m}{\text{V}}}\ (t_j\ (t + 1)\ \wedge\ w_{ij}) \right] \qquad \text{(revised 7.5)}$$

For all n places, the expression presented above together takes the form of (7.6).

$$N(t + 1) = N(t) \text{ V } (W \text{ o } T(t + 1)) \qquad (7.6)$$

where N = fuzzy belief vector of dimension (n $\times$ 1), the r-th component of which represents fuzzy belief of place p_r, W = fuzzy weight matrix from places to their input transition of dimension (m $\times$ n), and T = token vector for transitions with dimension (m $\times$ 1), such that the r-th component of T represents the fuzzy truth token of the r-th transition. The operator "o" represents fuzzy max-min composition operator, which is carried out analogous to conventional matrix multiplication algebra with the replacement of algebraic multiplication by fuzzy AND (Min) and addition by fuzzy OR (Max) operator.

7.3.2 State-Space Model for FTT Updating of Transitions

Applying De Morgan's law, the transition updating equation described by expression (7.4), can be re-modeled as follows:

$$t_q(t + 1) = \left(\underset{\exists k\ =\ 1}{\overset{n}{\text{V}}} (n_k)^c \right)^c \wedge\ u\left[\left(\underset{\exists k\ =\ 1}{\overset{n}{\text{V}}} (n_k)^c \right)^c - th_q \right] \qquad \text{(revised 7.4)}$$

where c above a parameter represents its one's complementation. Since all places are not connected to the input of transition tr_q, to represent connectivity from places to transitions we use the binary PTC matrix Q, with elements $q_{qk} \in \{0,1\}$. Thus we find:

$$
t_q(t+1) = \overset{n}{\underset{\forall k=1}{V}} \overset{c}{(n_k(t)} \wedge \overset{c}{q_{ik})} \wedge u \left[\left(\overset{n}{\underset{\forall k=1}{V}} \overset{c}{(n_k(t)} \wedge \overset{c}{q_{ik})} \right) - th_q \right]
$$

(revised 7.4)

For m transitions as in Figure 7.2, we will have m number of expressions as presented above. Combining all the m expressions we get the following state-space form for token updating at transitions.

$$
T(t+1) = (Q \circ \overset{c c}{N(t)}) \wedge U[(Q \circ \overset{c c}{N(t)}) - Th]
$$

(7.7)

where
Th =   threshold Vector of dimension (m × 1), the i-th component of which represents the threshold of transition tr_i,
U =   Unit step vector of dimension (m × 1) , and
$\overset{c}{N}$ =   Complement of the belief vector, the i-th, component of which is the complement of the i-th component of N.

The other parameters in expression (7.7) have the same meaning as defined earlier.

7.3.3 State-Space Model for Weights

The weight updating equation given by expression (7.3) can be written in vector-matrix form as presented below:

$$
W(t+1) = (1 - \alpha) W(t) + ((N(t) \cdot T^T(t)) \wedge P)
$$

(7.8)

Here, W = $[w_{ij}]$ is a weight matrix of dimension (n × m), P = binary TPC matrix of dimension (n × m) with element $p_{ij} \in \{0,1\}$, where p_{ij} represents connectivity from tr_j to p_i, T over a vector denotes its transposition. The other parameters of expression (7.8) have been defined earlier.

7.4 Stability Analysis of the Cognitive Model

The results of stability analysis are presented in theorems 7.1 through 7.3.

Theorem 7.1: *$n_i(t)$ in expression (7.5) always converges for $n_k(0)$ lying between 0 and 1, for all k.*

Proof: Since $n_k(0)$ for all k are bounded in [0, 1], and expressions (7.4) and (7.5) involve AND (Min) and OR (Max) operators, $n_i(t)$ at any time t remains bounded in [0, 1]. Further, from expression (7.5) it is evident that $n_i(t + 1) \geq n_i(t)$. Thus $n_i(t + 1)$ can never decrease with respect to $n_i(t)$, and thus $n_i(t + 1)$ cannot be oscillatory. Consequently, $n_i(t + 1)$ being bounded in [0, 1] and oscillation-free converges to a stable point after some finite number of iterations. Hence, the theorem holds good.

Theorem 7.2: *The fuzzy truth tokens $t_i(t)$ at transition tr_j for all j given by expression (7.4) converge to stable points for $n_k(0)$ lying in [0, 1] for all k.*

Proof: Since fuzzy beliefs at places in the cognitive net converge to stable points, therefore, by expression (7.4) the proof of the theorem is obvious [9].

Theorem 7.3: *The weights in the encoding model given by expression (3) satisfy the following conditions.*

> *stable when $0 < \alpha < 2$,*
> *limit cycle when $\alpha = 2$, and*
> *unstable when $\alpha > 2$.*

Proof: Rewriting expression (7.3) by extended difference operator E, we find

$$(E - 1 + \alpha) \, w_{ij}(t) = t_j(t) \cdot n_i(t) \tag{7.9}$$

Now, to prove the theorem we require to solve the above equation. The complementary function (CF) for the above equation is given in the expression

$$CF: w_{ij}(t) = A \, (1 - \alpha)^t \tag{7.10}$$

where A is a constant.

The particular integral (PI) for expression (7.9) is given by expression (7.11).

$$PI: w_{ij}(t) = \frac{t_j(t) \, n_i(t)}{E - 1 + \alpha} \tag{7.11}$$

Since PI represents the steady-state value of $w_{ij}(t)$, let us consider the steady state values of $t_j(t)$ and $n_i(t)$ in expression (7.8). Let $t_j(t)$ as $t \to \infty$ be called t_j^* and $n_i(t)$ as $t \to \infty$ be called n_i^*.

Replacing $t_j(t)$ and $n_i(t)$ with their steady-state values in expression (7.11), we get the numerator of expression (7.11) to be constant. Therefore, we set $E = 1$ in expression (7.11), which yields the resulting PI given by expression (7.12).

$$\text{PI:} \quad w_{ij}(t) = t_j^* \, n_i^* / \alpha \tag{7.12}$$

The complete solution of equation (7.9) is given by (7.13).

$$w_{ij}(t) = A(1 - \alpha)^t + t_j^* \, n_i^* / \alpha \tag{7.13}$$

Substituting $t = 0$ in expression (7.13) we find the value of A, which on further substitution in (7.13) yields the complete solution given by (7.14).

$$w_{ij}(t) = (w_{ij}(0) - t_j^* \, n_i^* / \alpha)\,(1 - \alpha)^t + t_j^* \, n_i^* / \alpha \tag{7.14}$$

The condition for stability, limit cycle, and instability, now, directly follows from expression (7.14).

Example 7.2: Consider the cognitive map for bird (X) given in Figure 7.4. The initial value of beliefs, weights, and thresholds are shown in the figure itself. The encoding and recall models given in expressions (7.6), (7.7), and (7.8) are repeated until limit cycles, steady-state, or instability is observed on the plots, vide Figure 7.5. The results obtained in these figures are consistent with Theorem 7.3.

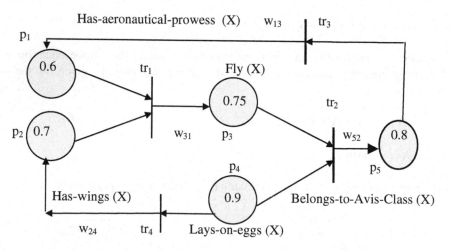

Initial weights: $w_{31}(o) = 0.85$, $w_{52}(0) = 0.7$, $w_{13}(0) = 0.85$, $w_{24}(0) = 0.8$.
Thresholds: $th_i = 0$ for all i.

Fig. 7.4: A cognitive map for birds built with FPN.

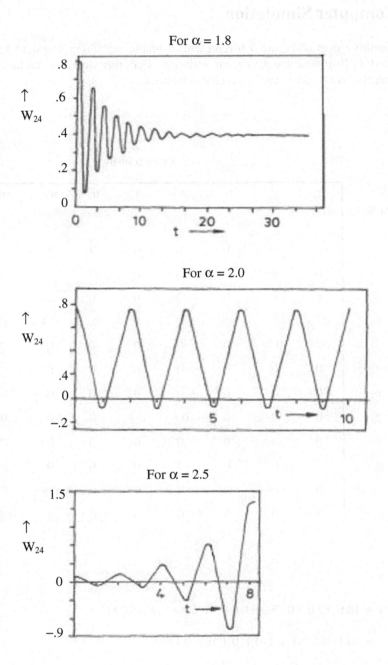

Fig.7.5: Effect of α on the dynamics of a sample weight w_{24} in Figure 7.4.

7.5 Computer Simulation

The cognitive map of Figure 7.6 represents a graph, signifying a system for automated Collision/Accident-free car driving. For this map, we initialize W(0) matrix, N(0) vector, and Th vector as follows.

W(0) =				From transition				
To place	tr_1	tr_2	tr_3	tr_4	tr_5	tr_6	tr_7	tr_8
p_1	0	0	0	0	0	0	0	0
p_2	0	0	0	0	0	0	0	0
p_3	0	0	0	0	0	0	0	0
p_4	0.95	0	0.85	0	0	0.75	0	0
p_5	0	0.7	0	0	0	0	0	0
p_6	0	0	0	0	0	0	0	0
p_7	0	0	0	0	0	0	0	0
p_8	0	0	0	0.8	0.4	0	0	0
p_9	0	0	0	0	0	0	0	0
p_{10}	0	0	0	0	0	0	0	0
p_{11}	0	0	0	0	0	0	0.9	0
p_{12}	0	0	0	0	0	0	0	0.85

$N(0) = [0.6\ 0.7\ 0.5\ 0.75\ 0.6\ 0.6\ 0.9\ 0.6\ 0.8\ 0.7\ 0.5\ 0.6]^T$

$Th = [0.1\ 0.1\ 0.1\ 0.1\ 0.1\ 0.1\ 0.1\ 0.1]^T$

The P and Q matrices for the above system are given on page no. 218.

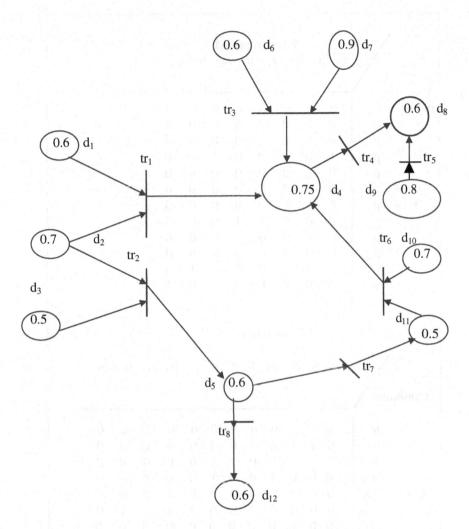

d_1 = Car behind side-car narrow in breadth, d_2 = Side-car of the front car too close, d_3 = Car behind side-car is wide, d_4 = Front car speed decreases, d_5 = Front car speed increases, d_6 = Passer-by changes his/her direction, d_7 = Passer-by crosses the road, d_8 = Rear car speed decreases, d_9 = Front car changes direction, d_{10} = Rear car changes direction, d_{11} = Rear car speed increases, d_{12} = Rear car keeps safe distance with respect to front car.

Fig. 7.6: Fuzzy cognitive map, representing a car driver's problem.

From transition

		tr_1	tr_2	tr_3	tr_4	tr_5	tr_6	tr_7	tr_8
To places									
	p_1	0	0	0	0	0	0	0	0
	p_2	0	0	0	0	0	0	0	0
	p_3	0	0	0	0	0	0	0	0
	p_4	1	0	1	0	0	1	0	0
	p_5	0	1	0	0	0	0	0	0
	p_6	0	0	0	0	0	0	0	0
$P =$	p_7	0	0	0	0	0	0	0	0
	p_8	0	0	0	1	1	0	0	0
	p_9	0	0	0	0	0	0	0	0
	p_{10}	0	0	0	0	0	0	0	0
	p_{11}	0	0	0	0	0	0	1	0
	p_{12}	0	0	0	0	0	0	0	1

From place

		p_1	p_2	p_3	p_4	p_5	p_6	p_7	p_8	p_9	p_{10}	p_{11}	p_{12}
To transitions													
	tr_1	1	1	0	0	0	0	0	0	0	0	0	0
	tr_2	0	1	1	0	0	0	0	0	0	0	0	0
	tr_3	0	0	0	0	0	1	1	0	0	0	0	0
	tr_4	0	0	0	1	0	0	0	0	0	0	0	0
$Q =$	tr_5	0	0	0	0	0	0	0	0	1	0	0	0
	tr_6	0	0	0	0	0	0	0	0	0	1	1	0
	tr_7	0	0	0	0	1	0	0	0	0	0	0	0
	tr_8	0	0	0	0	1	0	0	0	0	0	0	0

The fuzzy encoding and recall equations given by expressions (7.8) and (7.7) and (7.6) are updated with $\alpha = 1.8$ (<2) in order until a convergence in weights is attained. The steady-state value of weights, thus obtained, is saved by the cognitive system for application during recognition phase. Its value, as obtained through simulation, is found to be W^*, where

$w_{41}^* = 0.25$, $w_{43}^* = 0.25$, $w_{46}^* = 0.25$, $w_{56}^* = 0.17$, $w_{84}^* = 0.31$, w_{85}^* $= 0.33$, $w_{11,7}^* = 0.2$, $w_{12,8}^* = 0.2$, and $w_{ij} = 0$ for all other values of i, j.

Now, with a new $N(0)$ vector, collected from a source, as given below and the W^* matrix, the cognitive system can derive new steady-state inferences N^* by using expressions (7.7) and (7.6) only in order. The value of $N(0)$ and N^* are presented below:

$$N(0) = [0.2\ \ 0.3\ \ 0.4\ \ 0.0\ \ 0.0\ \ 0.3\ 0.35\ \ 0.0\ \ 0.4\ \ 0.3\ \ 0.0\ 0.0]^{\text{T}}$$

$$N^* = \quad [0.2\ \ 0.3\ \ 0.4\ \ 0.25\ \ 0.17\ \ 0.3\ \ 0.35\ \ 0.33\ \ 0.4\ \ 0.3\ \ 0.17\ 0.17]^{\text{T}}$$

Since among the concluding places $\{p_4\ ,\ p_5,\ p_8\ ,\ p_{11},\ p_{12}\}$, p_8 has the highest steady-state belief ($ = 0.33$), therefore d_8 has to be executed.

7.6 Implication of the Results

Stability analysis of a cognitive map, represented by fuzzy Petri net models, has been accomplished in detail in this chapter. The result of the stability analysis envisages that the encoding model is conditionally stable ($0 < \alpha < 2$), while the recall model is unconditionally stable. Further, the cognitive map never destroys any weights during the process of encoding. This is evident from the fact that when $w_{ij}(0) \neq 0$, w_{ij}^* too is nonzero for any i, j. This is an important issue because it ensures that the cognitive map never destroys its weights (connectivity) completely.

7.7 Knowledge Refinement by Hebbian Learning

This section presents a new method for automated estimation of certainty factors of knowledge from the proven and historical databases of a typical reasoning system. Certainty factors, here, have been modeled by weights in a special type of recurrent fuzzy neural Petri net. The beliefs of the propositions, collected from the historical databases, are mapped at places of a fuzzy Petri net and the weights of directed arcs from transitions to places are updated synchronously following the Hebbian learning principle until an equilibrium condition [15], [9] following which the weights no longer change further is reached. The model for weight adaptation has been chosen for maintaining consistency among the initial beliefs of the propositions and thus the derived steady-state weights represent a more accurate measure of certainty factors than those assigned by a human expert.

7.7.1 The Encoding Model

The process of encoding of weights consists of three basic steps, presented below:

Step I: A transition tr_i is enabled if all its input places possess tokens. An enabled transition is firable. On firing of a transition tr_i, its FTT t_i is updated using expression (7.15) [20], where places $p_k \in I$ (tr_i), n_k is the belief of proposition mapped at place p_k, and th_i is the threshold of transition tr_i.

$$t_i(t+1) = (\bigwedge_{1 \leq k \leq n} n_k(t)) \wedge u \, [(\bigwedge_{1 \leq k \leq n} n_k(t)) - th_i] \qquad (7.15)$$

Expression (7.15) reveals that if $\bigwedge_{1 \leq k \leq n} n_k > th_i$,

$$t_i(t+1) = \bigwedge_{1 \leq k \leq n} n_k(t)$$
$$= 0, \text{ otherwise.}$$

Step II: After the FTTs at all the transitions are updated synchronously, we revise the fuzzy beliefs at all places concurrently. The fuzzy belief n_j at place p_j is updated using expression 7.16 (a), where $p_i \in O(tr_j)$ and by using 7.16(b) when p_i is an axiom, having no input arc.

$$n_i(t+1) = \bigvee_{j=1}^{m} (t_j(t+1) \wedge w_{ij}(t)) \qquad (7.16(a))$$

$$= n_i(t), \text{ when } p_i \text{ is an axiom} \qquad (7.16(b))$$

Step III: Once the updating of fuzzy beliefs are over, the weights w_{ij} of the arc connected between transition tr_j and its output place p_i are updated following Hebbian learning [25] by expression (7.17).

$$w_{ij}(t+1) = t_j(t+1) \wedge n_i(t+1) \qquad (7.17)$$

The above three-step cycle for encoding is repeated until the weights become time-invariant. Such a time-invariant state is called equilibrium. The steady-state values of weights are saved for subsequent reasoning in analogous problems.

Theorem 7.4: *The encoding process of weights in a cognitive map realized with FPN is unconditionally stable.*

Proof: Since $n_k(0)$ for all k is bounded in [0, 1], and the encoding model (expressions (7.15-7.17)) involves only AND (min) and OR (max) operators, $n_k(t)$ will remain bounded at any time t. Further, from expression 7.16(a) we have

$$n_i(t+1) = \bigvee_{j=1}^{m} (t_j(t+1) \wedge w_{ij}(t)) \qquad (7.16(a) \text{ re-written})$$

$$= \bigvee_{j=1}^{m} \{t_j(t+1) \wedge t_j(t) \wedge n_i(t)\} \qquad \text{by (7.17)}$$

$$= \bigvee_{j=1}^{m} [\{t_j(t+1) \wedge t_j(t)\}] \wedge n_i(t) \qquad (7.18)$$

Thus, $n_i(t+1) \leq n_i(t)$. Consequently, $n_i(t+1)$ can only decrease with respect to its last value, and thus it cannot have an oscillation. Therefore, $n_i(t+1)$ being bounded and oscillation-free is stable. This holds for any i, $1 \leq i \leq n$, where n is the number of places in the FPN.

If $n_k(t+1)$ for all k is stable, then by (7.15) $t_j(t+1)$ too is stable. Finally, $n_i(t+1)$ and $t_j(t+1)$ being stable, $w_{ij}(y+1)$ by expression (7.17) is also stable. Hence, the encoding process of weights in a cognitive map by the proposed model is stable.

7.7.2 The Recall/Reasoning Model

The reasoning model of a recurrent FPN has been reported elsewhere [8-9]. During the reasoning phase, we can use any of these models, including the new model proposed below.

The reasoning/recall model in a FPN can be carried out in the same way as in the first two steps of the encoding model with the following exceptions.

- While initiating the reasoning process, the known fuzzy beliefs for the propositions of a problem are to be assigned to the appropriate places. It is to be noted that, in the encoding model, the fuzzy beliefs of propositions were submitted using proven case histories.

- The reasoning model should terminate when the fuzzy beliefs associated with all propositions reach steady-state values, i.e., when for all places,

$n_i^*(t+1) = n_i^*(t)$ at $t = min(t)$. The steady-state beliefs thus obtained are used to interpret the results of typical analogous reasoning problems. The execution of the reasoning model is referred to as belief revision cycle [8].

Theorem 7.5: *The recall process in a FPN unconditionally converges to stable points in belief space.*

Proof: Directly follows from the first part of the proof of Theorem 7.4 showing that $n_i(t+1)$ for all i is stable.

7.7.3 Case Study by Computer Simulation

In this study, we consider two proven case histories described by (Rule base I, Database I) and (Rule base II, Database II). The beliefs of each proposition in the FPNs (Fig. 7.7 and 7.8) for these two case histories are known. The encoding model for the cognitive map presented above has been used to estimate the CFs of the rules in either cases. In case the estimated CF of a rule, obtained from two case histories differs, we take the average of the estimated values as its CF.

Case History I

Rule base I :

PR1 :   Loves(x,y), Loves(y,x) → Lover(x,y)

PR2 :   Young(x), Young(y),
         Opposite-Sex(x,y) → Lover(x,y)

PR3 :   Lover(x,y), Male(x), Female(y) → Marries(x,y)

PR4 :   Marries(x,y) → Loves(x,y)

PR5 :   Marries(x,y), Husband(x,y) → Wife(y,x)

PR6 :   Father(x,z), Mother(y,z) → Wife(y,x)

Database I :

Loves(ram,sita), Loves(sita,ram), Young(ram),
Young(sita), Opposite-Sex(ram,sita),
Male(ram), Female(sita), Husband(ram,sita),
Father(ram, kush),
Mother(sita, kush)

Table 7.1: Parameters of case history I

Initial Weights w_{ij}	$w_{71} = 0.8$, $w_{72} = 0.7$, $w_{93} = 0.6$, $w_{14} = 0.9$, $w_{13,5} = 0.8$, $w_{13,6} = 0.5$
Initial Fuzzy Beliefs n_i	$n_1 = 0.2$, $n_2 = 0.8$, $n_3 = 0.75$, $n_4 = 0.9$, $n_5 = 0.6$, $n_6 = 0.75$, $n_7 = 0.35$, $n_8 = 0.85$, $n_9 = 0.45$, $n_{10} = 0.85$, $n_{11} = 0.7$, $n_{12} = 0.65$, $n_{13} = 0$.
Steady-state weights after 4 iterations	$w_{71} = 0.35$, $w_{72} = 0.60$, $w_{93} = 0.35$, $w_{14} = 0.35$, $w_{13,5} = 0.35$, $w_{13,6} = 0.50$
$th_j = 0$ for all transitions tr_j	

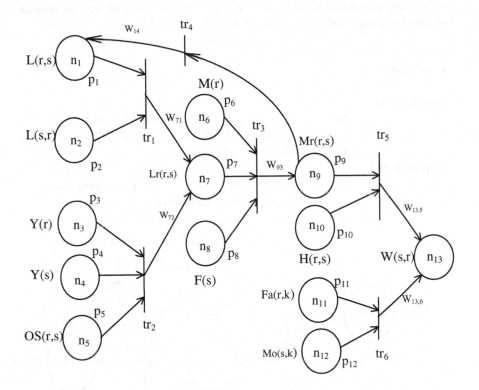

L = Loves, Y = Young, OS = Opposite-Sex, Lr = Lover, M = Male, F = Female,
Mr = Married, H = Husband, W = Wife, Fa = Father, Mo = Mother
r = Ram. s = Sita. k = Kush

Fig. 7.7: A FPN with initially assigned known beliefs and random weights.

While reasoning in analogous problems, the rules may be assigned with the estimated CFs, as obtained from the known histories. In this example, case III is a new test problem, whose knowledge base is the subset of the union of the knowledge bases for case I and case II. Thus, the CFs of the rules are known. In case III, the initial beliefs of the axioms only are presumed to be known. The aim is to estimate the steady-state belief of all propositions in the network. Since stability of the reasoning model is guaranteed, the belief revision process is continued until steady-state is reached. In fact, steady-state occurs in the reasoning model of case-III after five belief revision cycles. Once the steady-state condition is reached, the network may be used for generating new inferences.

The FPN, given in Figure 7.7, has been formed using the above rule-base and database from a typical case history. The fuzzy beliefs of the places in Figure 7.7 are found from proven historical database. The initial weights in the network are assigned arbitrarily and the model for encoding of weights is used for computing the steady-state value of weights (Table 7.1).

Case History II

Rule base II :

PR1 :   Wife(y,x), Loves(x,z), Female(z) → Hates(y,x)
PR2 :   Husband(x,y), Loves(x,y) → Wife(y,x)
PR3 :   Hates(z,x) → Loves(x,y)

Database II :

Husband(jadu,mala),   Loves(jadu,mala),
Loves(jadu,rita), Female(rita)

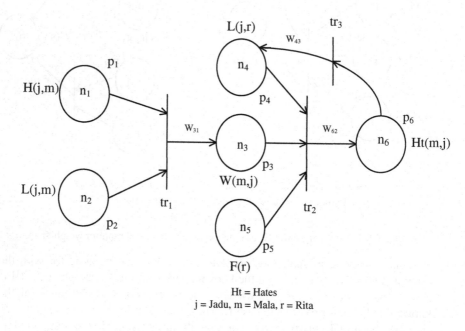

Ht = Hates
j = Jadu, m = Mala, r = Rita

Fig. 7.8: A second FPN with known initial beliefs and random weights.

The FPN of Figure 7.8 is formed with the rule base and database given above. The system parameters of the FPN of Figure 7.8 are presented in Table 7.2.

The Current Reasoning Problem

Now, to solve a typical reasoning problem, whose knowledge and databases are presented herewith, we need to assign the derived weights from the last two case histories. The reasoning model can be used in this example to compute the steady-state belief of the proposition: Hates(lata, ashoke), with the given initial beliefs of all the propositions. Table 7.3 shows the system parameters of the FPN given in Figure 7.9.

Table 7.2: Parameters of Case History 2

Initial Weights w_{ij}	$w_{31} = 0.75$, $w_{62} = 0.95$, $w_{43} = 0.8$
Initial Fuzzy Beliefs n_i	$n_1 = 0.8$, $n_2 = 0.7$, $n_3 = 0.1$, $n_4 = 0.2$, $n_5 = 0.9$, $n_6 = 0.3$
Steady-state weights after three iterations	$w_{31} = 0.7$, $w_{62} = 0.10$, $w_{43} = 0.10$
$th_j = 0$ for all transitions tr_j	

Current Rule Base:

PR1 :   Loves(x,y), Loves(y,x) → Lover(x,y)
PR2 :   Young(x),   Young(y),   OS(x,y)   →
        Lover(x,y)
PR3 :   Lover(x,y),   Male(x),   Female(y)   →
        Marries(x,y)
PR4 :   Marries(x,y) → Loves(x,y)
PR5 :   Marries(x,y), Husband(x,y) → Wife(y,x)
PR6 :   Father(x,z), Mother(y,z) → Wife(y,x)
PR7 :   Wife(y,x),   Loves(x,z),   Female(z)   →
        Hates(y,x)
PR8 :   Hates(z,x) → Loves(x,y)

Current Database:

Loves(ashoke,lata), Loves(lata,ashoke),
Young(ashoke), Young(lata), Opposite-
Sex(ashoke, lata), Male(ashoke),
Female(lata), Husband(ashoke,lata),
Father(ashoke, kamal), Mother(lata,
kamal), Loves(ashoke,tina),
Female(tina)

7.7.4 Implication of the Results

The analysis of stability envisages that both the encoding and the recall model of the fuzzy cognitive map are unconditionally stable. The time required for convergence of the proposed model is proportional to the number of transitions on the largest path (cascaded set of arcs) [11] in the network. The model could be used for determination of CF of rules in a KB by maintaining consistency among the beliefs of the propositions of known case histories.

Table 7.3: Parameters of the current reasoning problem

Initial weight w_{ij} taken from the steady-state CFs of corresponding rules from earlier case histories I and II shown in parenthesis	$w_{71} = w_{71}(I) = 0.35$, $w_{72} = w_{72}(I) = 0.60$, $w_{93} = w_{93}(I) = 0.35$, $w_{14} = w_{14}(I) = 0.35$, $w_{13,5} = w_{13,5}(I) = 0.35$, $w_{13,6} = w_{13,6}(I) = 0.50$, $w_{16,7} = w_{62}(II) = 0.10$, $w_{15,8} = w_{43}(II) = 0.10$
Initial Fuzzy Beliefs n_i	$n_1 = 0.4$, $n_2 = 0.8$, $n_3 = 0.75$, $n_4 = 0.85$, $n_5 = 0.65$, $n_6 = 0.9$, $n_7 = 0.3$, $n_8 = 0.7$, $n_9 = 0.3$, $n_{10} = 0.95$, $n_{11} = 0.65$, $n_{12} = 0.6$, $n_{13} = 0.25$, $n_{14} = 0.55$, $n_{15} = 0.35$, $n_{16} = 0.40$
Steady-state Belief at place P_{16} for proposition Hates(l,a)	**$n_{16} = 0.10$**
$th_j = 0$ for all transitions tr_j	

7.8 Conclusions

This chapter developed two models of fuzzy cognitive maps that are free from the limitations of existing models. The first model employed Hebbian learning with an additional forgetfulness factor α. The conditional convergence of the first model in encoding and unconditional convergence in recall phases has been proved. The second model has been proved to converge to stable points in both encoding and recall phases.

Both the models can be used for fusion of knowledge from multiple sources. In the case of, the first model, used for knowledge refinement, then α for each FPN representing the case histories should be equal. Selection of α also is an important problem. A large value of α (<2) causes an overdamped behavior of weights. On the other hand, a very small value of α (<0.2) requires significant time for the oscillations of weights to settle down to stable values. Typically, α should be chosen in the range [1, 1.2] to avoid quick memory refresh but stable learning.

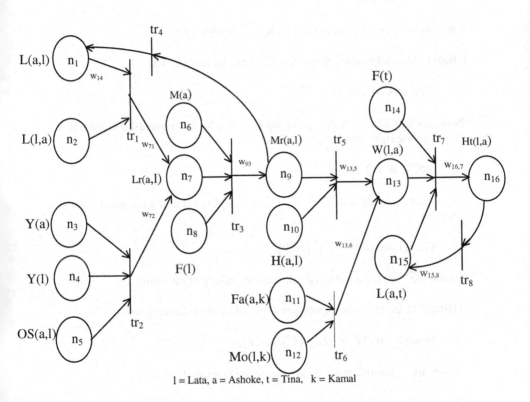

Fig. 7.9: A FPN used for estimating the belief of Ht(l,a) with known initial belief and CFs.

The second model does not include α, and thus can easily be employed for knowledge refinement from multiple sources. The inferences derived by the second model during the recall phase, however, are more approximate in comparison to those of the first model. In summary, the first model is a more accurate representation of the cognitive learning by human beings, and naturally it invites more difficulty in tuning α, which does not arise in the second model.

Exercises

1. Show that at steady state the model given by expression (7.6) satisfies the following inequality:

    $$N^* \geq W^* o\ T^*$$

    where the * above a parameter denotes its steady-state value.

    [**Hints:** At steady-state expression (7.6) can be rewritten as

    $$N^* = N^* \vee (W^* o\ T^*).$$

    Now, the left-hand side N^* can be equal to the N^* in the right-hand side, if the N^* in the right-hand side satisfies:

    $$N^* \geq W^* o\ T^*.]$$

2. Show that at steady state the weight matrix in the encoding equation (7.8) is given by

    $$W^* = (1/\alpha)\ [(N^* . (T^*)^T) \wedge P].$$

    where the * above a parameter denotes its steady-state value.

    [**Hints:** At steady state expression (7.8) takes the following form:

    $$W^* = (1 - \alpha)\ W^* + [(N^* . (T^*)^T) \wedge P],$$

    which after simplification yields the desired results.]

3. The solution of the following equation describes the steady state minimum value of weight matrix in the encoding cycle represented by the model (7.6-7.8).

    $$W^* = (1/\alpha)\ [\ ((W^* o\ T^*) . (T^*)^T)\ \wedge\ P]$$

    [**Hints:** The result follows by combining the results of the last two problems.]

References

[1] Bugarin, A. J. and Barro, S., "Fuzzy reasoning supported by Petri nets," *IEEE Trans. on Fuzzy systems*, vol. 2, no. 2, pp. 135-150, 1994.

[2] Cao, T. and Sanderson, A.C ., "Task sequence planing using fuzzy Petri nets," *IEEE Trans. on Systems, Man and Cybernetics*, vol. 25, no. 5, pp. 755-769, May 1995.

[3] Cao, T. and Sanderson, A. C., "A fuzzy Petri net approach to reasoning about uncertainty in robotic systems," *Proc. of IEEE Int. Conf. Robotics and Automation*, Atlanta, GA, pp. 317-322, 1993.

[4] Cao, T., "Variable reasoning and analysis about uncertainty with fuzzy Petri nets," *Lecture Notes in Computer Scince 691*, Marson, M. A. (Ed.), Springer-Verlag, New York, pp. 126-145, 1993.

[5] Cardoso, J., Valette, R., and Dubois, D., "Petri nets with uncertain markings," In *Advances in Petri Nets*, Lecture notes in computer science, Rozenberg, G. (Ed.), Springer-Verlag, New York, vol. 483, pp. 65-78, 1990.

[6] Daltrini, A. and Gomide, F., "An extension of fuzzy Petri nets and its applications," *IEEE Trans. on Systems, Man and Cybernetics*, 1993.

[7] Daltrini, A., *Modeling and Knowledge Processing Based on the Extended Fuzzy Petri Nets*, M.Sc. degree thesis, UNICAMP-FEE0DCA, May 1993.

[8] Garg, M. L., Ashon, S. I., and Gupta, P. V., "A fuzzy Petri net for knowledge representation and reasoning," *Information Processing Letters*, vol. 39, pp. 165-171, 1991.

[9] Konar, A., *Uncertainty Management in Expert System Using Fuzzy Petri Nets*, Ph.D. dissertation, Jadavpur University, India, 1994.

[10] Konar, A., *Artificial Intelligence and Soft Computing: Behavioral and Cognitive Modeling of the Human Brain*, CRC Press, Boca Raton, FL, 1999.

[11] Konar, A., *Computational Intelligence, Principles, Techniques and Applications*, Springer-Verlag, Heidelberg, December 2004.

[12] Konar, A. and Chakraborty, U. K.,   "Reasoning and unsupervised learning in a fuzzy cognitive map," *Information Sciences*, 2005 (in press).

[13] Konar, A. and Mandal, A.K., "Non-monotonic reasoning in expert systems using fuzzy logic," *AMSE Publication*, 1992.

[14] Konar, A. and Mandal A. K., "Uncertainty management in expert systems using fuzzy Petri nets," *IEEE Trans. on Knowledge and Data Engineering*, vol. 8, no. 1, pp. 96-105, February, 1996.

[15] Kosko, B., "Fuzzy cognitive maps," *International Journal of Man-Machine Studies*, vol. 24, pp. 65-75, 1986.

[16] Kosko, B., *Fuzzy Engineering*, Prentice-Hall, Englewood Cliffs, NJ, ch. 15, pp. 499-527, 1997.

[17] Lipp, H. P. and Gunther, G., "A fuzzy Petri net concept for complex decision making process in production control," *Proc. First European Congress on Fuzzy and Intelligent Technology (EUFIT '93)*, Aachen, Germany, volI, pp. 290–294, 1993.

[18] Looney, C. G., "Fuzzy Petri nets for rule-based decision making," *IEEE Trans. on Systems, Man, and Cybernetics*, vol. 18, no. 1, pp.178-183, 1988.

[19] Pal, S. and Konar, A., "Cognitive reasoning using fuzzy neural nets," *IEEE Trans. on Systems, Man and Cybernetics*, Part B, vol. 26, no. 4, 1996.

[20] Paul, B., Konar, A. and Mandal, A. K., "Estimation of certainty factor of rules by associative memory realized with fuzzy Petri nets," *Proc. IEEE Int. Conf. on Automation, Robotics and Computer Vision*, Singapore, 1996.

[21] Paul, B., Konar, A. and Mandal, A. K., "Estimation of certainty factor of knowledge using fuzzy Petri nets," presented in *FUZZ-IEEE '98 Int. Conf.*, Alaska, 1998.

[22] Pedrycz, W., *Fuzzy Sets Engineering*, CRC Press, FL, Boca Raton, 1995.

[22] Pedrycz, W. and Gomide, F., "A generalized fuzzy Petri net model," *IEEE Trans. on Fuzzy Systems*, vol. 2, no. 4, pp. 295-301, Nov 1994.

[24] Peters, J. F. and Sohi, N., "Coordination of multi-agent systems with fuzzy clocks," *Concurrent Engineering Research and Applications*, vol. IV, no. 1, March 1996.

[25] Scarpelli, H. and Gomide, F. "High level fuzzy Petri nets and backward reasoning," In *Fuzzy Logic and Soft Computing*, Bouchon-Meunier, B., Yager, R. R. and Zadeh, L. A. (Eds.), World Scientific, 1995.

[26] Scarpelli, H., Gomide, F. and Yager, R., "A reasoning algorithm for high level Fuzzy Petri nets," *IEEE Trans. on Fuzzy Systems* , vol. 4, no. 3, pp. 282-295, Aug. 1996.

[27] Simpson, P. K., *Artificial Neural System: Foundation, Paradigm, Application and Implementation*, Pergamon Press, 1990.

Chapter 8

Supervised Learning by a Fuzzy Petri Net

Feed-forward neural networks used for pattern classification usually have one input layer, one output layer, and several hidden layers. The hidden layers in these networks add extra nonlinearity for realization of precise functional mapping from the input to the output layer, but are unable to explain any semantic relations between them with their predecessor and successor layers. This chapter presents a novel scheme for supervised learning on a fuzzy Petri net that provides semantic justification of the hidden layers, and is capable of approximate reasoning and learning from noisy training instances. An algorithm for training a feed-forward fuzzy Petri net together with an analysis of its convergence has been presented in the chapter. The chapter also examines the scope of the learning algorithm in object recognition from their 2-D geometric views.

8.1 Introduction

There exists a vast literature on fuzzy Petri nets (FPNs) [1-7], [9-13], [18-19] that deal with knowledge representation and reasoning in presence of inexact data and knowledge bases. The credit of machine learning with fuzzy AND-OR neurons [8], [14], [16] in general and fuzzy Petri nets [17] in particular goes with Pedrycz. In [15], while modeling a specialized cognitive structure, Pedrycz examined the scope of the model in fuzzy pattern recognition. This chapter addresses an alternative scheme of supervised learning on fuzzy Petri nets to study its application in pattern recognition. Our method combines the power of high-level reasoning of FPN and the ease of training of the supervised feed-forward neural net compositely on a common platform.

The proposed model of fuzzy Petri net comprises fuzzy OR and AND neurons respectively represented by places and transitions of the network. Normally, a set of transitions followed by a set of places constitutes a layer. An l-layered fuzzy Petri net thus contains $(l - 1)$ layers of transitions followed by places with an additional input layer consisting of places only. The places in the last layer are called concluding places. Such a network has 2-fold representational benefits. First, it can represent inexact knowledge like conventional FPN. Secondly, the network can be trained with a set of input-output patterns as is done in a feed-forward neural net. Such a network used for object recognition from their fuzzy features, offers the benefits of both inexact reasoning and machine learning on a common platform.

The chapter starts with a formal description of the proposed model of fuzzy Petri net. An algorithm for adaptation of thresholds associated with the transitions from a set of training instances has been presented for applications in object recognition from their fuzzy features. An analysis of the proposed training algorithm has been presented to study its unconditional convergence.

This chapter is organized in six sections. The proposed model of fuzzy Petri nets is covered in Section 8.2. The algorithm for training the fuzzy Petri net with n sets of input-output training instances is presented in Section 8.3. The analysis for convergence of the algorithm is presented in Section 8.4. An illustrative case study, highlighting the training with 2-D geometric objects and their recognition, is covered in Section 8.5. Conclusions are listed in Section 8.6.

8.2 Proposed Model of Fuzzy Petri Nets

The principles of supervised learning to be introduced in this chapter have been developed based on the following model of FPN.

Definition 8.1: *A FPN is an 8-tuple, given by*

$FPN = <P, Tr, T, D, I, O, Th, n>$
where
  $P = \{p_1, p_2, \ldots, p_n\}$ *is a finite set of places;*
  $Tr = \{tr_1, tr_2, \ldots, tr_m\}$ *is a finite set of transitions;*
  $T = \{t_1, t_2, \ldots, t_m\}$ *is a set of fuzzy truth tokens in the interval [0,1] associated with the transitions $tr_1, tr_2, \ldots, tr_m$ respectively;*
  $D = \{d_1, d_2, \ldots, d_n\}$ *is a finite set of propositions, where proposition d_k corresponds to place p_k;*

> $P \cap Tr \cap D = \emptyset$; *cardinality of* $(P) = $ *Cardinality of* (D);
> *I:* $Tr \to P^{\infty}$ *is the input function, representing a mapping from transitions to bags of (their input) places;*
> *O:* $Tr \to P^{\infty}$ *is the output function, representing a mapping from transitions to bags of (their output) places;*
> $Th = \{th_1, th_2, \ldots, th_m\}$ *represents a set of threshold values in the interval [0,1] associated with transitions* $tr_1, tr_2, \ldots, tr_m$ *respectively;*
> *n:* $P \to [0,1]$ *is an association function hereafter called fuzzy belief, representing a mapping from places to real values between 0 and 1;* $n(p_i) = n_i,$ *say.*

A few more definitions, which will be referred to frequently while presenting the model, are in order.

Definition 8.2: *If* $p_i \in I$ (tr_a) *and* $p_i \in O$ (tr_b) *then* tr_a *is **immediately reachable** from* tr_b [4]. *Again, if* tr_a *is immediately reachable from* tr_b *and* tr_b *is immediately reachable from* tr_c, *then* tr_a *is **reachable** from* tr_c.

The reachability property is the reflexive, transitive closure of the immediate reachability property [4]. We would use IRS (tr_a) and RS (tr_a) operators to denote the set of transitions immediately reachable and reachable from transition tr_a, respectively.

Moreover, if $tr_a \in$ [IRS {IRS (IRS .... k-times (tr_b))}], denoted by IRS^k (tr_b), then tr_a is reachable from tr_b with a degree of reachability k. For reachability analysis, two connectivity matrices P and Q, defined below will be needed.

Definition 8.3: *A **place to transition connectivity (PTC)** matrix Q is a binary matrix whose elements* $q_{jk} = 1,$ *if for places* p_k *and transition* $tr_j,$ $p_k \in I$ (tr_j); *otherwise* $q_{jk} = 0.$ *If the FPN has n places and m transitions, then the Q matrix is of* ($m \times n$) *dimension* [9].

Definition 8.4: *A **transition to place connectivity (TPC)** matrix P is a binary matrix whose element* $p_{ij} = 1,$ *if for places* p_i *and transition* $tr_j,$ $p_i \in O$ (tr_j); *otherwise* $p_{ij} = 0,$ *With n places and m transitions in the FPN, the P matrix is of* ($n \times m$) *dimension* [10].

Definition 8.5: *If* $p_{ij},$ *the (i, j)th element of matrix P is 1, then place* p_i *is said to be reachable from transition* $tr_j.$ *We call this reachability **transition-to-place reachability** to distinguish it from transition-to-transition reachability,*

Since the max-min composition (Q o P) represents mapping from transitions to their immediately reachable transitions, the presence of a 1 in the matrix $M_1 = $ (Q o P) at position (j, i) represents that $tr_j \in$ IRS (tr_i). Analogously, a 1 at position (j, i) in matrix $M_r = $ (Q o P)^r for positive integer r, represents $tr_j \in$

IRS^r(tr$_j$), i.e., tr$_j$ is reachable from tr$_i$ with a degree of reachability r. Further, a 1 at position (i, j) in the matrix P o (Q o P)^r denotes that there is a place p$_i$ reachable from transition tr$_j$ with a *degree of reachability r from transitions to transitions* and a *degree of reachability one from transition to places*.

Definition 8.6: *A transition tr$_j$ is* **enabled** *if p$_i$ possesses fuzzy beliefs for* $\forall p_i \in$ *I(tr$_j$).*

An enabled transition **fires** *by generating a fuzzy truth token (FTT) at its output arc. The value of the FTTs is given by:*

$$\left. \begin{array}{l} t_j\,(t+1) = [\wedge\{\, n_i \mid p_i \in I(tr_j\,)\} - th_j\,],\ if\ \wedge\{\, n_i \mid p_i \in I(tr_j\,)\} > th_j \\ \quad\quad \forall i \quad\quad\quad\quad\quad\quad\quad\quad\quad\quad\quad\quad\quad \forall i \\ \quad\ = 0,\ otherwise. \end{array} \right\} \quad (8.1)$$

Since FTT computation at a transition involves taking fuzzy AND (Min) of the beliefs of its input places, the transitions may be regarded as AND neurons.

Definition 8.7: *After firing of tr$_j$'s the fuzzy belief of n$_k$ at the place p$_k$, where p$_k \in$ O(tr$_j$), $\forall j$, is given by:*

$$n_k(t+1) = n_k(t)\ V\ [\ V\{\ t_j\,(t+1)\}] \qquad\qquad\qquad (8.2)$$
$$\forall j$$

Since belief computation involves fuzzy OR (max) operation, the places may be regarded as **OR neurons**.

Definition 8.8: *A set of transitions {tr$_x$} and a set of places {p$_y$}, where $\forall$ place p $\in$ {p$_y$} and $\forall$ transition tr $\in$ {tr$_x$}, p $\in$ O (tr) constitutes a* **layer** *in the proposed FPN.*

Definition 8.9: *The* **input layer** *in a FPN is a special layer which consists of only input places, and there does not exist any tr$_j$ such that p $\in$ O (tr$_j$).*

8.2.1 State-Space Formulation

Let N$_t$ be the fuzzy belief vector, whose i-th component n$_i$ denotes the belief of proposition d$_i$ located in place p$_i$ at time t. Also assume T$_t$ to be the FTT vector, whose j-th component t$_j$ is the FTT of transition tr$_j$ at time t. Then the FTT update equation (8.1) for m transitions together can be described [10] in state-space form as follows:

$$T_{t+1} = (Q\ o\ N_t^c)^c - Th,\ \ if\ \ (Q\ o\ N_t^c)^c > Th \qquad\qquad (8.3)$$

where Th denotes the threshold vector of transitions, such that its i-th component th_i is the threshold of transition tr_i and c over a vector denotes its component-wise one's complement.

The belief updating expression (8.2) of the entire FPN can also be represented in state-space form as follows:

$$N_{t+1} = N_t \, V \, (P \circ T_{t+1})$$

$$= N_t \, V \, [P \circ \{ (Q \circ N_t^c)^c - Th \}], \text{ if } (Q \circ N_t^c)^c > Th \qquad (8.4)$$

For a FPN with l layers, one may recursively update expression N_{t+1} l-times in (8.4) to find the belief vector N_l in terms of the initial belief vector N_0 as follows:

$$N_l = N_0 \, V \, [P \circ \{ \overset{l}{\underset{i=1}{V}} \, \{ (Q \circ N_i^c)^c - Th \} \}] \qquad (8.5)$$

It may be noted that the components of the initial belief vector N_0 for the input layer are non-zero, while other components of N_0 are set to zero.

The vector N_l yields the fuzzy belief at all places in the NFPN at steady-state. The components of the steady-state belief vector N_l corresponding to the places in the last layer may now be considered for estimation of error vector E_k. If the estimation of the error vector after one forward pass in the neural FPN followed by the adjustment of thresholds of transitions in a pre-defined layer is termed a single *threshold adjustment cycle (TAC)*, then the error vector after the k-th TAC is given by

$$E_k = D - A \circ N_l(k) \qquad (8.6)$$

where D is the target vector, corresponding to the places at the last layer and N_l (k) is the N_l vector after the k-th TAC and A is a binary mask matrix of the following form :

$A =$

From / To		Places in all excluding the last layer				Places in the last layer			
		p_1	p_2	......	p_m	p_{m+1}	p_{m+2}	.....	p_{m+n}
Places in the last layer	p_{m+1}	0	0	......	0	1	0	.....	0
	p_{m+2}	0	0	......	0	0	1	.....	0
	⋮	⋮	⋮	⋮	⋮	⋮	⋮	⋮	⋮
	p_{m+n}	0	0	......	0	0	0	.....	1

$=$

From / To Places	Places	
Places	Φ	I

$$(8.7)$$

which on pre-multiplication with $N_l(k)$ yields the vector corresponding to the output layered places.

8.3 Algorithm for Training

Training in the proposed neural FPN refers to layer-wise adaptation of thresholds of the transitions. Unlike the classical back-propagation algorithm, the training in the neural FPN begins at the input layer and is continued layer-wise until the last layer. The training may further involve several phases of adaptation of thresholds in the entire network.

Before formally presenting the training algorithms, we here briefly outline the principle of training. Given an input-output training instance, the neural FPN first computes the output through a forward pass in the network and evaluates the outputs of the last layered AND-neurons. It then evaluates the error vector by taking the component-wise difference of the computed output vector from the prescribed target vector (output instance). Let th_j be the threshold of a transition tr_j in a given layer and e_u be the scalar error at the concluding place u in the output layer, such that p_u has connectivity from tr_j. By the phrase "connectivity" we mean the following: if a concluding place $p_u \in O(tr_k)$ then $tr_k \in RS\ (tr_j)$. Then the threshold th_j is adapted using

$$th_j \leftarrow th_j - \bigvee_{\forall u} (e_u) \cdot \tag{8.8}$$

After the thresholds in a given layer is adapted by the above principle, the network undergoes a forward pass for re-evaluation of the error vector at the output layer and adaptation of the thresholds for the transitions in the subsequent layer. The subsequent layer in the present context usually is the next layer, but it can be the first layer as well if the threshold adaptation is currently performed in the last layer. The whole scheme of training a neural FPN is represented in vector-matrix form as follows.

Redefining that a forward pass in the network for computing the error vector at the output layer followed by threshold adjustment of all the transitions in a given layer together constitutes a TAC, let

  k be the total number of TAC so far performed + 1, the last 1 is due to the current TAC, and

  r be the total number of training cycles, i.e., number of times training has so far been given to the entire network.

Considering l to be the number of layers in the network, we define z = k − r × l, where z denotes the layer currently selected for threshold adaptation. The definition of z follows easily from the definitions of k and r. For convenience, we can rewrite the last expression as k = r × l + z. Then for any positive integer k, z = k modulo l. For example, let k = 8 and l = 3, then z = 8 modulo 3 = 2. The above result indicates that if the current TAC is the 8-th cycle, and the FPN contains three layers, then the second layer is currently selected for threshold adaptation. It is indeed important to note that r can be defined as k div l, which in the present context is found to be 8 div 3 = 2, i.e., already 2 training cycles has elapsed.

In order to transfer the error vector E_k of the k-th TAC to the layer $(k - r \times l)$, we need to define a matrix M that denotes the connectivity from the transitions at layer $(k - r \times l)$ to the places at the output layer, where

$$M = P \text{ o } (Q \text{ o } P)^{(k-r \times l)}. \tag{8.9}$$

Consequently, M^T denotes connectivity from the places in the output layer to the transitions at layer $(k - r \times l)$. Now, to adapt the thresholds at the $(k - r \times l)$ layer, we need to construct a rotational partitioned matrix $W_{k - r \times l}$, where partition matrix I occupies the $(k - r \times l)$-th position out of l possible partitions in matrix $W_{k-r \times l}$ and the remaining $(l - 1)$ positions contain null matrices ϕ.

$$W_{(k-r \times l)} = \begin{bmatrix} [\Phi] \\ \overline{[I]} \\ [\Phi] \end{bmatrix}_{m \times s}$$

For example, for a 3-layered net, if $l = 3, k = 1, r = 0$, then $W_{(k-r \times l)}$ would be:

$$W_1 = \begin{bmatrix} [I] \\ [\Phi] \\ [\Phi] \end{bmatrix}.$$

Similarly, for $l = 3, k = 2, r = 0$, $\quad W_2 = \begin{bmatrix} [\Phi] \\ [I] \\ [\Phi] \end{bmatrix}$;

for $l = 3, k = 3, r = 0$, $\quad W_3 = \begin{bmatrix} [\Phi] \\ [\Phi] \\ [I] \end{bmatrix}$;

for $l = 3, k = 4, r = 1$, $\quad W_1 = \begin{bmatrix} [I] \\ [\Phi] \\ [\Phi] \end{bmatrix}$; and so on.

Adaptation of thresholds at layer $k - r \times l$ now can be done easily using the following expression:

$$Th := Th - W_{(k - r \times l)} \text{ o } M^T \text{ o } E_k \text{ where}$$

$$M = P \text{ o } (Q \text{ o } P)^{(k - r \times l)} \text{ and}$$

$$W_{(k-r \times l)} = \begin{bmatrix} [\Phi] \\ [I] \\ [\Phi] \end{bmatrix}_{m \times s}$$

where Th is the $(m \times 1)$ threshold vector.

The algorithm for training a multi-layered feed-forward FPN with a single input-output pattern is presented below.

Procedure Train-with-single-I/O-pattern (N_o, P, Q, D, Th_k);
Begin

    k:=1; error-sum:=1000; //large error-sum// r:=0;
    While (error-sum > pre-assigned-limit) AND (r < r_{max}) **do**
    Begin

$$N_l(k) := N_o \ V \left[Po \left\{ \overset{l}{\underset{i=1}{V}} \left\{ (Q o N_i^c)^c - Th \right\} \right\} \right] ;$$

$$E_k := D - Ao \ N_l(k);$$

$$\text{Error-sum} := \underset{\forall u}{\Sigma} (E_u)^{1/3} ; \ // \ E_u = \text{components of } E_k \text{ at}$$

concluding places p_u//

$$Th := Th - W_{(k-r \times l)} \ o \ M^T \ o \ E_k ;$$

where $M := P \ o \ (Q \ o \ P)^{(k-r \times l)}$

and $W_{(k-r \times l)} := \begin{bmatrix} [\Phi] \\ [I] \\ [\Phi] \end{bmatrix}_{m \times s} ;$

    If Mod(k/*l*)=0 then r:= r + 1;
    k:= k+ 1;
End While;
End.

The while loop in the above procedure first computes the output vector Ao $N_l(k)$ and determines the error vector E_k by taking the difference of Ao $N_l(k)$ from the target vector D. The E_k is then used to layer-wise adapt the thresholds starting from the first layer. It is indeed important to note that although the threshold vector Th includes the thresholds of all the transitions present in the network, the adaptation rule changes the thresholds at the $(k - r \times l)$-*th* layer only. A check of mod $(k / l) = 0$ is included at the end of the while loop to test whether thresholds of all layers in the FPN have been updated. If yes, then r is increased by one, indicating a fresh start of threshold adaptation beginning with the first layer. The TAC count k is increased after each threshold adaptation in the layers of the FPN.

The performance of learning in the above procedure depends greatly on an important attribute called error-sum, which is computed by summing the cube root of the errors at the concluding places. Naturally, the question arises: why cube root? Since the components of the error vector E_k are bounded in [0, 1], the cube roots increase the error level for small signals maintaining their signs. The threshold adaptation in layers of the FPN is continued until the error-sum does not converge within a permissible limit, or the number of complete training cycles r is within a prescribed limit r_{max}. The algorithm terminates when the condition specified in the WHILE-loop is no longer satisfied.

The training algorithm with n input-output patterns is presented below.

Procedure Train-FPN (training-patterns[N_o]$_i$, D_i, Th)
Begin
r :=0;
  Repeat
    $S_1 := 0$; $S_2 := 0$;...... $S_5 := 0$; Sum := 0;
    For i := 1 to n **do**
    Begin
      For k:= (r+1) to (r + no-of-layers) **do**
      Begin
      $[N_l(k)]_i := [N_0]_i$ V $[Po\{V\{Q$ o $N_i^c)^c - Th\}\}]$
      $[E_k]_i := D_i -$ Ao $[N_l(k)]_i$;
        For j := 1 to number-of-components-of $[Ek]_i$ **do**
        Begin
        $X_j := \{[E_k (j)]_i \}^{1/3}$;
        $S_j := S_j + X_j$; $e_j := [S_j]^3$; //e_j= j-th
                         component of composite
                         error vector $E_k^{/}$//
      End For ;
      End For;
    Th := Th $- W_{(k-r\times l)}$ o M^T o $E_k^{/}$
    where, M := P o $(Q$ o $P)^{(k-r\times l)}$

$$\text{and } W_{(k-r\times l)} := \begin{bmatrix} [\Phi] \\ \hline [I] \\ \hline [\Phi] \end{bmatrix}_{m\times s} ;$$

    End For;
    For j := 1 to number-of-components-of $E_k^{/}$ **do**
    Begin
      Sum := Sum + S_j ;

$PI_r := Sum; r := r + 1;$
  End For;
  Until $PI_{r-1} \leq PI_r;$
End.

For n sets of input-output patterns in procedure Train-FPN, we determine $[E_k]_i$ for $i = 1$ to n and take the effects of the individual error vector E_k to construct a new error vector E_k'. The j-th component of E_k' is evaluated by taking the cube of the sum of the cube-roots of the individual j-th components of $[E_k]_i$ for $i = 1$ to n. The E_k' is then used to adapt the thresholds by the previous procedure. A performance index (PI) that attempts to minimize the sum of the components of E_k' is employed as the termination criterion of the procedure.

8.4 Analysis of Convergence

One important aspect of a learning algorithm is its study of convergence. The algorithm proposed so far incidentally converges to a stable point at the origin. A proof of the convergence for the proposed algorithm is addressed in Theorem 8.1.

Theorem 1: *The error vector E_k unconditionally converges to the origin in error-space.*

Proof: The error vector E_k is given by :

$$E_k = D - A \text{ o } N_l(k)$$

$$= D - A \text{o } N_0 \text{ V} \left[\text{ P o} \left\{ \overset{l}{\underset{i=1}{V}} \left\{ (Q \text{ o } N_i^c)^c - Th_k \right\} \right] \right] \qquad (8.10)$$

where

$$Th_k = Th_{k-1} - W_{(k-r \times l)} \text{ o } M^T \text{ o } E_k \text{ with}$$
$$M = P \text{ o } (Q \text{ o } P)^{(k-r \times l)} \text{ and}$$

$$W_{(k-r \times l)} = \begin{bmatrix} [\Phi] \\ \hline [1] \\ \hline [\Phi] \end{bmatrix}_{m \times s}$$

and Th_k = threshold vector Th at a time, after updating thresholds at (k −1) number of layers, all of which need not be distinct.

$$E_k = D - A \text{o } \left[N_0 \text{ V} \left[\text{ P o} \left\{ \overset{l}{\underset{i=1}{V}} \left\{ (Q \text{ o } N_i^c)^c - Th_{k-1} + W_{(k-r \times l)} \text{ o } M^T \text{ o } E_k \right\} \right] \right] \right]$$

$$E_{k-1} = D - Ao\ [\ N_0\ V\ [\ P\ o\ \{\ \overset{l}{\underset{i=1}{V}}\ \{\ (\ Q\ o\ N_i^c)^c - Th_{k-1}\}]]\ [\text{by expression (8.10)}]$$

Now, $\Delta E_{k-1} = E_k - E_{k-1}$

$$= Ao\ [\ N_0\ V\ [\ P\ o\ \{\ \overset{l}{\underset{i=1}{V}}\ \{\ (\ Q\ o\ N_i^c)^c - Th_{k-1}\}]]$$

$$- Ao\ [N_0\ V\ [\ P\ o\ \{\ \overset{l}{\underset{i=1}{V}}\ \{\ (\ Q\ o\ N_i^c)^c - Th_{k-1} + W_{(k-r \times l)}\ o\ M^T\ o\ E_k\}]\ (8.11)$$

Let $\overset{l}{\underset{i=1}{V}}\ \{\ (\ Q\ o\ N_i^c)^c - Th_{k-1}\} = (\ Q\ o\ N_j^c)^c - Th_{k-1}$ \hfill (8.12)

Substituting (8.12) in (8.11) we have:

$$\Delta E_{k-1} = Ao\ [\ N_0\ V\ [\ P\ o\ \{(\ Q\ o\ N_j^c)^c - Th_{k-1}\}]]$$

$$- A\ o\ [\ N_0\ V\ [\ P\ o\ \{(\ Q\ o\ N_j^c)^c - Th_{k-1} + W_{(k-r \times l)}\ o\ M^T\ o\ E_k\}]]$$

$$= A\ o\ [\ N_0\ V\ (P\ o\ y)\ ] - A\ o\ [\ N_0\ V\ \{\ Po\ (y + W_{(k-r \times l)}\ o\ M^T\ o\ E_k)\}]$$

where $y = (\ Q\ o\ N_j^c)^c - Th_{k-1}$ \hfill (8.13)

Four possible cases are now considered.

Case I : $P\ o\ y \geq N_0$ \hfill (8.14)

$$\text{and}\quad P\ o\ (y + W_{(k-r \times l)}\ o\ M^T\ o\ E_k) \geq N_0 \hfill (8.15)$$

We now present two extreme cases of analysis.

Case I (a): Let the signs of all components of E_k be positive. Then

$$\Delta E_{k-1} = A\ o\ P\ o\ y - A\ o\ P\ o\ (y + W_{(k-r \times l)}\ o\ M^T\ o\ E_k)$$

$$\geq - A\ o\ Po\ W_{(k-r \times l)}\ o\ M^T\ o\ E_k\ [\text{Since}\ Xo\ Y + X\ o\ Z \geq X\ o\ (Y+Z)]$$

$$= - [\alpha_{ij}]\ o\ E_k,\ \text{say, where}\ [\alpha_{ij}] = A\ o\ P\ o\ W_{(k-r \times l)}\ o\ M^T$$

Since elements α_{ij} are bounded in the interval: $0 \leq \alpha_{ij} \leq 1$, for any component of $[\alpha_{ij}] \text{ o } E_k$ close to the positive maximum (one), the corresponding component of ΔE_{k-1} is ≥ -1, which corresponds to convergence.

Case I (b): Let the signs of all components of E_k be negative. Then

$$\Delta E_{k-1} \leq - \text{Ao Po } W_{(k-r \times l)} \text{ o } M^T \text{ o } E_k \quad [\text{Since } X \text{ o } Y + X \text{ o } Z \leq X \text{ o } (Y + Z)]$$

Now, for any component of $[\alpha_{ij}] \text{ o } E_k$ being the negative maximum $(-1 + \in)$, for a pre-assigned small positive number $\in$, the corresponding component of ΔE_{k-1} is $\leq 1 - \in$, which too forces the system to converge.

When E_k contains both positive and negative components, the analysis could be carried out easily by point-wise formulation of the expressions for change of error at the output places in the network.

Case II: When $N_0 < P \text{ o } y$ (8.16)
     and $N_0 > P \text{ o } (y + W_{(k-r \times l)} \text{ o } M^T \text{ o } E_k)$ (8.17)

$$\Delta E_{k-1} = A \text{ o } P \text{ o } y - A \text{ o } N_0 \geq \mathbf{0}.$$

Since all components of E_k are negative [refer expressions (8.16) & (8.17)], $\Delta E_{k-1} \geq \mathbf{0}$ proves convergence of E_k.

Case III: When $N_0 > P \text{ o } y$ (8.18)
     and $N_0 < P \text{ o } (y + W_{(k-r \times l)} \text{ o } M^T \text{ o } E_k)$ (8.19)

$$\Delta E_{k-1} = A \text{ o } N_0 - \text{Ao Po } (y + W_{(k-r \times l)} \text{ o } M^T \text{ o } E_k) \leq \mathbf{0} \; [\text{ by}(8.18) \text{ and } (8.19)]$$

Since $E_k >$ null vector and $\Delta E_{k-1} <$ null vector, therefore E_k converges.

Case IV: When $N_0 > P \text{ o } y$ (8.20)
     and $N_0 > P \text{ o } (y + W_{(k-r \times l)} \text{ o } M^T \text{ o } E_k)$ (8.21)

$$\Delta E_{k-1} = \mathbf{0}, \text{ and hence } E_k \text{ attains a constant value.}$$

But in view of cases I, II, and III, this constant value will be the null vector. Thus, E_k unconditionally converges to the origin from any initial value in the error space.

8.5 Application in Fuzzy Pattern Recognition

This section provides an application of the proposed learning algorithm in fuzzy pattern recognition. As a case study we consider the problem of recognizing 2-D objects such as circle, ring, ellipse, rectangle, and hexagon (Fig. 8.1) from their geometric features. The features we considered for the set of objects include area, perimeter, maximum length along the x-axis, maximum length along the y-axis, and inverse sphericity. The features have been selected from the point of view of their relative independence and totality, so that feature-count is as minimum as possible and features together is a sufficient descriptor of the objects. Since the measurements of these features may not be free from noise (measurement imprecision), we fuzzify the measurements and train a judiciously selected neural Petri net (Fig. 8.2) to recognize similar objects.

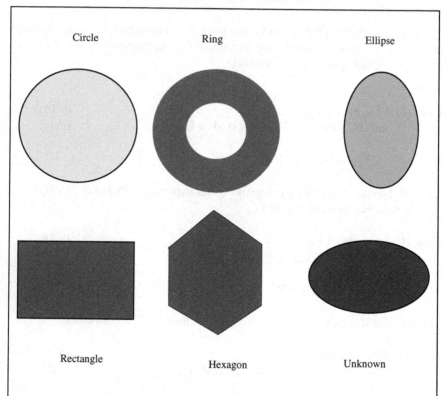

Circle   Ring   Ellipse

Rectangle   Hexagon   Unknown

Fig. 8.1: Various objects used for training the FPN along with the unknown Object.

For fuzzification of features we could take typical fuzzy membership functions, such as triangular or Gaussian-shaped functions. We, however, selected the following function, which is widely used in communication engineering for its inherent characteristics of mapping small variations in x to smaller variations in y, and very large variations in x to relatively smaller variations in y. Further, the function has a single control parameter that is used for scaling its x-range to make it suitable for specific applications.

$$y = \frac{\log_{10}(1 + 5x/\gamma)}{\log_{10}(1 + 5x)} \tag{8.22}$$

where x denotes the measured value of a feature, y represents the fuzzified membership value of x, and γ denotes a normalizing factor, selected based on the range of x.

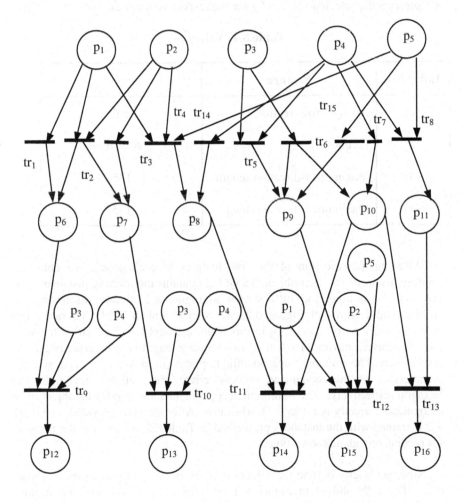

Fig. 8.2: The proposed fuzzy Petri net (multiple copies of input places are shown for neatness in the figure.)

The first four features: area of cross section, perimeter, maximum x-length and maximum y-length, are fuzzified using the above membership function. The fifth feature, inverse sphericity, is computed by taking the ratio of the perimeter of the object and square root of its approximate cross-sectional area, and is fuzzified by subtracting a value of 3.8 from its estimated value. The subtraction is needed to keep the membership in [0, 1].

The value of γ in expression (8.22) for the first four features has been determined experimentally to keep the range y of the function in [0, 1]. Table 8.1 presents the selected values of γ for the respective features.

Table 8.1: Values of γ

Index No.	Feature	γ
(1)	Area of cross-section	10^4
(2)	Perimeter	6×10^2
(3)	Maximum x-direction length	135
(4)	Maximum y-direction length	135

After the fuzzification of the five features of each object, we construct training instances for each object. Table 8.2 contains the training instances. It is prepared for training the FPN shown in Figure 8.2. Before discussing the architectural features of Figure 8.2, we briefly explain Table 8.2 here. The columns under p_1, p_2, . . ., p_5 in this table represent the 5 fuzzified features: area, perimeter, maximum x-length, maximum y-length, and inverse sphericity, respectively. The columns corresponding to places p_{12}, p_{13}, . . ., p_{16} denote the membership of the recognized objects: circle, ring, ellipse, rectangle and hexagon respectively. The minimum acceptable membership for an object to be recognized correctly is set as 0.41 arbitrarily. After the given network of Figure 8.2 is trained with the instances prescribed in Table 8.2, we can use the network for recognizing an unknown object.

Now, we briefly outline the selection of the places and transitions in Figure 8.2. The FPN shown in Figure 8.2 contains three layers: one input layer consisting of places p_1 through p_5, one hidden layer comprising of transitions tr_1, tr_2, ..., tr_8 followed by places p_6, p_7, . . ., p_{11} and one output layer comprising of transitions tr_9, tr_{10}, . . ., tr_{13} followed by concluding places p_{12}, p_{13}, . . ., p_{16}. The fuzzified features cross sectional area, perimeter, maximum x-length, maximum y-length and inverse sphericity are mapped at the input places p_1, p_2, . . . , p_5 respectively. The places p_{12}, p_{13}, ... , p_{16} in the output layer correspond to objects circle, ring, ellipse, rectangle, and hexagon, respectively.

The most important aspect in the construction of the FPN of Figure 8.2 is the selection of transitions and places in the hidden layer. The inputs of the transitions tr_1, tr_2, . . . , tr_8 in the hidden layer is selected by taking into account the logical joint occurrence of the features. For example, large area and perimeter should co-exist in circular patterns. So, both places p_1 and p_2 are considered as input places of transition tr_2. Further, to emphasize the importance of area (see Table 8.2, column under p_1), we use transition tr_1. To recognize circles, p_6 is considered as the output place of both tr_1 and tr_2. Recognition of circles thus can be accomplished by considering p_6, p_3, and p_4 as the input places of tr_9 and p_{12} as the output place of tr_8. Since inverse sphericity is small for circles (see Table 8.2), we do not use p_5 for recognition of circles.

In principle, both transitions and places in the hidden layers are selected by analyzing the influence of both co-existence and independence of features. For each set of joint features, we need one transition. The places in the hidden layer are selected to emphasize one independent feature (or joint feature) over one or more other features. The semantic interpretation of the places in the hidden layer is explained below for convenience.

p_6:   objects having higher priority on large area than the joint occurrence of large area and large perimeter.

p_7:   objects having higher priority on large perimeter than the joint occurrence of large area and large perimeter.

p_8:   objects having higher priority on maximum y-length than the joint occurrence of large area, large perimeter, and inverse sphericity.

p_9:   objects having higher priority on (inverse sphericity or maximum x-length) than the joint occurrence of maximum x-length and maximum y-length.

p_{10}:   objects having priority on maximum x-length and maximum y-length.

p_{11}:   objects having priority on maximum y-length and inverse sphericity.

Table 8.2: Training instances

Places	Input vector components at					Target vector components at				
Object	p_1	p_2	p_3	p_4	p_5	p_{12}	p_{13}	p_{14}	p_{15}	p_{16}
Circle	.77	.62	.74	.82	.38	0.41	10^{-4}	10^{-4}	10^{-4}	10^{-4}
Ring	.55	.92	.88	.88	.38	10^{-4}	0.41	10^{-4}	10^{-4}	10^{-4}
Ellipse	.65	.56	.53	.84	.49	10^{-4}	10^{-4}	0.41	10^{-4}	10^{-4}
Rectangle	.69	.72	.88	.44	.95	10^{-4}	10^{-4}	10^{-4}	0.41	10^{-4}
Hexagon	.53	.51	.44	.82	.68	10^{-4}	10^{-4}	10^{-4}	10^{-4}	0.41

The threshold values obtained following the training algorithm are presented in Table 8.3. The training cycle requires 157 iterations with an estimated error margin of 10^{-4}.

Table 8.3: Threshold values estimated for the transitions

tr_1	tr_2	tr_3	tr_4	tr_5	tr_6	tr_7	tr_8
0.06	0.21	0.09	0.05	0.00	0.33	0.02	0.00

tr_9	tr_{10}	tr_{11}	tr_{12}	tr_{13}	tr_{14}	tr_{15}
0.32	0.41	0.05	0.28	0.27	0.87	0.00

A plot of the performance index PI_r for the given three-layered network (Fig. 8.2) for 78 complete training cycles is presented in Figure 8.3.

After the training in the network is over, we can use the network for application in object recognition. We experimented with various unknown objects and obtained interesting results. For example, when the fuzzified feature vector for a regular pentagon is supplied at the input of the pretrained neural FPN, the network classifies it to its nearest class (hexagon). When the fuzzified feature vector of a 90° rotated ellipse (see the unknown sample in Fig. 8.1) is supplied as the input of the pretrained network, it correctly classifies the object as an ellipse (see Table 8.4). It is indeed important to note that the current ellipse pattern is definitely different from the training instance pattern of ellipse as these two patterns differ in their maximum x-length and maximum y-length features.

Table 8.4: Input/output vector components for an unknown object

Places	Input vector components at					Output vector components at				
Object	Area	Perimeter	Max. x-length	Max. y-length	Curvature	p_{12}	p_{13}	p_{14}	p_{15}	p_{16}
unknown object	0.60	0.54	0.45	0.86	0.66	0.13	0.03	0.43	0.26	0.38

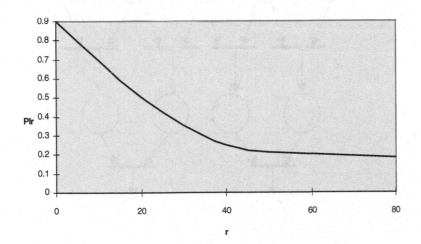

Fig. 8.3: PI_r versus training cycles r.

8.6 Conclusions

This chapter presented a new learning model of neural nets capable of representing the semantics of high-level knowledge. The unconditional convergence of the error states to the origin in error space has been proved and is an added advantage of the proposed learning model. The model has successfully been applied to a practical problem in fuzzy pattern recognition. The generic scheme of the model will find applications in both many-to-many fuzzy semantic function realization as well as recognition of objects from their fuzzy feature space.

Exercises

1. Suppose you need to classify oranges, coconuts, and guava from a mixture of the given three fruits. What features should you select to design an automatic classifier for the problem? Are the selected features independent of each other?

2. Consider the fuzzy Petri net given in Figure 8.4. Name the places and transitions in the net and hence determine P, Q, M and *l* for this network.

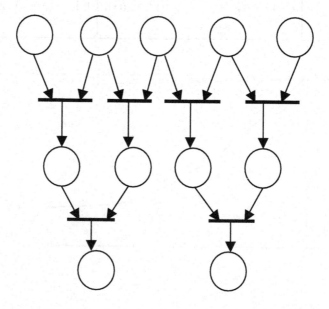

Fig. 8.4: A feed-forward fuzzy neural Petri net.

Consider one pair of input-output fuzzy training instances, and assign initial values of thresholds for the transitions. Verify that after one complete adaptation of thresholds for the transition, the error vector decreases appreciably.

3. For the same network, construct a set of four training instances, and assign initial values of thresholds arbitrarily. Using the algorithm for multiple I/O training instances, adapt the thresholds of the transitions and plot performance index in each learning cycles (epochs).

4. What type of fuzzifiers do you require to fuzzify the measurements into fuzzy sets? What characteristics of the fuzzifiers did you identify for their selection? Can you design a comparator to compare two fuzzy membership functions of a given variable (measurement) in a given fuzzy set?

5. Consider the FPN (Fig. 5.6) we considered in Chapter 5 in connection with a criminology problem. First eliminate the feedback connection from SR(r, s) to T(r, s) through transition tr_5 to make the network feed-forward. Now, consider three input/output training instances to prove s, r, and l as culprits, respectively. Randomly select initial thresholds and then train the network with these training instances.

References

[1] Bugarin, A. J. and Barro, S., "Fuzzy reasoning supported by Petri nets," *IEEE Trans. on Fuzzy systems*, vol. 2, no. 2, pp. 135-150,1994.

[2] Cao, T. and Sanderson, A. C., "Task sequence planning using Fuzzy Petri nets," *IEEE Trans. on Systems, Man and Cybernetics*, vol. 25, no. 5, pp. 755-769, May 1995.

[3] Cardoso, J., Valette, R., and Dubois, D., "Petri nets with uncertain markings," In *Advances in Petri Nets*, Lecture notes in computer science, Rozenberg, G. (Ed.), vol. 483, Springer-Verlag, New York, pp. 65-78, 1990.

[4] Chen, S. M., Ke, J. S., and Chang, J. F., "Knowledge representation using fuzzy Petri nets," *IEEE Trans. on Knowledge and Data Engineering*, vol. 2, no. 3, pp. 311-319, 1990.

[5] Chen, S. M., "Fuzzy backward reasoning using fuzzy Petri nets," *IEEE Trans. on Systems, Man and Cybernetics-Part B*, vol. 30, no. 6, December 2000.

[6] Daltrini, A., *Modeling and Knowledge Processing Based on the Extended Fuzzy Petri Nets*, M.Sc. degree thesis, UNICAMP-FEE0DCA, May 1993.

[7] Garg, M. L., Ashon, S. I., and Gupta, P. V., "A fuzzy Petri net for knowledge representation and reasoning," *Information Processing Letters*, vol. 39, pp. 165-171, 1991.

[8] Hirota, K. and Pedrycz, W.,"OR-AND neuron in modeling fuzzy set connectives," *IEEE Trans. on Fuzzy Systems*, vol. 2, no. 2, 1994.

[9] Konar, A., Chakraborty, U., and Wang, P. P., "Supervised learning by fuzzy Petri nets," *Information Systems*, Elsevier, 2005 (in press).

[10] Konar, A. and Mandal, A. K., "Uncertainty management in expert systems using fuzzy Petri nets," *IEEE Trans. on Knowledge and Data Engineering*, vol. 8, no. 1, pp. 96-105, February 1996.

[11] Lipp, H. P. and Gunther, G., "A fuzzy Petri net concept for complex decision making process in production control," *Proc. First European Congress on Fuzzy and Intelligent Technology (EUFIT '93)*, Aachen, Germany, vol. I, pp. 290 - 294.

[12] Looney, C. G., "Fuzzy Petri nets for rule-based decision making," *IEEE Trans. on Systems, Man, and Cybernetics*, vol. 18, no. 1, pp. 178-183, 1988.

[13] Pal, S. and Konar A., "Cognitive reasoning using fuzzy neural nets," *IEEE Trans. on Systems, Man and Cybernetics*, Part B, vol. 26, no. 4, 1996.

[14] Pedrycz, W., *Fuzzy Sets Engineering*, CRC Press, Boca Raton, FL, 1995.

[15] Pedrycz, W, "A fuzzy cognitive structure for pattern recognition," *Pattern Recognition Letters*, vol. 9, pp. 305-313, 1988.

[16] Pedrycz, W., "Logic–oriented fuzzy neural networks," *Int. J. of Hybrid Intelligent Systems*, vol. 1, no. 1-2, pp. 3-11, 2004.

[17] Pedrycz, W. and Gomide, F., "A generalized fuzzy Petri net model," *IEEE Trans. on Fuzzy systems*, vol. 2, no. 4, pp. 295-301, Nov 1994.

[18] Scarpelli, H. and Gomide, F., "High level fuzzy Petri nets and backward reasoning," In *Fuzzy Logic and Soft Computing*, Bouchon-

Meunier, B., Yager, R. R. and Zadeh, L. A. (Eds.), World Scientific, Singapore, 1995.

[19] Scarpelli, H., Gomide, F. and Yager, R., "A reasoning algorithm for high level Fuzzy Petri nets," *IEEE Trans. on Fuzzy systems*, vol. 4 , no. 3, pp. 282-295, Aug. 1996.

Chapter 9

Distributed Modeling of Abduction, Reciprocity, and Duality by Fuzzy Petri Nets

Chapters 3 and 4 introduced several forward and backward reasoning models of fuzzy Petri nets. These models aimed at computing the singleton membership/belief of a proposition from the supplied beliefs of the selected propositions in the network. This chapter extends the classical fuzzy Petri net models by considering membership/belief distributions of propositions instead of their singleton beliefs. The chapter begins with a formalization of the proposed extension, and then presents abductive reasoning, bi-directional reasoning, and duality issues in fuzzy Petri nets using the extended model. One interesting observation in bi-directional reasoning is restoration of belief distribution of propositions after n-forward (backward) followed by n-backward (forward) belief revisions in the network. This is called reciprocity. It has been shown that a network supporting reciprocity has an interesting relationship between its structural framework and the relational matrices associated with the transitions. Examples of diagnosis problems with simple electronic circuits are given to illustrate the scope of applications of the proposed models.

9.1 Introduction

Given a causal relation $P \rightarrow Q$ and an effect Q, the principle of abductive reasoning attempts to determine the possible binary valuation space of P. When the fact Q is contaminated with imprecision and/or the causal relation P → Q is not free from uncertainty [11], determination of the degree of truth of P is a complex problem of paramount importance. With multiple chaining of causal relationships, comprising many-to-many dependencies, the complexity of the problem increases to a great extent.

This chapter presents a unified approach to handling the above problem by fusing the structural characteristics of Petri nets with the computational power of incompleteness management of the logic of fuzzy sets [41]. The rules embedded in the knowledge base of a system are mapped to the transitions of a Petri net in a manner such that the antecedent parts of the rules correspond to the input places and the consequent parts correspond to the output places of the transitions. The causal relation describing the antecedent-consequent connectivity of each transition is represented by a fuzzy relational matrix.

For realizing abduction on a FPN, we need to compute the membership distributions of the antecedent clauses from the known membership distributions of the consequent clauses of a rule. In Petri net terminology, the membership distribution vectors of the antecedent clauses associated with the input places of the transitions are computed from the available distributions of the consequent clauses associated with the output places. For a transition with one input and one output place, the computation involves taking the max-min composition of the consequent clause's distribution with the pre-inverse of the relational matrix tagged with the transition. When the number of input/output places of a transition is more than one, the composition of the component-wise maximum [1] of the membership distribution of the consequent propositions with the pre-inverse of the relational matrices is essentially needed. The result is considered as the membership distribution of the input places for the selected transition.

Further, when the rules in the knowledge base are pipelined [15], an output place of one transition becomes the input place of another transition. As an example, let us consider that there are two transitions tr_1 and tr_2 such that p_1 is the input place of tr_1, p_2 is the output place of tr_1 but it is also the input place of tr_2, and p_3 is the output place of tr_2. Let the relational matrices associated with tr_1 and tr_2 be R_1 and R_2, respectively. To evaluate the membership distribution of the antecedent clause at place p_1 from the supplied distribution at place p_3, we need the following computation. First, we compose the membership distribution of the consequent at p_3 by the pre-inverse of the relational matrix R_2 and the resulting distribution thus obtained is again pre-composed with the pre-inverse of the relational matrix R_1. This principle has been applied in succession to compute the membership distribution of the initiating propositions (axioms) in the network. The computational paradigm employed in this chapter thus has a

---

[1] The result directly follows from the computation by state equations in backward time (refer to expression 9.17) on a fuzzy Petri net model.

similarity with backtracking in a tree or a directed graph with an aim to compute the distribution of the predecessors from the known distribution of the successors in the prescribed data structure.

The chapter is unique both in its theme and computational perspectives. The existing literature on causal models usually employs Bayesian reasoning on a network [27], the computational principle of which is attractive for its simplicity, but it is prohibitive too because of the computational need of a large number of conditional probability matrices associated with the link of the network. The fuzzy relational matrices that model the certainty of the rules of the knowledge base, however, can be constructed by a logical basis with minimal prior experience of the respective subject domain. The significance of fuzzy reasoning [20-21], [26], [28] in comparison to Bayesian reasoning in a causal network [27] is thus apparent.

There also exist quite an extensive work on fuzzy reasoning on Petri nets [1], [3-6], [8], [10], [15-19], [22-23], [34-35], [39-40], but none of these works refer to the particular problem handled in this chapter with fuzzy Petri net models [31-32].

The chapter has been subdivided into nine sections. The parameters of a fuzzy Petri net are introduced in Section 9.2. A state-space formulation of the fuzzy Petri net model and its special cases are discussed in Section 9.3. An analysis of stability of the proposed state-space model is presented in Section 9.4. An algorithm for forward reasoning in a Fuzzy Petri net is presented in Section 9.5. The concept of backward chaining in a fuzzy Petri net is formalized in Section 9.6. It needs mention here that the concept of backward chaining presented in this chapter has a considerable difference with backward reasoning addressed by other authors [7], [34]. An application of backward chaining in the diagnosis problem of electronic circuits has been considered in Section 9.6 of the chapter to justify the feasibility of backward chaining. Starting with the definition of bi-directional IFF-type reasoning, the condition of reciprocity is derived in Section 9.7. The concepts of duality in a FPN have been introduced in Section 9.8. The chapter concludes with a discussion on the pros and cons of the proposed techniques with respect to the existing methods.

9.2 Formal Definitions

Definition 9.1 *A **FPN** is a directed bipartite graph with 9 tuples denoted by*

$$FPN = \{P, D, N, Tr, T, Th, I, O, R_i\}$$

where

$P = \{p_1, p_2, \ldots, p_n\}$ *is a finite set of places;*

$D = \{d_1, d_2, \ldots, d_n\}$ *is a finite set of predicates, each d_i having a correspondence to each p_i for $1 \leq i \leq n$;*

$N = \{n_1, n_2, \ldots, n_n\}$ *is a finite set of discrete fuzzy membership distributions, each distribution n_i having correspondence to each predicate d_i;*

$Tr = \{tr_1, tr_2, \ldots, tr_m\}$ *is a finite set of transitions: $P \cap Tr \cap D = \varnothing$;*

T: $Tr \rightarrow [0,1]$ represents a mapping from transitions to real numbers between 0 and 1. Let $T = \{t_1, t_2, \ldots, t_m\}$ where $t_i \in [0, 1]$. Then t_i denotes the fuzzy truth token (FTT) of transition tr_i;

Th: $Tr \rightarrow [0,1]$ represents a mapping from transitions to real numbers between 0 and 1. Let $Th = \{th_1, th_2, \ldots, th_m\}$ where $th_i \in [0, 1]$. Here, th_i denotes the threshold associated with transition tr_i;

I: $Tr \rightarrow P^\infty$ represents a mapping from transitions tr_i to the bags of its input places $1 \leq \forall i \leq m$;

O: $Tr \rightarrow P^\infty$ represents a mapping from transitions tr_i to the bags of its output places $1 \leq \forall i \leq m$;

R: $Tr \rightarrow [0, 1] \times [0, 1]$ represents a mapping from each transition to a matrix. For instance, R_i is a fuzzy relational matrix associated with transition tr_i.

Example 9.1: Let us consider the FPN shown in Figure 9.1. Here,
$P = \{p_1, p_2, p_3, p_4\}$ is the set of places,
$D = \{d_1, d_2, d_3, d_4\}$ are the predicates, where d_1 = Tall (ram), d_2 = Stout (ram), d_3 = Fast-runner (ram), d_4 = Has-nominal-pulse-rate (ram).
$N = \{n_1, n_2, n_3, n_4\}$ is the set of fuzzy membership distributions associated with predicates d_1, d_2, d_3, d_4, respectively. Here, $n_1 = [0.6 \; 0.8 \; 0.9 \; 0.4]^T$, $n_2 = [0.2 \; 0.9 \; 0.6 \; 0.3]^T$, $n_3 = n_4$ = null vector.

It may be added here that these fuzzy membership distribution are discretized (sampled) and denoted by vectors. They are assigned at time t = 0 and may be updated in each membership updating cycle. It is therefore convenient to include the notion of time in the argument of n_i for $1 \leq i \leq 9$. As an instance, we can refer to n_i at t = 0 by $n_i(0)$.

$Tr = \{tr_1, tr_2\};$

$T = \{t_1, t_2\}$ where t_1 and t_2 are FTT vectors associated with tr_1 and tr_2, respectively. Like n_i's, t_j's are also time-varying quantities and are denoted by $t_j(t)$. $R = \{R_1, R_2\}$ where R_1 and R_2 are relational matrices associated with tr_1 and tr_2, respectively. These matrices need to be supplied by respective experts of their domain. $I(tr_1) = \{p_1, p_2\}$, $I(tr_2) = \{p_3\}$, $O(tr_1) = \{p_3\}$, and $O(tr_2) = \{p_2, p_4\}$.

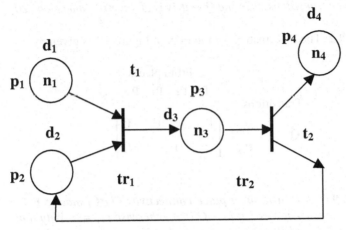

$d_1 =$ Tall (ram), $d_2 =$ Stout (ram), $d_3 =$ Fast runner (ram),
$d_4 =$ Has-nominal-pulse-rate (ram)

Fig. 9.1: An illustrative FPN.

It needs mention here that Figure 9.1 corresponds to the fuzzy rule-based system given by the following two production rules [2], [38]:

> *PR1: Tall(x), Stout (x) → Fast-runner (x)*
> *PR2: Fast-runner (x) → Has-nominal-pulse-rate (x), Stout (x).*

In the PR above, Tall (x), Stout (x), etc. denote predicates and the commas in the left- and right-hand sides of the implication sign (→) denote AND and OR operations, respectively. The elements of the relational matrices [30] supplied with the rules denote the certainty factor [35] of the respective rule for each discrete membership value of the antecedent and the consequent clauses.

Definition 9.2: *If $p_i \in I(tr_a)$ and $p_j \in O(tr_a)$ then p_j is **immediately reachable** from p_i. Again, if p_j is immediately reachable from p_i and p_k is immediately reachable from p_j, then p_k is **reachable** from p_i. The reachability property is the reflexive, transitive closure of the immediate reachability property [9]. We would use IRS (p_i) and RS (p_i) operators to denote the set of places immediately reachable and reachable from the place p_i, respectively.*

Moreover, if $p_j \in [IRS\{IRS(IRS. \ldots k\text{-times } (p_i)\}]$, denoted by $IRS^k (p_i)$, then p_j is reachable from p_i with a **degree of reachability k**. The following two connectivity matrices are used throughout the chapter.

Definition 9.3: *A place to transition connectivity (PTC) matrix Q is a binary matrix whose elements $q_{jk} = 1$ if $p_k \in I (tr_j)$, otherwise $q_{jk} = 0$. If the FPN has n places and m transitions, then the Q matrix is of $(m \times n)$ dimension [20].*

Example 9.2: The PTC matrix for the FPN of Figure 9.1 is given by

$$
Q = \begin{matrix} & & \text{From places} \\ \text{To} & & p_1 \;\; p_2 \;\; p_3 \;\; p_4 \\ \text{Transitions} & & \\ \begin{matrix} tr_1 \\ tr_2 \end{matrix} & \begin{bmatrix} 1 & 1 & 0 & 0 \\ 0 & 0 & 1 & 0 \end{bmatrix} \end{matrix}
$$

Definition 9.4: *A transition to place connectivity (TPC) matrix P is a binary matrix whose element $p_{ij} = 1$ if $p_i \in O (tr_j)$, otherwise $p_{ij} = 0$. With n places and m transitions in the FPN, the P matrix is of $(n \times m)$ dimension [19].*

Example 9.3: The TPC matrix corresponding to Figure 9.1 is given by

$$
P = \begin{matrix} & & \text{From transitions} \\ \text{To} & & tr_1 \;\;\;\; tr_2 \\ \text{places} & & \\ \begin{matrix} p_1 \\ p_2 \\ p_3 \\ p_4 \end{matrix} & \begin{bmatrix} 0 & 0 \\ 0 & 1 \\ 1 & 0 \\ 0 & 1 \end{bmatrix} \end{matrix}
$$

9.3 State-Space Formulation of the Proposed FPN Model

The dynamic behavior of an FPN is modeled by updating FTTs at transitions and memberships at places. In fact, the enabling condition of transitions is first checked. All enabled transitions are fireable; on firing of a transition, the FTT distribution at its outgoing arcs [16] is estimated based on the membership distribution of its input places and the relational matrix associated with the transition. It may be noted that on firing of a transition, the membership distribution of its input places is not destroyed as with conventional Petri nets [26]. After the FTTs at all transitions are updated concurrently, the membership distribution at the places is also updated concurrently. The revised membership distribution at a place p_j is a function of the FTT of those transitions whose output place is p_j. The concurrent updating of FTT distribution at transitions followed by concurrent updating of membership distribution at places is termed as a **membership revision cycle**.

9.3.1 The Behavioral Model of FPN

Let us consider a transition tr_i, where $I(tr_i) = \{p_k, p_m\}$ and $O(tr_i) = \{p_u, p_v\}$. Assume that th_i is the threshold vector, associated with the transition tr_i. The transition tr_i is enabled if

$$\mathbf{R_i} \; o \; (\mathbf{n_k} \wedge \mathbf{n_m}) \geq \mathbf{th_i}.$$

An enabled transition fires, resulting in a change in the FTT vectors at its output arcs. It is to be noted that the FTT vectors at all the output arcs of a transition are always equal. In case the transition tr_i is not enabled, the FTT distribution at its output arcs is set to null vector. The extended model of FPN, designed after Looney [24], that satisfies the above constraints is formally presented below:

$$\mathbf{t_i}(t + 1) = \mathbf{t_i}(t) \wedge [\mathbf{R_i} \; o \; (\mathbf{n_k}(t) \wedge \mathbf{n_m}(t))] \wedge \mathbf{U}[\mathbf{R_i} \; o \; (\mathbf{n_k}(t) \wedge \mathbf{n_m}(t)) - \mathbf{th_i}] \quad (9.1)$$

In expression (9.1), **U** denotes a unit step vector, each component of which is unity when its corresponding argument is non-negative and, otherwise is zero. In fact, the enabling condition of the transition tr_i is tested using this vector. Moreover, the $\wedge$ operation between two vectors is done component-wise like column vector addition in conventional matrix algebra. It may be noted that if tr_i has m input places $p_1, p_2, \ldots, p_m$ and k output places $p_{m+1}, p_{m+2}, \ldots, p_{m+k}$ (Fig. 9.2), then expression (9.1) can be modified with the replacement of

$$\mathbf{n_k}(t) \wedge \mathbf{n_m}(t) \quad \text{by} \quad \bigwedge_{w=1}^{m} \mathbf{n_w}(t).$$

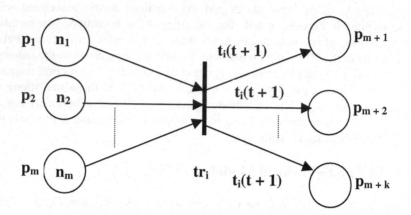

Fig. 9.2: A transition tr_i with m input and k output places.

After the FTT distribution at all the transitions in the FPN are updated concurrently, the membership distribution at all places can be updated in parallel following expression (9.2). Let us consider a place p_j such that $p_j \in [O(tr_1) \cap O(tr_2) \cap \ldots \cap O(tr_s)]$ (Fig. 9.3). The updating of membership distribution n_j at place p_j is given by

$$n_j(t+1) = n_j(t) \ \mathbf{V} \ [t_1(t+1) \ \mathbf{V} \ t_2(t+1) \mathbf{V} \ldots . \mathbf{V} \ t_s(t+1)]$$

$$= \ n_j(t) \ \mathbf{V} \ (\overset{s}{\underset{r=1}{\mathbf{V}}} \ t_r(t+1) \ ). \tag{9.2}$$

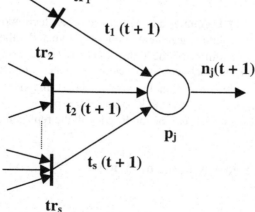

Fig. 9.3: A place p_j that belongs to output places of s transitions.

9.3.2 State-Space Formulation of the Model

For convenience of analysis and simplicity of realization of FPN by arrays, instead of linked list structures, the dynamic behavior of the entire FPN is represented by a single vector-matrix (state-space) equation. For the formulation of the state-space model, we first construct a membership distribution vector N (t) of dimension (z. n) × 1, such that

$$N(t) = [\, n_1(t) \; n_2(t) \; . \; . \; . \; . \; n_n(t)]^T,$$

where each membership distribution vector $n_i(t)$ has z components. Analogously, we construct a FTT vector T and threshold vector Th of dimension (z. m) × 1, such that

$$T(t) = [\, t_1(t) \; t_2(t) \; . \; . \; . \; . \; t_m(t) \,] \quad \text{and} \quad Th = [\, th_1 \; th_2 \; . \; . \; . \; . \; th_m \,]$$

where each t_j (t) and th_i have z components. We also form a relational matrix R, given by

$$R = \begin{bmatrix} R_1 & \Phi & \Phi & ... & \Phi \\ \Phi & R_2 & \Phi & ... & \Phi \\ \Phi & \Phi & R_3 & ... & \Phi \\ ... & .... & ... & ... & ... \\ \Phi & \Phi & \Phi & ... & R_m \end{bmatrix}$$

where R_i for $1 \le i \le m$ is the relational matrix associated with transition tr_i and Φ denotes a null matrix of dimension equal to that of R_i's, with all elements = 0. It may be noted that position of a given R_i on the diagonal in matrix R is fixed. Moreover, extended P and Q matrices denoted by P' and Q', respectively, may be formed by replacing each unity and zero element in P and Q by identity and square null matrices, respectively, of dimensions equal to the number of components of t_i and n_j, respectively. Now consider a FPN with n places and m transitions. Omitting the U vector for brevity, the FTT updating equation at a given transition tr_i may now be described by expression (9.3).

$$t_i\,(t+1) = t_i(t) \; \Lambda \; \underset{\exists \, w = 1}{\overset{n}{[R_i \; o \; (\Lambda \; n_w\,(t))]}} , \text{ where } p_w \in I \; (tr_i) \tag{9.3}$$

$$= t_i(t) \; \Lambda \; \underset{\forall w = 1}{\overset{n}{[\, R_i \; o \; \{q_w \; \Lambda \; (\Lambda \; n_w\,(t))\}]}}, \tag{9.3 a}$$

where element of $\mathbf{q}_w = 1$ when $p_w \in I$ (tr_i) otherwise $= 0$. Now, applying De Morgan's law in (9.3a) we obtain:

$$t_i(t+1) = t_i(t) \wedge [\mathbf{R}_i \circ \{ \mathbf{q}_w' \vee \{ \overset{n}{\underset{\forall w\,=\,1}{\mathbf{V}}} \mathbf{n}^c{}_w(t) \} \}^c]$$
(9.3 b)

where the elements $\mathbf{q}_w' \in \{0, 1\}$ and c over a vector denotes the one's complements of its corresponding elements.

Combining the FTT updating equations for m transitions, we have

$$\mathbf{T}(t+1) = \mathbf{T}(t) \wedge [\mathbf{R} \circ (\mathbf{Q'} \circ \mathbf{N}^c(t))^c].$$
(9.4)

Similarly the membership updating equations for n places can now be combined using expression (9.2) as follows:

$$\mathbf{N}(t+1) = \mathbf{N}(t) \vee [\mathbf{P'} \circ \mathbf{T}(t+1)].$$
(9.5)

Combining expressions (9.4) and (9.5) yields

$$\mathbf{N}(t+1) = \mathbf{N}(t) \vee [\mathbf{P'} \circ \{\mathbf{T}(t) \wedge \{\mathbf{R} \circ (\mathbf{Q'} \circ \mathbf{N}^c(t))^c \}\}].$$
(9.6)

Including the $\mathbf{U}$ vector in the expression (9.6), we have

$$\mathbf{N}(t+1) = \mathbf{N}(t) \vee \mathbf{P'} \circ [\{\mathbf{T}(t) \wedge \{\mathbf{R} \circ (\mathbf{Q'} \circ \mathbf{N}^c(t))^c \}\}$$
$$\wedge \mathbf{U} \{\mathbf{R} \circ (\mathbf{Q'} \circ \mathbf{N}^c(t))^c - \mathbf{Th}\}].$$
(9.7)

Estimation of $\mathbf{N}(r)$ for $r > 1$ from $\mathbf{N}(0)$ can be performed by updating $\mathbf{N}(r)$ iteratively r times using expression (9.7) .

9.3.3 Special Cases of the Model

In this section, two special cases of the above model, obtained by eliminating state feedback [21] from FTT and membership distributions, are considered. This renders the membership at any given place be influenced by not only its parents [20], as in the general model, but its global predecessors also [27].

Case I: In expression (9.1), the current value of FTT distribution is used for estimation of its next value. Consequently, the FTT distribution $t_i(t+1)$ at a given time $(t+1)$ depends on the initial value of $t_i(0)$. Since $t_i(0)$ for any arbitrary transition tr_i in the FPN is not always available, it is reasonable to keep $t_i(t+1)$ free from $t_i(t)$. If $t_i(t)$ is dropped from expression (9.1), the modified state-space equation can be obtained by setting all components of $\mathbf{T}(t) = 1$ in the expressions (9.6) and (9.7). The revised form of expression (9.7), which will be referred to frequently, is rewritten as expression (9.8).

$$N(t + 1) = N(t) \ V \ [\mathbf{P}' \ o \ \{\mathbf{R} \ o \ (\mathbf{Q}' \ o \ \mathbf{N}^c \ (t))^c \ \} \\ \Lambda \ \{\mathbf{U} \ (\mathbf{R} \ o \ (\mathbf{Q}' \ o \ \mathbf{N}^c \ (t))^c - \mathbf{Th})\}]. \tag{9.8}$$

Case II: A second alternative is to keep both $t_i(t + 1)$ and $n_j(t + 1)$, $\forall i,j$ independent of their last values. However, when $n_j(t)$ is dropped from expression (9.2), places with no input arcs, called axioms [35], cannot restore their membership distribution, since $t_r(t)$ for $r = 1$ to s in expression (9.2) are absent and hence zero. In order to set $n_j(t + 1) = n_j(t)$ for axioms, we consider self-loop around each axiom through a virtual transition tr_k, such that the $\mathbf{R}_k$ and $\mathbf{th}_k$ are set to identity matrix and null vector, respectively. The $\mathbf{P}'$, $\mathbf{Q}'$, and $\mathbf{R}$ matrices are thus modified and denoted by $\mathbf{P}'_m$, $\mathbf{Q}'_m$, and $\mathbf{R}_m$ respectively. The state-space model for case II without U is thus given by

$$N(t + 1) = \mathbf{P}'_m \ o \ \{ \mathbf{R}_m \ o \ (\mathbf{Q}'_m \ o \ \mathbf{N}^c \ (t))^c \ \} \tag{9.9}$$

Example 9.4 In this example, the formation of $\mathbf{P}'$, $\mathbf{Q}'$, and $\mathbf{R}$ matrices (Fig. 9.4) are demonstrated for the model represented by expression (9.7).

The $\mathbf{P}$ and $\mathbf{Q}$ matrices for the FPN of Figure 9.4 are given by

$$
\mathbf{P} = \begin{array}{cc} & \begin{array}{ccc} \textbf{From} \\ tr_1 & tr_2 & tr_3 \end{array} \\ \textbf{To} & \\ \begin{array}{c} p_1 \\ p_2 \\ p_3 \end{array} & \left(\begin{array}{ccc} 0 & 0 & 1 \\ 0 & 0 & 0 \\ 1 & 1 & 0 \end{array} \right) \end{array}
\qquad
\mathbf{Q} = \begin{array}{cc} & \begin{array}{ccc} \textbf{From} \\ p_1 & p_2 & p_3 \end{array} \\ \textbf{To} & \\ \begin{array}{c} tr_1 \\ tr_2 \\ tr_3 \end{array} & \left(\begin{array}{ccc} 0 & 1 & 0 \\ 1 & 1 & 0 \\ 0 & 0 & 1 \end{array} \right) \end{array}
$$

Assuming the n_j and t_i vectors of dimension (3×1) we construct the $\mathbf{P}'$ and $\mathbf{Q}'$ matrix

$$
\mathbf{P}' = \begin{bmatrix} \Phi & \Phi & I \\ \Phi & \Phi & \Phi \\ I & I & \Phi \end{bmatrix}, \quad
\mathbf{Q}' = \begin{bmatrix} \Phi & I & \Phi \\ I & I & \Phi \\ \Phi & \Phi & I \end{bmatrix}
$$

where Φ and I denote null and identity matrices each of dimension (3×3).

The relational matrix $\mathbf{R}$ in the present context is given by

$$R = \begin{bmatrix} R_1 & \Phi & \Phi \\ \Phi & R_2 & \Phi \\ \Phi & \Phi & R_3 \end{bmatrix}$$

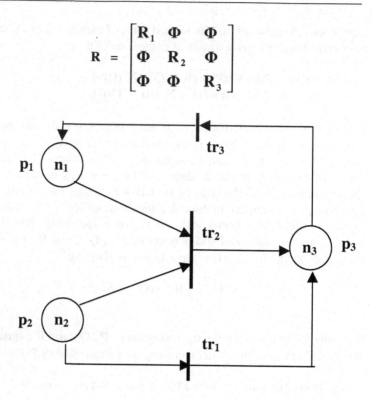

Fig. 9.4: An FPN for illustrating the formation of **P′**, **Q′**, and **R** matrices.

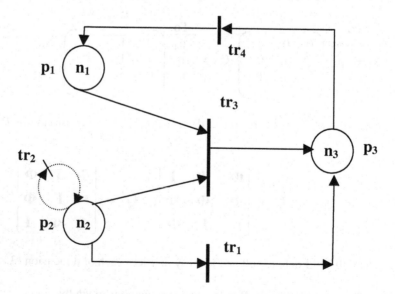

Fig. 9.5: The modified form of Figure 9.4 with self-loop around place p_2 and renamed transitions.

Further, $N = [n_1 \ n_2 \ n_3 \ ]^T$, $T = [t_1 \ t_2 \ t_3]^T$, $Th = [th_1 \ th_2 \ th_3 \ ]^T$. Expression (9.7) can be used for updating N with the above parameters. For updating N with expression (9.9), we, however, redraw the FPN with a virtual self-loop around place p_2 (vide Fig. 9.5) and reconstruct P, Q, and consequently P'_m, Q'_m, and R_m matrices. It may be noted that the virtual transitions around place p_j should be named tr_j ($j = 2$, here) to satisfy equation (9.9) and other transitions should be renamed distinctively.

9.4 Stability Analysis

In this section, the analysis of the dynamic behavior of the proposed model will be presented. A few definitions, which are used to understand the analysis, are in order.

Definition 9.5: *An FPN is said to have reached an* **equilibrium state (steady-state)** *when* $N(t^* + 1) = N(t^*)$ *for some time* $t = t^*$, *where* t^* *is the minimum time when the equality of the vectors is first attained. The* t^* *is called the equilibrium time.*

Definition 9.6: *A FPN is said to have* **limit cycles** *if the fuzzy memberships* n_j *of at least one place* p_j *in the network exhibits periodic oscillations, described by* $n_j(t + k) = n_j(t)$ *for some positive integer* $k > 1$ *and sufficiently large* t, *numerically greater than the number of transitions in the FPN.*

The results of stability analysis of the proposed models are presented in the Theorems 9.1 through 9.9.

Theorem 9.1: *The model represented by expression (9.7) is unconditionally stable and the steady state for the model is attained only after one membership revision cycle in the network.*

Proof: For an oscillatory response, the membership at a place should increase as well as decrease with respect to time. However, since $N(t + 1) \geq N(t)$ for all integer $t \geq 0$ (vide expression (9.7)), hence, the components of $N(t)$ cannot exhibit oscillatory response. Further, as components of $N(0)$, $T(0)$, Th, and R are bounded between 0 and 1, and $N(t + 1)$ is derived by using fuzzy AND/OR operators over R, $N(t)$, and $T(t)$, hence, components of $N(t)$ remain bounded. Thus, fuzzy membership at places being bounded and free from oscillation exhibit stable behavior. This completes the first part of the theorem.

Now to prove the second part, we first rewrite expression (9.7) in the following form:

$$T(t + 1) = T(t) \wedge [R \ o \ (Q' \ o \ N^c(t))^c] \wedge U [R \ o \ (Q' \ o \ N^c(t))^c - Th] \quad (9.10)$$

and $\qquad N(t + 1) = N(t) \ V \ (P' \ o \ T(t + 1)) \qquad\qquad$ [(9.5) rewritten]

Now, we compute $\mathbf{T}(1)$ and $\mathbf{T}(0)$ and $\mathbf{N}(1)$ and $\mathbf{N}(0)$ by expressions (9.10) and (9.5), respectively, and find

$$\mathbf{N}(1) = \mathbf{N}(0) \ \vee \ (\mathbf{P'} \circ \mathbf{T}(1)) \qquad\qquad (9.11)$$

Now, computing $\mathbf{T}(2)$ and $\mathbf{N}(2)$ by expressions (9.10) and (9.5) it is found

$$\mathbf{T}(2) \le \mathbf{T}(1) \qquad\qquad (9.12)$$

and
$$
\begin{aligned}
\mathbf{N}(2) &= \mathbf{N}(1) \ \vee \ (\mathbf{P'} \circ \mathbf{T}(2)) & \\
&= \mathbf{N}(0) \vee (\mathbf{P'} \circ \mathbf{T}(1)) \vee (\mathbf{P'} \circ \mathbf{T}(2)) & [\text{ by } (9.11)] \\
&= \mathbf{N}(0) \ \vee (\mathbf{P'} \circ \mathbf{T}(1)) & [\text{ by } (9.12)] \\
&= \mathbf{N}(1)
\end{aligned}
$$

Thus, it can be easily shown that $\mathbf{N}(t + 1) = \mathbf{N}(t)$ for all integer $t \ge 1$. Therefore, steady state is reached after only one membership revision step. Hence, the theorem follows.

Theorem 9.2: *The model represented by expression (9.8) is unconditionally stable and the non-zero steady state membership vector N* satisfies the inequality (9.13).*

$$\mathbf{N^*} \ge \mathbf{P'} \circ \{ \ \mathbf{R} \circ (\mathbf{Q'} \circ \mathbf{N^*}^{\,c})^{\,c} \},$$

when $\mathbf{R} \circ (\mathbf{Q'} \ \circ \ \mathbf{N^c}(t))^c \ \ge \ \mathbf{Th}, \forall \ t \ge 0.$ \qquad (9.13)

Proof: Unconditional stability of the model can be proved following the steps analogous to the proof of Theorem 9.1.

To prove the second part of the Theorem, let us assume that the equilibrium condition is reached at time $t = t^*$. Thus, by Theorem 9.1

$$\mathbf{N}(t^* + 1) \ = \ \mathbf{N}(t^*) \ \ (= \mathbf{N^*}, \text{ by statement of the theorem }) \qquad (9.14)$$

Now, expression (9.8) satisfies expression (9.14) when

$$\mathbf{N^*} \ge \mathbf{P'} \circ [\{\mathbf{R} \circ (\ \mathbf{Q'} \circ \mathbf{N^{*c}})^{\,c}\} \wedge \mathbf{U} \ \{ \ \mathbf{R} \circ (\ \mathbf{Q'} \circ \mathbf{N^{*c}})^c - \mathbf{Th}\}]. \qquad (9.15)$$

Further, if $\mathbf{R} \circ (\mathbf{Q'} \circ \mathbf{N^{*c}})^c \ge \mathbf{Th}$, all the components of $\mathbf{U}$ vector being unity, it can be dropped from expression (9.15). Thus, we get expression (9.13).

The following definitions will facilitate the analysis of the model represented by expression (9.9).

Definition 9.7: *An arc* $tr_i \times p_j$ *is called* **dominant** *at time* τ *if for* $p_j \varepsilon (\exists k \cap O(tr_k))$, $t_i(\tau) > t_k(\tau)$; *alternatively, an arc* $p_x \times tr_v$ *at time* τ *is* **dominant** *if* $\forall w, \; p_w \varepsilon I(tr_v), \; n_x(\tau) < n_w(\tau)$, *provided* $R_v \; o \; (\forall w, \wedge \; n_w) > Th_v$.

Definition 9.8: *An arc is called* **permanently dominant** *if, after becoming dominant at time* $t = \tau$, *it remains so for all time* $t > \tau$.

The limit cycle behavior of the model represented by expression (9.9) is stated in Theorem 9.3.

Theorem 9.3: *If* *all the n number of arcs on any of the cycles of an FPN remains dominant from r_1-th to r_2-th membership revision step, by using the model represented by expression (9.9), then each component of the fuzzy membership distribution at each place on the cycle would exhibit*

> *i) at least "a" number of periodic oscillations, where a = integer part of $\{(r_2 - r_1)/n\}$ and*
> *ii) limit cycles with $r_2 \to \infty$.*

Proof: Proof is similar to the proof of Theorem 4.5 (see Chapter 4).

The model represented by expression (9.9) also yields an equilibrium condition if none of the cycles have all their arcs permanently dominant. The number of membership revision steps required to reach the equilibrium condition for this model is estimated below.

Let $l_1 =$ the worst number of membership revision steps required on the FPN for transfer of fuzzy membership distribution from the axioms to all the places on the cycles, which are directly connected to the axioms through arcs lying outside the cycle,

$l_2 =$ the worst number of membership revision steps required for the transfer of fuzzy membership distribution from the places on the cycles to the terminal places in the network,

$n =$ number of transitions on the largest cycle,

$l_3 =$ the worst number of membership revision steps required for the transfer of fuzzy membership distribution from the axioms to all the terminal places through the paths, which do not touch the cycles.

Theorem 9.4: *In the case when steady state is reached in an FPN by using the model represented by the expression (9.9), the total number of membership revision steps required in the worst case to reach steady state is given by*

$$T_{worst} = Max\{l_3, (l_1 + l_2 + n - 1)\}. \qquad (9.16)$$

Proof: Proof of the theorem is similar to the proof of Theorem 4.7 (see Chapter 4).

It may be added that the number of membership revision steps required for the model, represented by (9.8), is the same as computed above.

9.5 Forward Reasoning in FPN

Forward reasoning is generally carried out in fuzzy logic by extending the principle of generalized modus ponens (GMP) [36]. For illustration, consider the following rule having fuzzy quantifiers and the observed antecedent:

> **Rule:** *if x-is-A AND y-is-B Then z-is-C*
> **Observed antecedent:** *x-is-A' AND y-is-B'*
>
> ---
>
> **Conclusion:** *z-is-C'*

The conclusion z-is-C' is inferred by the reasoning system based on the observed level of quantifiers A' and B'. While representing the above problem using FPN, we consider that two discrete membership distributions are mapped at places p_1 and p_2 with proposition $d_1 = $ x-is-A and $d_2 = $ y-is-B respectively. Further, let $p_1, p_2 \in I(tr_i)$, then p_3 which corresponds to $d_3 = $ z-is-C is an element of $O(tr_i)$. Here, the membership distribution of z-is-C' may be estimated using the distribution of x-is-A' and y-is-B'.

Further, for representing chained modus ponens, a Petri net is an ideal tool. For example, consider the second rule z-is-C $\rightarrow$ w-is-D and the observed antecedent z-is-C'. We subsequently infer w-is-D'. This too can be realized by adding one transition tr_j and a place p_4 such that $p_3 \in I(tr_j)$ and $p_4 \in O(tr_j)$.

The most complex and yet unsolved problem of forward reasoning, perhaps, is reasoning under self-reference. This problem too can be easily modeled and solved by using FPN. We now present an algorithm for forward reasoning that is applicable to all the above kinds of problems independent of their structures of the FPNs. Procedure forward reasoning is described below based on the state space equation (9.8), which is always stable.

Procedure forward-reasoning (FPN, R, P',Q', N(0), Th)
Begin
  N(t): = N(0) ;
  While N(t+1) $\neq$ N(t)
      temp := **R** o (**Q** 'o N^c (t))^c ;
      N(t + 1) := N(t) V [(**P** 'o **temp**) $\wedge$ U (**temp** – Th)];

      $N(t) := N(t + 1)$;
End While;
End.

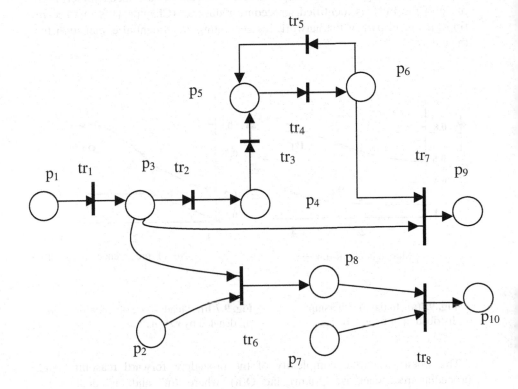

Predicate Definitions: L = Loves, T = Tortures, SR = Strained-relationships-between, HPM = Has precedence of murder, PTM = Proposes to marry, AMO = Accepts marriage offer from, Ht = hates, M = Murders,.

Persons involved: r = Ram, s = Sita, l = Lata

Database: d_1 = L(r, l), d_2 = L(l, r), d_3 = PTM (r, l), d_4 = HT (s, r), d_5 = T(r, s), d_6 = SR(r, s), d_7 = HPM(l), d_8 = AMO(l, r), d_9 = M(r, s), d_{10} = M(l, s).

Knowledge base: Rule 1: $d_1 \rightarrow d_3$, Rule 2: d_2, $d_3 \rightarrow d_8$, Rule 3: d_7, $d_8 \rightarrow d_{10}$, Rule 4: $d_3 \rightarrow d_4$, Rule 5: $d_4 \rightarrow d_5$, Rule 6: $d_5 \rightarrow d_6$, Rule 7: d_3, $d_6 \rightarrow d_9$, Rule 8: $d_6 \rightarrow d_5$.

Fig. 9.6: An FPN representing a murder history of a housewife "s" where the husband "r" and the girl friend "l" of "r" are the suspects.

The procedure forward reasoning is used to compute the steady state membership of $N(t)$ from its initial value $N(0)$. In an application like criminal investigation [20], these steady-state values of the predicates are used to identify the culprit from a given set of suspects. After the culprit, described by a terminal place of the FPN, is identified, procedure reducenet (Chapter 4) is invoked to find the useful part of the network for generating an evidential explanation for the culprit.

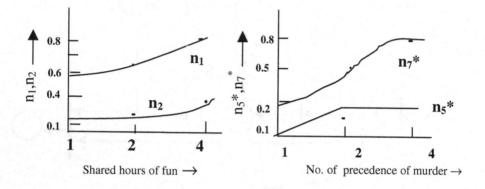

Fig. 9.7(a): Initial membership distribution of n_1 and n_2.

Fig. 9.7(b): Steady-state distribution of n_5 and n_7, denoted by n_5^*, n_7^*.

The worst case time complexity of the procedure forward reasoning and procedure reducenet are $O(a.m)$ and $O(n)$, where "m" and "n" denote the number of transitions, and number of places before reduction of the network, respectively, and "a" denotes number of axioms.

Example 9.5: In the FPN in Figure 9.6, the fuzzy membership distributions corresponding to places p_1 and p_2 are shown in Figure 9.7(a). The initial membership distributions of all other places are null vectors. Further, we assumed $R = I$. The steady-state membership distribution at all places in the entire FPN is obtained after five iterations using a forward reasoning algorithm, and their distributions at places p_5 and p_7 are shown in Figure 9.7(b). Since for all components, n_7 is larger than n_5, p_7 is marked as the concluding place and then procedure reducenet is invoked for tracing the explanation for the problem.

9.6 Abductive Reasoning in FPN

"Abductive reasoning" [33] or backward chaining [18] in fuzzy logic is concerned with inferring the membership distribution of the antecedent clauses, when the if-then rule and the observed distribution of the consequents are available. For example, given the rule and the observed consequent clause, the inference follows:

> **Rule:** *If x-is-A AND y-is-B THEN z-is-C*
> **Observed evidence:** *z-is-C'*
> _____
> **Inferred:** *x-is-A' AND y-is-B'*

In the above example, A, B, and C are three fuzzy quantifiers. C' is an observed quantifier of z and A' and B' are the inferred quantifiers of x and y, respectively. Here, given the membership (belief) distribution of z-is-C', one has to estimate the distribution x-is-A' and y-is-B'.

The classical problem of fuzzy backward chaining may be extended for application in cycle-free FPNs. In this chapter, we consider the model described by expression (9.9). Given the observed distribution of the clauses, corresponding to terminal places, the task is to estimate the membership distribution of the predicates for axioms.

For solving the above problem, we have to estimate the inverse of fuzzy matrices with respect to fuzzy AND-OR composition operators. Before describing the algorithm for estimation of fuzzy inverse matrices [34], let us first highlight its significance. Given that

$$N(t + 1) = P'_{fm} \text{ o } [\ R_{fm} \text{ o } (Q'_{fm} \text{ o } N^c(t))^c\ ] \tag{9.9a}$$

where the suffix "f" represents that the model corresponds to forward reasoning.

Premultiplying both sides of the above equation by the fuzzy inverse (pre-inverse to be specific [34]) of P'_{fm}, denoted by $P'_{fm}{}^{-1}$, we find

$$P'_{fm}{}^{-1} \text{ o } N(t + 1) = R_{fm} \text{ o } (Q'_{fm} \text{ o } N^c(t))^c.$$

After some elementary fuzzy algebra, we get

$$N(t) = [\ Q'_{fm}{}^{-1} \text{ o } \{R_{fm}{}^{-1} \text{ o } (P'_{fm}{}^{-1} \text{ o } N(t + 1))\}^c\ ]^c. \tag{9.17}$$

For estimation of the membership distribution of the axiom predicates, from the known membership distribution of the concluding predicates, the following steps are to be carried out in sequence:

i)   All of the concluding predicates at terminal places should have self-loops through virtual transitions. This helps maintain the initial membership distribution at these places. The threshold and the relational matrices for the virtual transitions should be set to null vector and identity matrices, respectively, to satisfy the above requirement.

ii) The $N(t + 1)$ vector in expression (9.17) is to be initialized by assigning non-zero vectors at the concluding places and null vectors at all other places. Call this membership vector N_{ini}.

iii) The expression (9.17) should be updated recursively in backward time until $N(t) = N(t + 1)$.

The algorithm for abductive reasoning may be formally stated as follows:

Procedure backward-reasoning (FPN, P'_{fm}, R_{fm}, Q'_{fm}, N_{ini})
Begin
  t: = m // m = no. of transitions in the FPN //;
  $N(t + 1) : = N_{ini}$;
  While $N(t) \neq N(t + 1)$ **do Begin**
   $N(t) : = [Q'_{fm}{}^{-1} \; o \; \{ R_{fm}{}^{-1} \; o \; (P'_{fm}{}^{-1} \; o \; N(t + 1))\}^c \;]^c$;
   $N(t + 1) : = N(t)$;
  End While;
End.

The worst case time complexity of the procedure backward-reasoning is proportional to $O(m)$, where m denotes the number of transitions in the FPN.

  It is apparent from the above procedure that for estimating $N(t)$ from $N(t + 1)$, the inverse of the fuzzy/binary matrices with respect to fuzzy composition operators is to be computed. Saha and Konar [31-33] have formulated the inverse problem as a constraint satisfaction problem [33] and established with the help of a heuristic function that transpose is always one of the inverses of a fuzzy/binary relational matrix with respect to the max-min composition operator. In the inverse computations highlighted above, we thus use the transpose of a fuzzy matrix as its inverse

Example 9.6: Consider the problem of diagnosis of a two-diode full-wave rectifier circuit. The expected rectifier output voltage is 12 volts, when the system operates properly. Under defective conditions, the output could be close to 0 volts or 10 volts depending on the number of defective diodes. The knowledge base of the diagnosis problem is extended into an FPN (Figs. 9.8). The task here is to identify the possible defects: defective (transformer) or defective (rectifier). Given the membership distribution of the predicate close-to (rectifier-out, 0 v) and more-or-less (rectifier-out, 10 v) (Figs. 9.9 and 9.10), one has to estimate the membership distribution of the predicate: defective (transformer) and defective (rectifier). However, for this estimation, one should have knowledge about the relational matrices and thresholds associated with the transitions. Let us assume for the sake of simplicity that the thresholds are zero and the relational matrices associated with transition tr_1 through tr_7 are equal. So, for $1 \le i \le 7$ let

$$\mathbf{R}_i = \begin{bmatrix} 0.3 & 0.5 & 0.6 \\ 0.4 & 0.6 & 0.9 \\ 0.8 & 0.7 & 0.2 \end{bmatrix}$$

be the same matrix, as chosen in the example. The $\mathbf{R}_i$ for $8 \le i \le 9$ will be the identity matrix.

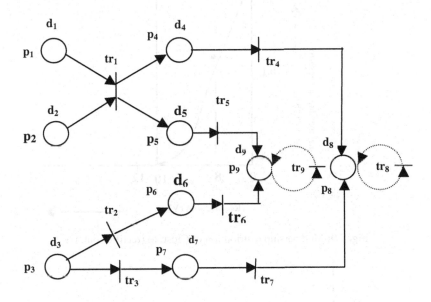

$d_1, d_2, \ldots, d_9$ (not shown) denote the predicates defined below:

d_1 = defective (transformer), d_2 = close-to (primary, 230), d_3 = defective (rectifier), d_4 = Close-to (trans-out, 0V), d_5 = Open (one half-of-secondary-coil), d_6 = Defective (one-diode), d_7 = Defective (two-diodes), d_8 = Close-to (rectifier-out, 0V), d_9 = More-or-less (rectifier-out, 0V).

Rule 1: $d_1, d_2 \leftrightarrow d_4, d_5$; Rule 2: $d_3 \leftrightarrow d_6$; Rule 3: $d_3 \leftrightarrow d_7$; Rule 4: $d_4 \leftrightarrow d_8$; Rule 5: $d_5 \leftrightarrow d_9$; Rule 6: $d_6 \leftrightarrow d_9$; Rule 7: $d_7 \leftrightarrow d_8$; where the $\leftrightarrow$ denotes the bi-directional iff relation.

Fig. 9.8: An FPN representing diagnostic knowledge of a two- diode full -wave rectifier.

The $\mathbf{R}_{fm}$ in the present context will be a (27×27) matrix, whose diagonal blocks will be occupied by $\mathbf{R}_i$. Further, since all the nondiagonal block matrices are null matrix, the $\mathbf{R}_f^{-1}$ can be constructed by substituting $\mathbf{R}_i$ in $\mathbf{R}_{fm}$ by $\mathbf{R}_i^{-1}$. The $\mathbf{P}'_{fm}$ and $\mathbf{Q}'_{fm}$ in the present context are also (27×27) matrices. $\mathbf{N}_{ini}$ is a (27×1) initial membership distribution vector, each consecutive three components (separated by semicolons) of which correspond the membership distribution of each proposition.

$$\mathbf{N}_{ini} = [n_1(0); n_2(0); n_3(0); n_4(0); n_5(0); n_6(0); n_7(0); n_8(0); n_9(0)]^T$$
$$= [000; \ 000; \ 000; \ 000; \ 000; \ 000; \ 000; \ 0.2 \ 0.1 \ 0.0; \ \ 0.4 \ 0.5 \ 0.6]^T_{27 \times 1}$$

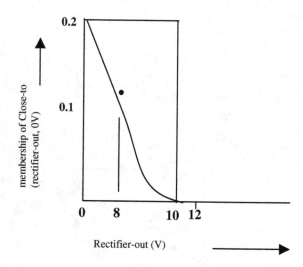

Fig. 9.9: Membership distribution of Close-to (rectifier-out, 0V).

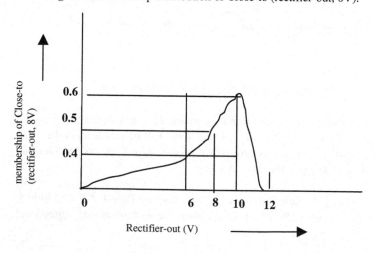

Fig. 9.10: Membership distribution of More-or-Less (rectifier-out, 10V).

Taking $\mathbf{P'}_{fm}^{-1} = (\mathbf{P'}_{fm})^T$ and $\mathbf{Q'}_{fm}^{-1} = (\mathbf{Q'}_{fm})^T$ and N_{ini} as given above, the algorithm for abductive reasoning is invoked. The algorithm terminates after three iterations and the steady-state membership distribution thus obtained is given by is given by

$$N_{S.S} = [\, n_1^*; \ n_2^*; \ n_3^*; \ n_4^*; \ n_5^*; \ n_6^*; \ n_7^*; \ n_8^*; \ n_9^* \,]$$

$$= [0.5 \ 0.4 \ 0.6; \ 0.5 \ 0.4 \ 0.6; \ 0.2 \ 0.2 \ 0.2; \ 0.2 \ 0.2 \ 0.2; \ 0.6 \ 0.4 \ 0.5;$$

$$0.6 \ 0.4 \ 0.5; \ 0.2 \ 0.2 \ 0.2; \ 0.2 \ 0.1 \ 0.0; \ 0.4 \ 0.5 \ 0.6\,]^T$$

It is evident from the above distribution that n_1^*, the steady-state values of each component of n_1, is larger than the corresponding components of n_3^*. The reasoning system thus infers predicate d_1: *defective (rectifier)* as the possible defect for the system.

9.7 Bi-directional Reasoning in an FPN

In a classical bi-directional if and only if (IFF) [15] type reasoning, the consequent part can be inferred when the antecedent part of the rule is observed and vice versa. In a fuzzy IFF type reasoning, one can infer the membership distribution of the one part of the rule, when the distribution of the other part is given. This principle has been extended in this section to estimate the membership distribution of all the predicates in the FPN, when the membership distribution of the predicates corresponding to the terminal places or the intermediate (nonterminal) places [16] are available. The bi-directional IFF type reasoning in FPN has, therefore, a pragmatic significance.

Like abductive reasoning, bi-directional IFF type reasoning too has been modeled in this chapter for acyclic FPNs [20] only by using the expression (9.9). For convenience of analysis, let us reformulate two basic fragments of expression (9.9), as

$$\mathbf{T}_f(t+1) = \mathbf{R}_{fm} \ o \ (\mathbf{Q'}_{fm} \ o \ \mathbf{N}_f^{\,c}(t))^c \qquad (9.18)$$

$$\mathbf{N}_f(t+1) = \mathbf{P'}_{fm} \ o \ \mathbf{T}_f(t+1). \qquad (9.19)$$

The above two expressions together represent the forward reasoning model of the IFF relation. An extra suffix (f) is attached to denote the "forward" direction of reasoning. For backward (back-directed) reasoning using IFF relation, however, one has to reverse the direction of the arrowheads in the FPN and then update $\mathbf{T}_b$ and $\mathbf{N}_b$ by using expression (9.12) and (9.13) in sequence until convergence [20] in $\mathbf{N}_b$ is attained.

$$T_b(t + 1) = R_{bm} \; o \; (Q'_{bm} \; o \; N_b{}^c(t))^c \qquad\qquad (9.20)$$

$$N_b(t + 1) = P'_{bm} \; o \; T_b(t + 1). \qquad\qquad (9.21)$$

The suffix "b" in expressions (9.20) and (9.21) stands for the backward direction of reasoning. It may be noted that, once P'_{fm} and Q'_{fm} are known, P'_{bm} and Q'_{bm} may be obtained using Theorem 9.1. It may be added here that for bi-directional IFF type reasoning in an FPN, a self-loop through virtual transitions with a relational matrix I and threshold vector zero have to be considered for all the concluding (terminal) places in the network. Without such augmentation in the network, the steady-state membership in the network can never be attained under abductive reasoning following expression (9.20) and (9.21). This augmentation should be reflected in the construction of P'_{bm}, Q'_{fm}, Q'_{bm} and P'_{fm} matrices.

Theorem 9.5: *The connectivity matrices* P'_{bm}, Q'_{fm}, Q'_{bm} *and* P'_{fm} *satisfy the following relationships:* $P'_{bm} = (Q'_{fm})^T$ *and* $Q'_{bm} = (P'_{fm})^T$.

Proof: Q'_{fm}, by definition, is a TPC matrix used in forward reasoning under IFF implication relationship, with elements $q_{ij} = 1$ when $p_i \in O(tr_j)$ and $q_{ij} = 0$, otherwise. P'_{bm}, on the other hand, is a PTC matrix used in back-directed reasoning under IFF implication relationship with elements $p_{ji} = 1$ when $p_i \in I(tr_j)$, otherwise $p_{ji} = 0$. Thus for all i, j, the relation $P'_{bm} = (Q'_{fm})^T$ holds. Analogously, $Q'_{bm} = (P'_{fm})^T$ can also be proved.

When the membership distributions of the axiom predicates are given, one has to use the forward reasoning model. On the other hand, when the membership distribution of the predicates for the concluding places is known, one should use the back-directed reasoning model of the IFF relation. Moreover, when the membership distributions of the predicates at the nonterminal places are available, one has to use both forward and back-directed reasoning models to estimate the membership distribution of the predicates corresponding to respective predecessors and successors of the given nonterminal places. However, this case, the estimated memberships of the predicates may not be consistent. In other words, after obtaining steady-state memberships at all places, if one recomputes memberships of the nonaxiom predicates with the known memberships of the axiom predicates, the computed memberships may not tally with their initial values. In order to overcome this problem, one requires a special relationship, called reciprocity [26]. It may be noted that in an FPN that holds (perfect) reciprocity property, n successive steps of forward (backward) reasoning followed by n successive steps of backward (forward) reasoning restores the value of the membership vector $N(t)$.

Definition 9.9: *An FPN is said to hold **reciprocity property** if updating FTT (membership) vector in the forward direction followed by updating of FTT (membership) vector in the backward direction restores the value of the FTT (membership) vector.*

Formally, we estimate $T_f(t + 1)$ from given $N_f(t)$ and $N_f(t)$ from $T_f(t+1)$ in succession,

$$\text{i.e., } T_f(t + 1) = R_{fm} \text{ o } (Q'_{fm} \text{ o } N_f{}^c(t))^c \tag{9.22}$$

$$\text{and } N_f(t) = P'_{bm} \text{ o } T_f(t + 1). \tag{9.23}$$

Combining expressions (9.22) and (9.23), we have

$$N_f(t) = P'_{bm} \text{ o } R_{fm} \text{ o } (Q'_{fm} \text{ o } N_f{}^c(t))^c$$

$$= (Q'_{fm})^T \text{ o } R_{fm} \text{ o } (Q'_{fm} \text{ o } N_f{}^c(t))^c \text{ [by Theorem 9.1]} \tag{9.24}$$

Suppose that the fuzzy FTT vector $T_f(t + 1)$ is known, and we want to compute $N_f(t + 1)$, i.e. the fuzzy distributions at the output places of the transitions. Now, with the known membership distribution at the places we want to retrieve the FTT vector associated with the transitions. The above two phenomena can be satisfied by executing the following two steps of computations in order.

$$N_f(t + 1) = P'_{fm} \text{ o } T_f(t + 1) \tag{9.25}$$

$$\text{and } T_f(t + 1) = R_{bm} \text{ o } (Q'_{bm} \text{ o } N_f{}^c(t+1))^c. \tag{9.26}$$

Replacing Q'_{bm} by $P'_{fm}{}^T$ by Theorem 9.1 and then substituting $N_f(t + 1)$ by the right hand side of (9.25) in expression (9.26) we have

$$T_f(t + 1) = R_{bm} \text{ o } (Q'_{bm} \text{ o } N_f{}^c(t+1))^c$$

$$= R_{bm} \text{ o } ((P'_{fm})^T \text{ o } N_f{}^c(t + 1))^c$$

$$= R_{bm} \text{ o } [(P'_{fm})^T \text{ o } (P'_{fm} \text{ o } T_f(t + 1))^c]^c. \tag{9.27}$$

Expressions (9.24) and (9.27), which are identities of N_f and T_f, respectively, taken together are called a reciprocity relation. For testing reciprocity conditions, however, one has to use the results of Theorem 9.6.

Theorem 9.6: *The condition of reciprocity in an FPN is given by*

$$(\mathbf{Q}'_{fm})^{T} \; o \; \mathbf{R}_{fm} \; o \; (\mathbf{Q}'_{fm} \; o \; \mathbf{I}^{c})^{c} \; = \; \mathbf{I} \tag{9.28(a)}$$

and $\qquad \mathbf{R}_{bm} \; o \; [(\mathbf{P}'_{fm})^{T} \; o \; (\mathbf{P}'_{fm})^{c}]^{c} \; = \; \mathbf{I} \tag{9.28(b)}$

Proof: We will use lemmas 9.1 and 9.2 to prove this theorem.

Lemma 9.1: *The distributive law of product holds good with respect to fuzzy composition operator, i.e.,*

$$\mathbf{A} \; o \; [\mathbf{B} \vee \mathbf{C}] \; = \; (\mathbf{A} \; o \; \mathbf{B}) \vee (\mathbf{A} \; o \; \mathbf{C}), \tag{9.29}$$

where A is a (n × m) fuzzy matrix and B and C are either fuzzy matrix or vectors of compatible dimensions.

Lemma 9.2: *De Morgan's laws hold good for any two fuzzy matrices A and B of respective dimensions (m × n) and (n × r), i.e.,*

$$(\mathbf{A} \; \ominus \; \mathbf{B}) \; = \; (\mathbf{A}^{c} \; o \; \mathbf{B}^{c})^{c} \tag{9.30(a)}$$

$$and \quad \mathbf{A} \; o \; \mathbf{B} \; = \; (\mathbf{A}^{c} \; \ominus \; \mathbf{B}^{c})^{c} \tag{9.30b)}$$

where $\ominus$ denotes fuzzy OR-AND composition operator, which plays the role of AND-OR composition operator, with the replacement of AND by OR and OR by AND operators.

Proof: The lemma can easily be proved by extending De Morgan's law from scalar fuzzy singleton membership to discrete distributions represented by matrices.

Now, let us consider the reciprocity relations given by expressions (9.24) and (9.27). Since expression (9.24) is nothing but an identity of $\mathbf{N}_{f}$, it is valid for any arbitrary fuzzy vector $\mathbf{N}_{f}$. Assume $\mathbf{N}_{f}$ to be a vector with only one element equal to 1 and the rest are zero. Further, to keep the proof brief, let us consider that $\mathbf{N}_{f}$ is of (3 × 1) dimension.

Thus, we get

$$\begin{bmatrix} 1 \\ 0 \\ 0 \end{bmatrix} = \mathbf{Q}'_{fm}{}^{T} \; o \; \mathbf{R}_{fm} \; o \; \left[\mathbf{Q}'_{fm} \; o \begin{bmatrix} 0 \\ 1 \\ 1 \end{bmatrix} \right]^{c}$$

$$= \mathbf{Q'}_{fm}{}^T \circ \mathbf{R}_{fm} \circ \begin{bmatrix} q_{12}^c \wedge q_{13}^c \\ q_{22}^c \wedge q_{23}^c \\ q_{32}^c \wedge q_{33}^c \end{bmatrix} \qquad (9.31)$$

Analogously,

$$\begin{bmatrix} 0 \\ 1 \\ 0 \end{bmatrix} = \mathbf{Q'}_{fm}{}^T \circ \mathbf{R}_{fm} \circ \begin{bmatrix} q_{11}^c \wedge q_{13}^c \\ q_{21}^c \wedge q_{23}^c \\ q_{31}^c \wedge q_{33}^c \end{bmatrix} \qquad (9.32)$$

and

$$\begin{bmatrix} 0 \\ 0 \\ 1 \end{bmatrix} = \mathbf{Q'}_{fm}{}^T \circ \mathbf{R}_{fm} \circ \begin{bmatrix} q_{11}^c \wedge q_{12}^c \\ q_{21}^c \wedge q_{22}^c \\ q_{31}^c \wedge q_{32}^c \end{bmatrix} \qquad (9.33)$$

where q_{ij} are the elements of $\mathbf{Q'}_{fm}$ matrix. Now, combining (9.31), (9.32), and (9.33) we find

$$\begin{bmatrix} 1 & 0 & 0 \\ 0 & 0 & 0 \\ 0 & 0 & 0 \end{bmatrix} \vee \begin{bmatrix} 0 & 0 & 0 \\ 0 & 1 & 0 \\ 0 & 0 & 0 \end{bmatrix} \vee \begin{bmatrix} 0 & 0 & 0 \\ 0 & 0 & 0 \\ 0 & 0 & 1 \end{bmatrix}$$

$$= \mathbf{Q'}_{fm}{}^T \circ \mathbf{R}_{fm} \circ \{ \begin{bmatrix} q_{12}^c \wedge q_{13}^c & : 0 & 0 \\ q_{22}^c \wedge q_{23}^c & : 0 & 0 \\ q_{32}^c \wedge q_{33}^c & : 0 & 0 \end{bmatrix} \vee$$

$$\begin{bmatrix} 0 : & q_{11}^c \wedge q_{13}^c & : 0 \\ 0 : & q_{21}^c \wedge q_{23}^c & : 0 \\ 0 : & q_{31}^c \wedge q_{33}^c & : 0 \end{bmatrix} \vee \begin{bmatrix} 0 & 0 : & q_{11}^c \wedge q_{12}^c \\ 0 & 0 : & q_{21}^c \wedge q_{22}^c \\ 0 & 0 : & q_{31}^c \wedge q_{32}^c \end{bmatrix} \}$$

(by Lemma 9.1)

where the column of M's partitions the matrix into blocks.

$$\Rightarrow I = Q'_{fm}{}^T \circ R_{fm} \circ \begin{bmatrix} q_{12}^c \wedge q_{13}^c & \vdots & q_{11}^c \wedge q_{13}^c & \vdots & q_{11}^c \wedge q_{12}^c \\ q_{22}^c \wedge q_{23}^c & \vdots & q_{21}^c \wedge q_{23}^c & \vdots & q_{21}^c \wedge q_{22}^c \\ q_{32}^c \wedge q_{33}^c & \vdots & q_{31}^c \wedge q_{33}^c & \vdots & q_{31}^c \wedge q_{32}^c \end{bmatrix}$$

$$\Rightarrow I = Q'_{fm}{}^T \circ R_{fm} \circ (Q'_{fm}{}^c \ominus I)$$

$$\Rightarrow I = Q'_{fm}{}^T \circ R_{fm} \circ (Q'_{fm} \circ I^c)^c \qquad \text{(by lemma 9.2)}$$

Thus, $Q'_{fm}{}^T \circ R_{fm} \circ (Q'_{fm} \circ I^c)^c = I$.

Now, extending the above operations for an $((n \cdot z) \times 1)$ N_f vector, the same results can be easily obtained. Considering expression (9.27), an identity of T_f, expression (9.28(b)) can be proved analogously.

Example 9.7: Consider the FPN given in Figure 9.11. Given $R_{fm} = I$ and $R_{bm} = I$, we want to test the reciprocity property of the FPN.

Here, $\quad P'_{fm} = \begin{bmatrix} \Phi & \Phi & I \\ I & \Phi & \Phi \\ \Phi & I & \Phi \end{bmatrix}$ and $Q'_{fm} = \begin{bmatrix} I & \Phi & \Phi \\ \Phi & I & \Phi \\ \Phi & \Phi & I \end{bmatrix}$

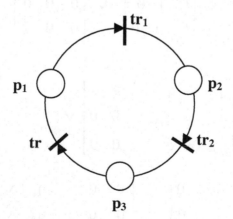

Fig. 9.11: An FPN used to illustrate the reciprocity property.

where Φ and **I** denote null and identity matrices of dimension (3×3), respectively. With these values of $\mathbf{P'}_{fm}$ and $\mathbf{Q'}_{fm}$, we found that the reciprocity conditions 9.28(a) and (b) hold good.

It is clear from expressions 9.28(a) and 9.28(b) that the condition of reciprocity implies a relationship between the structure of the FPN and the relational matrices associated with the transitions. Thus, for a given structure of a FPN, identification of the relational matrices $(\mathbf{R}_{fm}, \mathbf{R}_{bm})$ satisfying the reciprocity conditions is a design problem. In fact, rearranging expression 9.28(a) and 9.28(b), we find $\mathbf{R}_{fm}$ and $\mathbf{R}_{bm}$ as follows

$$\mathbf{R}_{fm} = [(\mathbf{Q'}_{fm})^T]^{-1}_{pre} \ o \ [(\mathbf{Q'}_{fm} \ o \ \mathbf{I}^c)^c]^{-1}_{post} \tag{9.34}$$

$$\mathbf{R}_{bm} = [\{(\mathbf{P'}_{fm})^T \ o \ (\mathbf{P'}_{fm})^c\}^c]^{-1}_{post} \tag{9.35}$$

where the suffixes "pre" and "post" denote pre-inverse and post-inverse of the matrices. Pedrycz and Gomide in a recent book [30] presented a scheme for computing inverse fuzzy relation using neural nets. Saha and Konar designed an alternative method [31-33] to the computation of the pre- and post-inverses. It can be shown from their work that a matrix can have a number of pre- and post-inverses, and the transpose should essentially be one of them [32]. Since computation of all the inverses is time-consuming, the transpose of the matrix can be used as its both pre- and post-inverse.

Since some choice of $\mathbf{R}_{fm}$ and $\mathbf{R}_{bm}$ by (9.34) and (9.35) satisfy the reciprocity condition, it is expected that the membership distribution at a given place of the FPN would retrieve its original value after n-forward (backward) steps followed by n-backward (forward) steps of reasoning in the network. Consequently, the steady-state [16] membership distribution at all places in the FPN will be consistent independent of the order of forward and backward computation. This, in fact, is useful when the initial membership distribution of the intermediate [12] places only in the FPN is known.

Example 9.8: Consider the problem of diagnosis of a two-diode full-wave rectifier circuit. The expected rectifier output voltage is 12 volts, when the system operates properly. Under defective conditions, the output could be close to 0 volts or 10 volts depending on the number of defective diodes. The knowledge base of the diagnosis problem is extended into a FPN (vide Fig. 9.12).

Let us now assume that the bi-directional IFF relationship exists between the predicates corresponding to input-output place pairs of the transitions in the network of Figure 9.12. We also assume that the membership distribution at places p_4 and p_5 only is known and one has to estimate the consistent memberships at all places in the network.

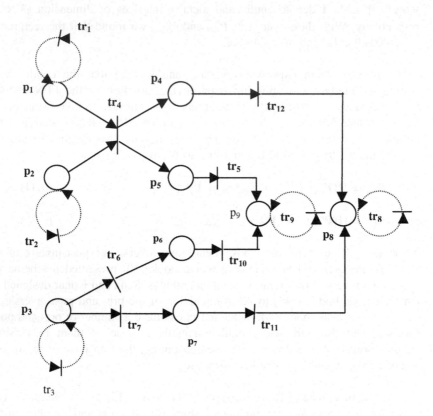

$d_1, d_2, \ldots, d_9$ (not shown) denote the predicates defined below:

d_1 = defective (transformer), d_2 = close-to (primary, 230), d_3 = defective (rectifier), d_4 = Close-to (trans-out, 0V), d_5 = Open (one half-of-secondary-coil), d_6 = Defective (one-diode), d_7 = Defective (two-diodes), d_8 = Close-to (rectifier-out, 0V), d_9 = More-or-less (rectifier-out, 0V).

Rule 1: $d_1, d_2 \leftrightarrow d_4, d_5$; Rule 2: $d_3 \leftrightarrow d_6$; Rule 3: $d_3 \leftrightarrow d_7$; Rule 4: $d_4 \leftrightarrow d_8$; Rule 5: $d_5 \leftrightarrow d_9$; Rule 6: $d_6 \leftrightarrow d_9$ Rule 7: $d_7 \leftrightarrow d_8$, where the $\leftrightarrow$ denotes the bi-directional iff relation.

Fig. 9.12: The FPN with self-loops through virtual transitions at the axioms and concluding places used to illustrate the reciprocity characteristics.

In the present context, we first construct $\mathbf{P'_{fm}}$ and $\mathbf{Q'_{fm}}$ from Figure 9.12 then determine $\mathbf{R_{fm}}$ and $\mathbf{R_{bm}}$ presented below by using expressions (9.34) and (9.35), and finally carry out one step forward reasoning followed by two steps back-directed reasoning using expressions (9.18) through (9.21). It has been checked that the membership vector, thus obtained, is unique and remains invariant (steady [16]) even if someone carries out one step back-directed reasoning followed by two steps forward and two steps back-directed reasoning.

$$
\mathbf{R}_{fm} =
\begin{bmatrix}
I & \Phi & \Phi & \Phi & \Phi & \Phi & \Phi & \Phi & \Phi & \Phi & \Phi \\
\Phi & I & \Phi & \Phi & \Phi & \Phi & \Phi & \Phi & \Phi & \Phi & \Phi \\
\Phi & \Phi & I & \Phi & \Phi & I & I & \Phi & \Phi & \Phi & \Phi \\
I & I & \Phi & \Phi & \Phi & \Phi & \Phi & \Phi & \Phi & \Phi & \Phi \\
\Phi & \Phi & \Phi & \Phi & I & \Phi & \Phi & \Phi & \Phi & \Phi & \Phi \\
\Phi & \Phi & I & \Phi & \Phi & I & I & \Phi & \Phi & \Phi & \Phi \\
\Phi & \Phi & I & \Phi & \Phi & I & I & \Phi & \Phi & \Phi & \Phi \\
\Phi & \Phi & \Phi & \Phi & \Phi & \Phi & \Phi & I & \Phi & \Phi & \Phi \\
\Phi & \Phi & \Phi & \Phi & \Phi & \Phi & \Phi & \Phi & I & \Phi & \Phi \\
\Phi & \Phi & \Phi & \Phi & \Phi & \Phi & \Phi & \Phi & \Phi & I & \Phi \\
\Phi & \Phi & \Phi & \Phi & \Phi & \Phi & \Phi & \Phi & \Phi & \Phi & I
\end{bmatrix}
\quad \text{and}
$$

$$
\mathbf{R}_{bm} =
\begin{bmatrix}
I & \Phi & \Phi & \Phi & \Phi & \Phi & \Phi & \Phi & \Phi & \Phi & \Phi & \Phi \\
\Phi & I & \Phi & \Phi & \Phi & \Phi & \Phi & \Phi & \Phi & \Phi & \Phi & \Phi \\
\Phi & \Phi & I & \Phi & \Phi & \Phi & \Phi & \Phi & \Phi & \Phi & \Phi & \Phi \\
\Phi & \Phi & \Phi & I & \Phi & \Phi & \Phi & \Phi & \Phi & \Phi & \Phi & \Phi \\
\Phi & \Phi & \Phi & \Phi & I & \Phi & \Phi & \Phi & I & I & \Phi & \Phi \\
\Phi & \Phi & \Phi & \Phi & \Phi & I & \Phi & \Phi & \Phi & \Phi & \Phi & \Phi \\
\Phi & \Phi & \Phi & \Phi & \Phi & \Phi & I & \Phi & \Phi & \Phi & \Phi & \Phi \\
\Phi & \Phi & \Phi & \Phi & \Phi & \Phi & \Phi & I & \Phi & \Phi & \Phi & \Phi \\
\Phi & \Phi & \Phi & \Phi & I & \Phi & \Phi & \Phi & I & I & \Phi & \Phi \\
\Phi & \Phi & \Phi & \Phi & I & \Phi & \Phi & \Phi & I & I & \Phi & \Phi \\
\Phi & \Phi & \Phi & \Phi & \Phi & \Phi & \Phi & I & \Phi & \Phi & I & I \\
\Phi & \Phi & \Phi & \Phi & \Phi & \Phi & \Phi & I & \Phi & \Phi & I & I
\end{bmatrix}
$$

9.8 Fuzzy Modus Tollens and Duality

In classical modus tollens [15], for predicates A and B, given the rule $A \rightarrow B$ and the observed evidence $\neg B$, then the derived inference is $\neg A$. Thus, the contrapositive rule: $(A \rightarrow B) \Leftrightarrow (\neg B \rightarrow \neg A)$ follows. It is known that in fuzzy logic the sum of the membership of an evidence and its contradiction is greater than or equal to one [22]. So, if the membership of an evidence is known, the membership of its contradiction cannot be easily ascertained. However, in many real-world problems, the membership of nonoccurrence of an evidence is to be estimated when the membership of nonoccurrence of its causal evidences is known. To tackle such problems, the concept of classical modus tollens of predicate logic is extended here to fuzzy logic for applications in FPN. Before formulation of the problem, let us first show that implication relations $(A \rightarrow B)$ and $(\neg B \rightarrow \neg A)$ are identical in the fuzzy domain, under the closure of Lukasiewicz implication function. Formally let a_i, $1 \le i \le m$ and b_j, $1 \le j \le n$ be the membership distribution of predicates A and B, respectively. Then the (i, j)th element of the relational matrix $\mathbf{R_1}$ for the rule $A \rightarrow B$ by Lukasiewicz implication function [34] is given by

$$\mathbf{R_1}(i, j) = \text{Min} \{1, (1 - a_i + b_j)\} \tag{9.36}$$

Again, the (i, j)th element of the relational matrix $\mathbf{R_2}$ for the rule $\neg B \rightarrow \neg A$ using Lukasiewicz implication function is given by

$$\mathbf{R_2}(i, j) = \text{Min}[1, \{1 - (1 - b_j) + (1 - a_i)\}] = \text{Min}\{1, (1 - a_i + b_j)\} \tag{9.37}$$

Thus, it is clear from expressions (9.36) and (9.37) that the two relational matrices $\mathbf{R_1}$ and $\mathbf{R_2}$ are equal. So, classical modus tollens can be extended to fuzzy logic under Lukasiewicz implication relation. Let us consider an FPN (vide Fig. 9.13(a)), referred to as the primal net, that is framed with the following knowledge base:

> *rule 1:* $d_1, d_2 \rightarrow d_3$
> *rule 2:* $d_2 \rightarrow d_4$
> *rule 3:* $d_3, d_4 \rightarrow d_1$

The dual of this net can be constructed by reformulating the above knowledge base using the contrapositive rules as follows:

> *rule 1:* $\neg d_3 \rightarrow \neg d_1, \neg d_2$
> *rule 2:* $\neg d_4 \rightarrow \neg d_2$
> *rule 3:* $\neg d_1 \rightarrow \neg d_3, \neg d_4$

Here, the comma on the right-hand side of the if-then operator in the above rules represents OR operation. It is evident from the reformulated knowledge base that the dual FPN can be easily constructed by replacing each predicate d_i by its negation and reversing the directivity in the network. The dual FPN of Figure 9.13(a) is given in Figure 9.13(b).

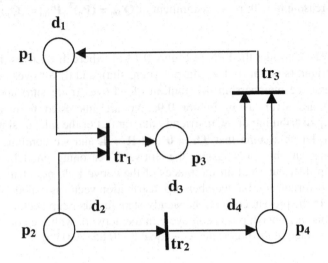

Fig. 9.13(a): The primal fuzzy Petri net.

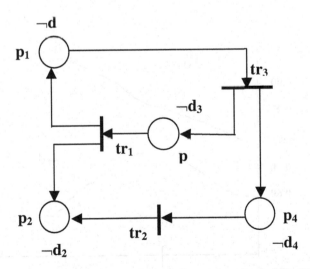

Fig. 9.13(b): The dual fuzzy Petri net corresponding to the primal of Fig. 9.13(a).

Reasoning in the primal model of FPN can be performed using the forward reasoning procedure. Let $R = R_p$, $P' = P_p$, and $Q' = Q_p$ denote the matrices for the primal model in expression (9.8). Then, for forward reasoning in the dual FPN, one should initiate $P' = (Q_p)^T$ and $Q' = (P_p)^T$ (vide Theorem 9.5), and $R =$

R_p prior to invoking the procedure forward reasoning. If the membership distributions at the concluding places are available, the membership distribution at other places of the dual FPN may be estimated by invoking the procedure backward-reasoning with prior assignment of $Q'_{fm} = (P_p)^T$, $P'_{fm} = (Q_p)^T$ and $R_{fm} = R_p$.

Example 9.9: Consider the FPN of Figure 9.13(a), where $d_1 \equiv$ Loves (ram, sita), $d_2 \equiv$ Girl-friend (sita, ram), $d_3 \equiv$ Marries (ram, sita), and $d_4 \equiv$ Loves (sita, ram). Suppose that the membership distribution of ¬Loves (ram, sita) and ¬Loves (sita, ram) are given as in Figure 9.9. We are interested to estimate the membership distribution of ¬Girl-friend (sita, ram). For the sake of simplicity in calculation, let us assume that **Th = 0** and **R = I,** and we continue updating memberships in the FPN using the forward reasoning model, until the membership distribution at all the places of the network do not change further. Such time-invariance of the membership distribution vector is called the **steady-state** [16]. In the present context, the steady-state membership vector is obtained only after one membership revision cycle in the network with the corresponding steady-state value of the predicate ¬ d_2 equal to $[0.85\ 0.9\ 0.95]^T$.

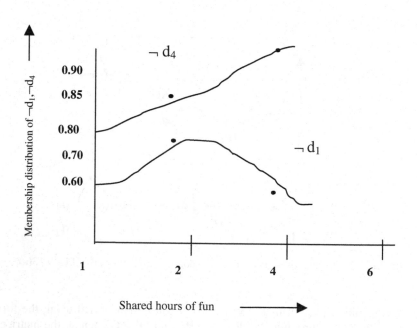

Fig. 9.14: Membership distribution of ¬ Loves (ram, sita) and ¬ Loves (sita, ram).

9.9 Conclusions

The principle of abductive reasoning has been extended in this chapter under the framework of a fuzzy relational system using Petri nets. The extension has been established following a new paradigm of inverse formulation of a fuzzy relational matrix. It has been shown elsewhere [33] that the transpose is one of the possible inverses of a fuzzy relational matrix. Since the evaluation of the fuzzy inverses is a NP- complete problem, instead of computing the best among all the inverses we employed the transpose as the inverse in the computations involved in abductive reasoning.

The most important aspect of the abductive reasoning formulation presented in the chapter lies in its extensive applications especially in diagnostic problems. The logic of predicates is unable to recast modus ponens for abductive reasoning. Default and non-monotonic logic find their own ways to handle the above problem but get stuck in limit cycles [24] in the presence of noisy measurement data. Pearl's belief revision model, which is quite competent to reason with imprecise measurement data, however, requires quite a large number of conditional probabilities, which are not always available in the respective problem domains. The logic of fuzzy sets thus remains the only tool to handle the said problem.

The uncertainty of knowledge has been modeled in the present context by the fuzzy relational matrices. It should be added here that the fuzzy relational matrices are usually constructed based on the subjective judgment of the experts and, thus, in most cases, no measurements (unlike in the case of conditional probabilities) are needed. It thus makes sense to design a fuzzy relational system to handle the said problem. The structural benefit of the Petri net in complex knowledge representation together with the advantage of the max-min inverse computation makes the abductive reasoning system much more robust and amenable to diagnostic applications.

Reciprocity, which is still an unexplored area in the domain of fuzzy sets, has great potential in the next-generation knowledge-based systems. The condition of reciprocity establishes a relationship between the relational matrices of the rules and the structure of the FPN. Thus, for a given structure of an FPN, the relational matrices of the system can easily be determined. But what happens when the relational matrices are known and we want to determine the structure to satisfy reciprocity. Because of the multiple solutions of the P'_{bm} and Q'_{fm} matrices for known R_{fm} and R_{bm}, a number of structures may come up as possible solutions. Such issues have interesting applications in knowledge clustering and hence in data mining. The matter remains an open problem.

Another interesting feature of the chapter is its application to determining the membership distributions of negated predicates in an FPN. This too is new in the realm of machine intelligence. This feature is important in fuzzy systems, as, unlike in probability, the sum of membership of a predicate and its negation need not be always one. Determination of the distribution for negated predicates will open up a new vista of research in criminology applications. For instance, consider A, B, and C to be three suspects in a criminology problem. Now, the grades of singleton membership of A and B not being suspects are, suppose, 0.7 and 0.8, respectively. This unfortunately does not help in finding the grade of membership of C not being a suspect. The existing models of non-monotonic logic and truth maintenance system [18] cannot handle such problem. Our dual FPN model, however, can answer such queries with a formal logical basis. This chapter thus brings an end to the long controversy [11] concerning the importance of fuzzy logic vis-à-vis classical non-monotonic logic.

Exercises

1. Consider the FPN shown in Figure 9.8. Determine its $\mathbf{P}$, $\mathbf{Q}$, $\mathbf{P}'_{fm}$, $\mathbf{Q}'_{fm}$. Select R_i for each transitions as given on page 277. Verify the results obtained for $N_{s.s}$ vector.

2. The max-min inverse-computing problem [33] can be formulated as follows.

   Evaluate Q such that

   $$Q \circ R = I'$$

   where $m = \sum_i \sum_j (I_{ij} - I'_{ij})^2$ is minimum.

   To solve this problem, we construct a heuristic function $h(q_{ij})$ that maximizes the min terms $(q_{ij} \wedge r_{jk})$, having contribution to the one in the resulting I matrix, and minimizes the min terms $(q_{ij} \wedge r_{jk})$, having contribution to the zeros in the I matrix. Next search q_{ij} in [0, 1] that satisfies the heuristic function. Using this principle, find the max-min inverse matrices of the following R.

   $$R = \begin{pmatrix} 0.2 & 0.6 & 0.9 \\ 0.3 & 0.8 & 0.4 \\ 0.5 & 0.7 & 0.2 \end{pmatrix}.$$

If there are multiple solutions, find the best R that minimizes m.

3. Prove that in a purely cyclic FPN, $(\mathbf{P} \ o \ \mathbf{Q})^k$ where k = number of transitions in the cycle.

4. Given that pre- and post-inverse matrix of **I** is **I,** show that the reciprocity relation holds for purely cyclic nets. Also show that R_{fm} and R_{bm} for such net = I.

5. Prove that a dual net can always be constructed by reversing the arrowheads on the primal net.

References

[1]   Bugarin, A. J., and Barro, S., "Fuzzy reasoning supported by Petri nets," *IEEE Trans. on Fuzzy Systems,* vol. 2, no. 2, pp. 135-150, 1999.

[2]   Buchanan, B. G. and Shortliffe, E. H., *Rule Based Expert Systems: The MYCIN Experiment of the Stanford University,* Addison-Wesley, Reading, MA, 1989.

[3]   Cao, T. and Sanderson, A. C., "A fuzzy Petri net approach to reasoning about uncertainty in robotic systems," *Proc. IEEE Int. Conf. Robotics and Automation,* Atlanta, GA, pp. 317-322, May 1993.

[4]   Cao, T., "Variable reasoning and analysis about uncertainty with fuzzy Petri nets," *Lecture Notes in Computer Science,* vol. 691, Marson, M. A. (Ed.), Springer-Verlag, New York, pp. 126-145, 1993.

[5]   Cao, T. and Sanderson, A. C., "Task sequence planning using fuzzy Petri nets," *IEEE Trans. on Systems, Man and Cybernetics,* vol. 25, no.5, pp. 755-769, May 1995.

[6]   Cardoso, J., Valette, R., and Dubois, D., "Petri nets with uncertain markings," In *Advances in Petri Nets,* Lecture Notes in Computer Science, Rozenberg, G. (Ed.), vol. 483, Springer-Verlag, New York, pp. 65-78, 1990.

[7]   Chen, S. M., "Fuzzy backward reasoning using fuzzy Petri nets," *IEEE Trans. on Systems, Man and Cybernetics, Part B: Cybernetics, vol. 30, no. 6, 2000.*

[8]   Daltrini, A., "Modeling and knowledge processing based on the extended fuzzy Petri nets," *M.Sc. degree book*, UNICAMP-FEE0DCA, May 1993.

[9]   Doyle, J., "Truth maintenance systems," *Artificial Intelligence*, vol. 12, 1979.

[10]   Garg, M. L., Ashon, S. I., and Gupta, P. V., "A fuzzy Petri net for knowledge representation and reasoning," *Information Processing Letters*, vol. 39, pp. 165-171, 1991.

[11]   Graham, I. and Jones, P. L., *Expert Systems: Knowledge, Uncertainty and Decision*, Chapman and Hall, London, 1988.

[12]   Hirota, K. and Pedrycz, W., "OR-AND neuron in modeling fuzzy set connectives," *IEEE Trans. on Fuzzy Systems*, vol. 2, no. 2, May 1999.

[13]   Hutchinson, S. A. and Kak, A. C., "Planning sensing strategies in a robot workcell with multisensor capabilities," *IEEE Trans. Robotics and Automation*, vol. 5, no. 6, pp. 765-783, 1989.

[14]   Jackson, P., *Introduction to Expert Systems*, Addison-Wesley, Reading, MA, 1988.

[15]   Konar, A. and Mandal, A. K., "Uncertainty management in expert systems using fuzzy Petri nets," *IEEE Trans. on Knowledge and Data Engineering*, vol. 8, no. 1, pp. 96-105, February 1996.

[16]   Konar, A., *Building an Intelligent Decision Support System for Criminal Investigation*, Report no. 1/2001/ETCE/J.U., submitted to All India Council for Technical Education as the completion report for the Career Award for Young Teachers, 2001.

[17]   Konar, A. and Mandal, A. K., "Non-monotonic reasoning in expert systems using fuzzy Petri nets," *Advances in Modeling and Analysis, B*, AMSE Press, vol. 23, no. 1, pp. 51-63, 1992.

[18]   Konar, A., *Artificial Intelligence and Soft Computing: Behavioral and Cognitive Modeling of the Human Brain*, CRC Press, Boca Raton, FL, 1999.

[19]   Konar, A., *"Uncertainty Management in Expert System Using Fuzzy Petri Nets,"* Ph. D. dissertation, Jadavpur University, India, 1999.

[20]   Konar, A. and Pal, S., "Modeling cognition with fuzzy neural nets," In *Fuzzy Theory Systems: Techniques and Applications,* Leondes, C. T. (Ed.), Academic Press, New York, 1999.

[21]   Kosko, B., *Neural Networks and Fuzzy Systems*, Prentice-Hall, Englewood Cliffs, NJ, 1999.

[22]   Lipp, H. P. and Gunther, G., "A fuzzy Petri net concept for complex decision making process in production control," In *Proc. of First European Congress on Fuzzy and Intelligent Technology (EUFIT '93)*, Aachen, Germany, vol. I, pp. 290 – 294, 1993.

[23]   Looney, C. G., "Fuzzy Petri nets for rule-based decision making," *IEEE Trans. on Systems, Man, and Cybernetics*, vol. 18, no. 1, pp. 178-183, 1988.

[24]   McDermott, V. and Doyle, J., "Non-monotonic logic I," *Artificial Intelligence*, vol. 13 (1-2), pp. 41-72, 1980.

[25]   Murata, T., "Petri nets: properties, analysis and applications," *Proceedings of the IEEE*, vol. 77, no. 4, pp. 541-580, 1989.

[26]   Pal, S. and Konar, A., "Cognitive reasoning using fuzzy neural nets," *IEEE Trans. on Systems, Man and Cybernetics*, August 1996.

[27]   Pearl, J., "Distributed revision of composite beliefs," *Artificial Intelligence*, vol. 33, 1987.

[28]   Pedrycz, W. and Gomide, F., "A generalized fuzzy Petri net model," *IEEE Trans. on Fuzzy Systems*, vol . 2, no. 4, pp. 295-301, Nov. 1999.

[29]   Pedrycz, W, *Fuzzy Sets Engineering*, CRC Press, Boca Raton, FL, 1995.

[30]   Pedrycz, W. and Gomide, F., *An Introduction to Fuzzy Sets: Analysis and Design*, MIT Press, Cambridge, MA, pp. 85-126, 1998.

[31]   Saha, P. and Konar, A., "Backward reasoning with inverse fuzzy relational matrices," *Proc. of Int. Conf. on Control, Automation, Robotics and Computer Vision,* Singapore, 1996.

[32]   Saha, P. and Konar, A., "Reciprocity and duality in a fuzzy network model," *Int. J. of Modelling and Simulation*, vol. 24, no. 3, pp. 168-178, 2004.

[33]   Saha, P. and Konar, A., "A heuristic algorithm for computing the Max-Min inverse fuzzy relation," *Int. J. of Approximate Reasoning,* vol. 30, pp. 131-137, 2002.

[34]   Scarpelli, H. and Gomide, F., "High level fuzzy Petri nets and backward reasoning," In *Fuzzy Logic and Soft Computing,* Bouchon-Meunier, B., Yager, R. R. and Zadeh L. A. (Eds.), World Scientific, Singapore, 1995.

[35]   Sil, J. and Konar, A., "Approximate reasoning using probabilistic predicate transition net model," *Int. J. of Modeling and Simulation,* vol. 21, no. 2, pp. 155-168, 2001.

[36]   Scarpelli, H., Gomide, F. and Yager, R., "A reasoning algorithm for high level fuzzy Petri nets," *IEEE Trans. on Fuzzy Systems,* vol. 4, no. 3, pp. 282-295, Aug. 1996.

[37]   Shafer, G., *A Mathematical Theory of Evidence,* Princeton University Press, Princeton, NJ, 1976.

[38]   Waterman, D. A. and Hayes-Roth, F., *Pattern Directed Inference Systems,* Academic Press, New York, 1977.

[39]   Yu, S. K., "Comments on 'Knowledge representation using fuzzy Petri nets'," *IEEE Trans. on Knowledge and Data Engineering,* vol. 7, no.1, pp. 190-191, Feb. 1995.

[40]   Yu, S. K., "Knowledge representation and reasoning using fuzzy Pr\T net-systems," *Fuzzy Sets and Systems,* vol. 75, pp. 33-45, 1995.

[41]   Zadeh, L. A. "The role of fuzzy logic in the management of uncertainty in expert system," *Fuzzy Sets and Systems,* vol. 11, pp. 199-227, 1983.

Chapter 10

Human Mood Detection and Control: A Cybernetic Approach

This chapter examines the scope of fuzzy relational models in human mood detection and control. Selected facial attributes such eye-opening, mouth-opening, and the length of eyebrow-constriction are first extracted from a given facial expression. These features are fuzzified and then mapped onto a mood space by employing a Mamdani-type relational model. The control of the human mood requires construction of a model of the human mind, which on being excited with positive/negative influences may undergo state transitions in the mood space. The chapter considers music, video clips, and messages as different inputs to the mind model to study the effect of positive/negative influences on the mood transition dynamics. A fuzzy relational model is then employed to select the appropriate music or video/audio clips to control the mood from a given state to a desired state. Experimental results show good accuracy in both detection and control of the human mood.

10.1 Introduction

Human-machine interaction is currently regarded as an important field of study in cognitive engineering. In the early 1960s, much emphasis was given to the human behavioral factors, including hand and body movements and speed of work, which nowadays are studied in ergonomics. Unfortunately, the composite interactions of the humans and the machines were overlooked at that time. Only in recent times has human-machine interaction been considered a closed-loop system, where the humans read the control panels of the machine and accordingly generate control commands for the machines to cause transition in system states. Driving a car, for instance, is a human-machine interactive system, where the cognitive skill of the driver is used to control the motion of the car from its indicators. In this chapter, human mood detection and control is taken up as a case study of human-machine interactive systems.

Human mood detection is currently gaining importance for its increasing applications in computerized psychological counseling and therapy and also in the detection of criminal and antisocial motives. Identification of human moods by a machine is a complex problem for the following reasons. First, identification of the exact facial expression from a blurred facial image is not easily amenable. Secondly, segmentation of a facial image into regions of interest sometimes is difficult, when the regions do not have significant difference in their imaging attributes. Thirdly, unlike humans, machines usually do not have the capability to map facial expressions into moods.

Very few works on human mood detection have so far been reported in the current literature on machine intelligence. Ekman and Friesen [4] proposed a scheme for recognition of facial expressions from the movements of cheek, chin, and wrinkles. They have reported that there exist many basic movements of human eyes, eyebrows, and mouth, which have direct co-relation with facial expressions. Kobayashi and Hara [9-11] designed a scheme for recognition of human facial expressions using the well-known back-propagation neural algorithms [7], [14-16]. Their scheme is capable of recognizing six common facial expressions: happy, sad, fear, angry, surprised, and disgusted. Among the well-known methods of determining human emotions, Fourier descriptor [15], template-matching [2], neural network models [15], and fuzzy integral [6] techniques need special mention. Yamada in one of his recent papers [17] proposed a new method for recognizing moods through classification of visual information. Fernandez–Dols et al. proposed a scheme for decoding emotions from facial expression and content [5]. Carroll and Russel [3] in a recent book chapter analyzed in detail the scope of emotion modeling from facial expressions.

The chapter provides an alternative scheme for human mood detection from facial imageries. It employs i) fuzzy C-Means clustering algorithm [1] for segmentation of a facial image into important regions of interest and ii) fuzzy reasoning to map facial expressions into moods. The scheme is both robust and insensitive to noise, because of the nonlinear mapping of image-attributes to moods in the fuzzy domain. Experimental results demonstrate that the detection accuracy of moods by the proposed scheme is as high as 98%. The later part of the chapter proposes a new model of human mood control using a fuzzy relational approach.

The chapter is organized into 10 sections. Section 10.2 provides new techniques for segmentation and localization of important components in a human facial image. In Section 10.3, a set of image attributes including eye-pening, mouth-opening, and the length of eyebrow constriction is determined on-line in the segmented images. In Section 10.4, we fuzzify the measurements of imaging attributes into three distinct fuzzy sets: HIGH, MEDIUM, and LOW. Principles of a fuzzy relational scheme for human mood detection are also stated in this Section. Validation of the proposed scheme has been undertaken in Section 10.5. The scheme attempts to tune the membership distributions so as to improve the performance of the overall system to a great extent.

A basic scheme of human mood control is presented in Section 10.6. In Section 10.7, we present a model for human mood transition dynamics. Some interesting properties on mood controllability and convergence have also been addressed in this section. A proportional control of human moods is given in Section 10.8. A scheme for human mood control using Mamdani's relational technique has been presented in Section 10.9. The principles used for ranking the audio/video clips and music are narrated in Section 10.10. Experimental results are included in Section 10.11 and conclusions are listed in Section 10.12.

10.2 Filtering, Segmentation, and Localization of Facial Components

Identification of facial expressions by pixel-wise analysis of images is both tedious and time consuming. This chapter attempts to extract significant components of facial expressions through segmentation of the image. Because of the difference in the regional profiles on an image, simple segmentation algorithms, such as histogram-based thresholding technique, do not always yield good results. After several experiments, it has been observed that for the segmentation of the mouth region, a color-sensitive segmentation algorithm is most appropriate. Further, because of apparent nonuniformity in the lip color profile, a fuzzy segmentation algorithm is preferred. Taking into account the above viewpoints, a color-sensitive Fuzzy-C-Means clustering algorithm has been selected for the segmentation of the mouth region.

Segmentation of eye regions, however, in most images has been performed successfully by the traditional thresholding method. The hair region in human face can also be easily segmented by thresholding technique. Segmentation of the mouth and the eye regions is required for subsequent determination of mouth-opening and eye-opening, respectively. Segmentation of the eyebrow region is equally useful to determine the length of eyebrow constriction. Details of the segmentation techniques of different regions are presented below.

10.2.1 Segmentation of the Mouth Region

Before segmenting the mouth region, we first represent the image in the L*a*b space from its conventional RGB space. The L*a*b system has the additional benefit of representing a perceptually uniform color space. It defines a uniform matrix space representation of color so that a perceptual color difference is represented by the Euclidean distance. The color information, however, is not adequate to identify the lip region. The position information of pixels, along with their color together is a good feature to segment the lip region from the face. The fuzzy-C Means (FCM) Clustering algorithm that we would employ to detect the human lip region is supplied with both color and pixel-position information of the image.

The FCM clustering algorithm is a well-known technique for pattern classification. But its use in image segmentation in general and lip region segmentation in particular is a virgin area of research until now. The FCM clustering algorithm is available in any book on fuzzy pattern recognition [1], [8], [17]. In this chapter, we just demonstrate how to use the FCM clustering algorithm in the present application.

A pixel in this chapter is denoted by five attributes: three for color information and two for pixel position information. The three for 3 color information are L*a*b and the two for pixel position information are (x, y). The objective of this clustering algorithm is to classify the above set of five dimensional data points into two classes/ partitions namely the lip region and non-lip region. Initial membership values are assigned to each five dimensional pixel data such that sum of membership in the lip and the non-lip region is equal to one. Mathematically for the kth pixel x_k,

$$\mu_L(x_k) + \mu_{NL}(x_k) = 1, \tag{10.1}$$

where μ_L and μ_{NL} denote the membership of x_k to fall in the lip and the non-lip region, respectively. Given the initial membership value of $\mu_L(x_k)$ and $\mu_{NL}(x_k)$ for k = 1 to n^2 (assuming that the image is of n × n size). With these initial memberships for all n^2 pixels, following the FCM algorithm, we first determine the cluster center of the lip and the non-lip regions by the following formulae.

$$V_L = \sum_{k=1}^{n^2} [\mu_L(x_K)]^m x_K / \sum [\mu_L(x_k)]^m \tag{10.2}$$

and

$$V_{NL} = \sum_{k=1}^{n^2} [\mu_{NL}(x_K)]^m x_K / \sum [\mu_{NL}(x_k)]^m \tag{10.3}$$

Here m (>1) is any real number that influences the membership grade. The membership value of each pixel in the image to fall both in the lip and the non-lip region is evaluated by the following formulae.

$$\mu_L(x_K) = \left(\sum_{j=1}^{2} \{ \|x_k - v_L\|^2 / \|x_k - v_j\|^2 \}^{1/(m-1)} \right)^{-1} \tag{10.4}$$

$$\text{and} \quad \mu_{NL}(x_K) = \left(\sum_{j=1}^{2} \{ \|x_k - v_{NL}\|^2 / \|x_k - v_j\|^2 \}^{1/(m-1)} \right)^{-1} \tag{10.5}$$

Determination of the cluster centers by expression (10.2) and (10.3) and membership evaluation by (10.4) and (10.5) are repeated several times following the FCM algorithm until the position of the cluster centers do not change further.

Figure 10.1 presents a section of a facial image with a large mouth opening. This image is passed through a median filter and the resulting image is shown in Figure 10.2. Application of FCM algorithm on Figure 10.2 yields Figure 10.3. In Fig. 10.4, we demonstrate the computation of mouth opening.

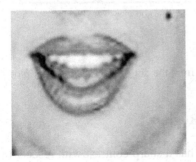

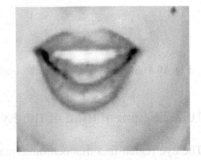

Fig. 10.1: The original face image.

Fig. 10.2: The median-filtered image.

Fig. 10.3: The image after the FCM clustering algorithm.

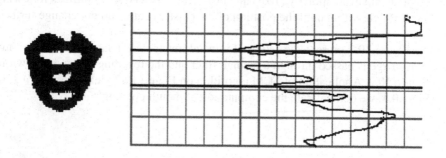

Fig. 10.4: Measurement of mouth-opening from the dips in average intensity plot.

10.2.2 Segmentation of the Eye Region

The eye region in a monochrome image has a sharp contrast with respect to the rest of the face. Consequently, the thresholding method can be employed to segment the eye region from the image. Images grabbed at poor illumination conditions have a very low average intensity value. Segmentation of eye region in these images is difficult because of the presence of dark eyebrows in the neighborhood of the eye region. To overcome this problem, we consider the images grabbed under good illuminating condition. After segmentation in the image is over, we need to localize the left and the right eyes on the image. In this chapter, we use template-matching scheme to localize the eyes. The eye template we used looks like Figure 10.5. The template-matching scheme we used here is taken from our previous works [2], [12]. It attempts to minimize the Euclidean distance between a fuzzy descriptor of the template with the fuzzy descriptor of the part of the image where the template is located. It needs mention here that even when the template is not a part of the image, the nearest matched location of the template in the image can be traced.

Fig. 10.5: A synthetic eye template.

10.2.3 Segmentation of Eyebrow Constriction

In a facial image, eyebrows are the second dark region after the hair region. The hair region is easily segmented by setting a very low threshold in the histogram-based thresholding algorithm. The eye regions are also segmented by the method discussed earlier. Naturally, a search for a dark narrow template can easily localize the eyebrows. It is indeed important to note that localization of eyebrows is essential for determining its length. This will be undertaken in the next section.

10.3 Determination of Facial Attributes

In this section, we present a scheme for on-line measurements of facial extracts such as mouth-opening (MO), eye-opening (EO), and the length of eyebrow constriction (EBC).

10.3.1 Determination of the Mouth-Opening

After segmentation of the mouth region, we plot the average intensity profile against the mouth-opening. Determination of MO in a black-and-white image becomes easier because of the presence of the white teeth. A plot of average intensity profile against the MO reveals that the curve will have several minima, out of which the first and the third correspond to one the inner region of the top lip and the inner region of the bottom lip, respectively. The difference along the MO axis (Y-axis) of the above two measures gives a measurement of the MO. An experimental instance of MO is shown in Figure 10.4. In this figure, the pixel count between the thick horizontal lines gives a measure of MO.

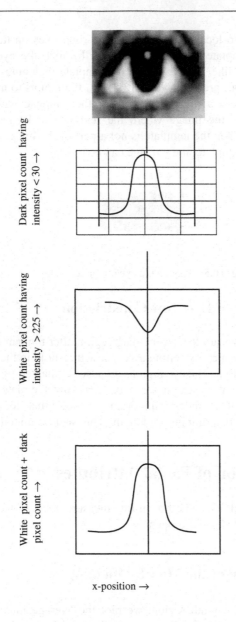

Fig. 10.6: Determination of the eye-opening.

10.3.2 Determination of the Eye-Opening

After the localization of the eyes, the count of dark pixels (having intensity <30) plus the count of the white pixels (having intensity > 225) is plotted against the x-position (Fig. 10.6). Suppose that the peak of this plot occurs at x = a. Then the ordinate at x = a gives a measure of the eye-opening.

10.3.3 Determination of the Length of Eyebrow Constriction

Constriction in the forehead region can be explained as a collection of white and dark patches called hilly and valley regions, respectively. The valley regions usually are darker than the hilly regions. Usually the width of the patches is around 10-15 pixels for a given facial image of (512×512) pixels.

Let I_{av} be the average intensity in a selected rectangular profile on the forehead and I_{ij} be the intensity of pixel (i, j). To determine the length of eyebrow constriction on the forehead region, we scan for variation in intensity along the x-axis of the selected rectangular region. The maximum x-width that includes variation in intensity is defined as the length of *eyebrow constriction*. The length of the eyebrow constriction has been measured in Figure 10.7 by using the above principle. An algorithm for eyebrow constriction is presented below.

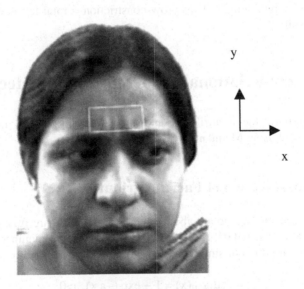

Fig 10.7: Determination of eyebrow constriction in the selected rectangular patch, identified by image segmentation.

Main steps:

1. Take a narrow strip over the eyebrow region with a thickness two-thirds of the eye-opening.

2. The length l of the strip is determined by identifying its intersection with the hair regions at both ends. Determine the center of the strip, and select a window of x-length $2l/3$ symmetric with respect to the center.

3. For x-positions central to window-right-end do

    a) Select nine vertical lines in the window and compute average intensity on each line;
    b) Take variance of the nine average intensity values;
    c) If the variance is below a threshold, stop;
        Else shift one pixel right;

4. Determine the total right shift.

5. Similar to step 3, determine the total left shift.

6. Compute length of eyebrow-constriction = total left shift + total right shift.

10.4 Fuzzy Relational Model for Mood Detection

Fuzzification of facial attributes and their mapping to the mood space is discussed here using Mamdani-type implication relation.

10.4.1 Fuzzification of Facial Attributes

The measurements we obtain about MO, EO, and EBC are fuzzified into three distinct fuzzy sets: HIGH, LOW, MODERATE. Typical membership functions that we have used in our simulation are presented below. For any real feature x

$$\mu_{HIGH}(x) = 1 - \exp(-a\,x),\ a>0,$$
$$\mu_{LOW}(x) = \exp(-b\,x),\ b>0,$$
$$\mu_{MODERATE}(x) = \exp[-(x - x_{mean})^2 / 2\sigma^2]$$

where x_{mean} and σ^2 are the mean and variance of the parameter x.

For the best performance we need to determine the optimal values of a, b, and σ. Details of these will be discussed in Section 10.5.

10.4.2 The Fuzzy Relational Model for Mood Detection

Mapping from fuzzified measurements to moods has been undertaken in this chapter by using a fuzzy relational scheme. To outline the scheme, we first consider a few sample rules.

Rule 1: IF (*eye-opening* is LARGE) &
  (*mouth-opening* is LARGE) &
  (*eyebrow-constriction* is SMALL)
THEN mood is VERY-MUCH-SURPRISED.

Rule 2: IF (*eye-opening* is LARGE) &
  (*mouth-opening* is SMALL/ MODERATE) &
  (*eyebrow-constriction* is SMALL)
THEN mood is VERY-MUCH-HAPPY.

Since each rule contains antecedent clauses of three fuzzy variables, their conjunctive effect is taken into account to determine the fuzzy relational matrix. The general formulation of a production rule with an antecedent clause of three linguistic variables and one consequent clause of a single linguistic variable is discussed below. Consider, for instance, a fuzzy rule:

If x is A and y is B and z is C
Then w is D.

Let $\mu_A(x)$, $\mu_B(y)$, $\mu_C(z)$ and $\mu_D(w)$ be the membership distribution of linguistic variables x, y, z, w belonging to A, B, C, and D, respectively. Then the membership distribution of the clause "x is A and y is B and z is C" is given by $t(\mu_A(x), \mu_B(y), \mu_C(z))$ where t denotes the fuzzy t-norm operator. Using Mamdani implication operator, the relation between the antecedent and consequent clauses for the given rule is described by

$$\mu_R(x, y, z; w) = \text{Min} \left[t \left(\mu_A(x), \mu_B(y), \mu_C(z) \right), \mu_D(w) \right] \qquad (10.6)$$

Taking Min as the t-norm, the above expression can be rewritten as

$$\mu_R(x, y, z; w) = \text{Min} \left[\text{Min} \left(\mu_A(x), \mu_B(y), \mu_C(z) \right), \mu_D(w) \right]$$

$$= \text{Min} \left[\mu_A(x), \mu_B(y), \mu_C(z), \mu_D(w) \right]. \qquad (10.7)$$

Now, given an unknown distribution of $(\mu_A'(x), \mu_B'(y), \mu_C'(z))$ then we can evaluate $\mu_D'(w)$ by the following fuzzy relational equation:

$$\mu_D'(w) = \text{Min} \left[(\mu_A'(x), \mu_B'(y), \mu_C'(z)) \right] \circ \mu_R(x, y, z; w). \quad (10.8)$$

For discrete systems, the relation μ_R (x, y, z; w) is represented by a matrix (Fig. 10.8) of the following form:

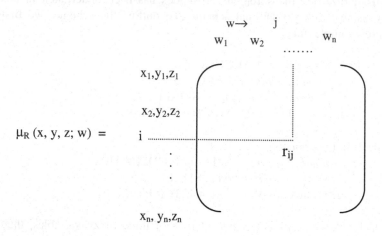

$$\mu_R \, (x, y, z; w) \; = \; i$$

Fig. 10.8: Structure of a fuzzy relational matrix.

where x_i, y_i, and z_i and w_i denote the specific instances of the variables x, y, z and w, respectively.

In our proposed application, the row index of the relational matrix is represented by conjunctive sets of values of mouth-opening, eye-opening, and eyebrow constriction. The column index of the relational matrix denotes the possible values of moods: happy, angry, sad, surprised, and anxious.

For determining the mood of a person, we define two vectors namely fuzzy descriptor vector F and mood vector M. The structural forms of the two vectors are given below.

$$F = [\mu_S \, (eo) \; \mu_M \, (eo) \; \mu_L(eo) \; \mu_S \, (mo) \, \mu_M \, (mo) \; \mu_L \, (mo) \; \mu_S(ebc) \; \mu_M(ebc)$$
$$\mu_L(ebc) \,] \qquad\qquad (10.9)$$

where suffices S, M and L stand for SMALL, MEDIUM, and LARGE, respectively.

Mood vector M
$$= [\mu_{VH} \, (mood) \; \mu_{MH} \, (mood) \; \mu_{N\text{-}So\text{-}H} \, (mood) \; \mu_{VA}(\, mood) \; \mu_{MA} \, (mood)$$
$$\mu_{N\text{-}So\text{-}A} \, (mood) \; \mu_{VS}(mood) \; \mu_{MS}(mood) \; \mu_{N\text{-}So\text{-}S}(mood) \; \mu_{Vaf}(mood)$$
$$\mu_{Maf} \, (mood) \; \mu_{N\text{-}So\text{-}Af}(mood) \; \mu_{VSr}(mood) \; \mu_{MSr}(mood) \; \mu_{N\text{-}So\text{-}Sr}(mood)$$
$$\mu_{Vag}(mood) \; \mu_{Mag}(mood) \; \mu_{N\text{-}So\text{-}Ag} \, (mood)]$$
$$(10.10)$$

where

> V, M, N-SO in suffices denote VERY, MODERATELY and NOT-SO,
> and   H, A, S, Af, Sr, and Ag in suffices denote HAPPY, ANXIOUS,
> SAD, AFRAID, SURPRISED, and ANGRY, respectively.

The relational equation used for the proposed system is given below:

$$M = F \circ R_{FM}, \tag{10.11}$$

where R_{FM} is the fuzzy relational matrix with the row and column indices as described above.

   Given an F vector and the relational matrix R_{FM}, we can easily compute the fuzzified mood vector M using the above relational equation. Finally, to determine the membership of the moods from their fuzzy memberships we need to defuzzify them using the following type of defuzzification rule:

$$\mu_{happy}(mood) = \frac{\mu_{VH}(mood)*w_1 + \mu_{MH}(mood)*w_2 + \mu_{N\text{-}SO\text{-}H}(mood)*w_3}{\mu_{VH}(mood) + \mu_{MH}(mood) + \mu_{N\text{-}SO\text{-}H}(mood)} \tag{10.12}$$

where w_1, w_2, and w_3 denote weights, which in the present context have been set equal to 0.33 arbitrarily.

10.5 Validation of System Performance

After a prototype design of an intelligent system is complete, we need to validate its performance. The term "validation" here refers to building the right system that truly resembles the system one intended to build. In other words, validation corresponds to the system and suggests reformulation of the problem characteristic and concepts based on the deviation of its performance from that of the desired (ideal) system [11]. It has been observed experimentally that performance of the proposed system greatly depends on the parameters of the fuzzifiers. To determine optimal settings of the parameters, a scheme for validation of the system's performance is proposed in Figure 10.9.

   In Figure 10.9, we tune parameter a, b, x_{mean}, and σ of the fuzzifiers by a supervised learning algorithm, so as to generate the desired mood from the given measurements of the facial extract. The back-propagation algorithm has been employed to experimentally determine the parameters a, b, m, σ. For a sample space of approximately 100 moods of different persons, the experiment was conducted and the following values of the parameters a = 2.2, b = 1.9, x_{mean} = 2.0, and σ = 0.17 have been found to yield the best results.

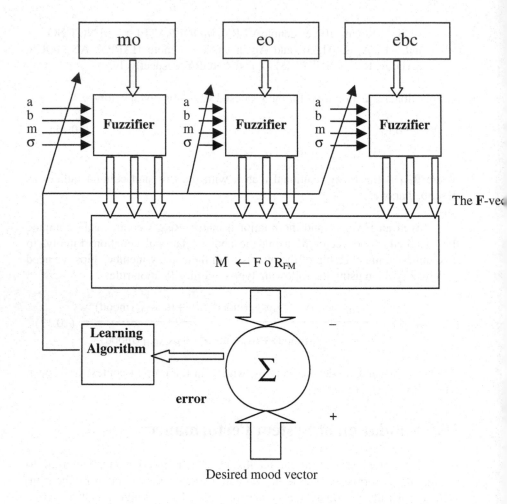

Fig. 10.9: Validation of the proposed system by tuning the parameters a, b, m, σ of the fuzzifiers.

10.6 A Basic Scheme of Human Mood Control

Human mood detection has already been undertaken as a significant research to study the psychological processes of the human mind. Various tools and techniques on human mood detection have already emerged, but very little work on human mood control has so far been reported. A research group, known as "Silent Sounds" led by Oliver Lowery of Georgia proposed a new means of controlling the human mind by music and ultrasonic signals.

According to a research bulletin reported by Silent Sounds, Inc. [13], the electroencephalograph (EEG) patterns/clusters of the human brain have a strong co-relation with the moods possessed by the subjects. Naturally, the EEG patterns may be used as a feedback to control the human mood towards the desired mood. A schematic view of the basic control scheme of human mood is presented in Figure 10.10.

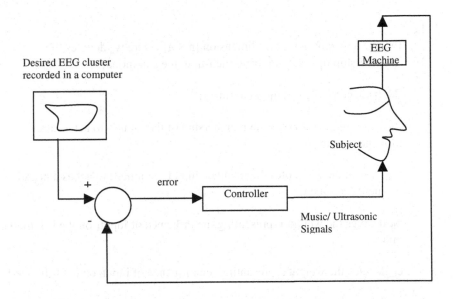

Fig. 10.10: A simple scheme to control the human emotion.

In this chapter we present an alternative scheme to control the human moods. Unlike the EEG clusters, we present a scheme that uses facial expression of the subject for the detection of his/her moods. Any classical method of human mood detection could have been embedded in our scheme. The preference to the scheme, however, depends on its suitability for real-time implementation. The fuzzy relational scheme for human mood detection, being very robust and amenable for real-time implementation, has been included in the proposed mood control scheme.

10.7 A Simple Model of Human Mood Transition Dynamics

Neuropsychologists and behavior therapists are of the opinion that the mood of a person at a given time is determined by the current sate of the his/her mind. The current state of the human mind is mainly controlled by the positive and the negative influences of input sensory data, including voice, video clips, ultrasonic signals, and music, on the human mind. In the proposed model, we consider that the mood at time t + 1 depends on the human mood at time t and the positive/negative influences that have been applied as stimulation of human mind at time t.

10.7.1 The Model

Let

[w_{ij}] be a weight matrix of dimension [n × n] where w_{ij} denotes the membership of transition from the i-th to the j-th mood;

M_i (t) denotes the i-th mood at time t;

$\mu_{POS\text{-}IN}$ (input$_k$, t) denotes the membership of the input k to act as a positive influence at time t;

$\mu_{NEG\text{-}IN}$ (input$_l$, t) denotes the membership of the input l to act as a negative influence at time t;

b_{ik} denotes the weight representing the influence of input$_k$ on the i-th mood; and

c_{il} denotes the weight representing the influence of input$_l$ on the i-th mood.

We can express $m_i(t + 1)$ as a function of $m_i(t)$ and positive as well as negative influences supplied to the human mind at time t.

$$m_i (t + 1) = \sum_{\forall j} w_{i,j}(t) \cdot m_j(t) + \sum_{\forall k} b_{i,k} \cdot \mu_{POS\text{-}IN} (input_k, t)$$

$$-\sum_{\forall l} c_{i,l} \cdot \mu_{NEG\text{-}IN} (input_l, t) \qquad (10.13)$$

It is indeed important to note that the first two terms in the right-hand side of equation 10.13 have a positive sign while the third term has a negative sign. The positive sign indicates that with growth of m_j and $\mu_{POS\text{-}IN}$ (input$_k$, t), $m_i(t + 1)$ also increases. The negative sign in the third term signifies that with growth in $\mu_{NEG\text{-}IN}$(input$_l$, t) $m_i(t + 1)$ decreases.

A question naturally appears in our mind: what does an increase or decrease in $m_i(t + 1)$ mean?

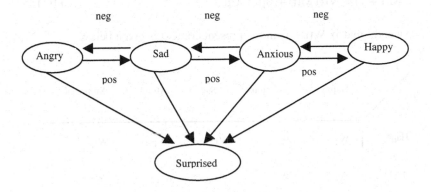

Fig. 10.11: The proposed mood transition graph.

Figure 10.11 provides a clear picture of transition of moods from state j to state i, where each state denotes a specific mood. It is clear from this Figure that with application of positive influence there is a transition of mood from angry state to the sad state or from the sad state to the anxious state or from the anxious state towards the happy state. An application of negative influence, on the other hand, attempts to control the mood in the reverse direction:

$$\text{Happy} \rightarrow \text{Anxious} \rightarrow \text{Sad} \rightarrow \text{Angry}.$$

It is to be noted further that the state of being surprised can be attained from any other states; and we in the mood control scheme, therefore, do not make any attempts to control the mood towards the surprised state.

Let

$M = [m_i]$ be the mood vector of dimension $(n \times 1)$;

$$\mu = \left(\mu_{\text{NEG-IN}} (\text{input}_k, t) \right) \quad \text{be the positive influence membership}$$

vector of dimension $n \times 1$;

$$\mu' = \left(\mu_{\text{NEG-IN}} (\text{input}_l, t) \right) \quad \text{be the negative influence membership}$$

vector of dimension $n \times 1$;

$B = [b_{ij}]$ be a $m \times n$ companion matrix to μ vector; and

$C = [c_{ij}]$ be a $m \times n$ companion matrix to μ' vector.

Iterating (10.13) for i = 1 to n we obtain:

$$M(t + 1) = W(t)\, M(t) + B\mu - C\mu'. \tag{10.14}$$

The weight matrix $W(t + 1)$ in the present context is given below.

$$W =$$

From To	Happy	Anxious	Sad	Afraid	Surprised
Happy	$W_{1,1}$	$W_{1,2}$	$W_{1,3}$	$W_{1,4}$	$W_{1,5}$
Anxious	$W_{2,1}$	$W_{2,2}$	$W_{2,3}$	$W_{2,4}$	$W_{2,5}$
Sad	$W_{3,1}$	$W_{3,2}$	$W_{3,3}$	$W_{3,4}$	$W_{3,5}$
Afraid	$W_{4,1}$	$W_{4,2}$	$W_{4,3}$	$W_{4,4}$	$W_{4,5}$
Surprised	$W_{5,1}$	$W_{5,2}$	$W_{5,3}$	$W_{5,4}$	$W_{5,5}$

$$\tag{10.15}$$

The equation (10.14) is represented diagrammatically in Figure 10.12.

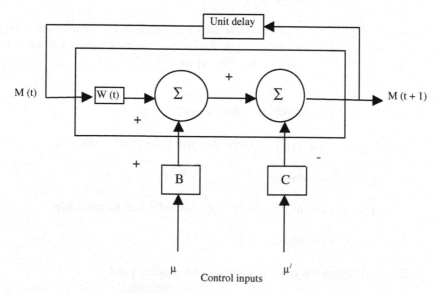

Fig. 10.12: The mood transition dynamics.

10.7.2 Properties of the Model

In an autonomous system, the system states change without application of any control inputs. The mood transition dynamics can be compared with an autonomous system with a setting of $\mu = \mu' = 0$. The limit cyclic behavior of an autonomous mood transition dynamics is given in Theorem 10.1 below:

Theorem 10.1: *If* $\mu = \mu' = 0$, *the mood vector M(t) exhibits limit cyclic behavior after every k-iterations if*

$$\prod_{i=0}^{k} w(i) = I. \tag{10.16}$$

Proof: Since $\mu = \mu' = 0$, we can rewrite expression (10.14) by

$$M(t + 1) = W(t)\, M(t) \tag{10.17}$$

Iterating $t = 0$ to $(n - 1)$, we have

$$M(1) = W(0)\, M(0) \tag{10.18}$$
$$M(2) = W(1)\, M(1) = W(1)\, (W(0)\, M(0)) \tag{10.19}$$

$$\vdots$$

$$M(n) = W(n-1)\, M(n-1)$$
$$= W(n-1)\, W(n-2)\, \ldots\ldots W(0)\, M(0)$$

$$= \left(\prod_{i=0}^{(n-1)} W(i) \right) . M(0) \tag{10.20}$$

Since the mood vector exhibits limit cyclic behavior after every k-iterations, then

$$M(k) = \left(\prod_{i=0}^{k-1} W(i) \right) . M(0)$$

$$= M(0) \text{ holds,} \tag{10.21}$$

which in turn requires

$$\prod_{i=0}^{k-1} W(i) = I, \text{ the identity matrix.}$$

The steady-state behavior of the mood transition dynamics with control input $\mu = \mu' = 0$ is given below in Theorem 10.2.

Theorem 10.2: *If $\mu = \mu' = 0$, $M(t + 1)$ vector reaches steady state M^* at time $t = n$, when $M(t + 1) = M(t)$. The weight matrix under this case is given by $W(n) = I$.*

Proof: Since $\mu = \mu' = 0$

$$M(t + 1) = W(t)\, M(t)$$

At $t = n$,

$$M(n + 1) = W(n)\, M(n)\ = M(n)\ \text{(given)} \tag{10.22}$$

$$\therefore W(n) = I.$$

Thus, the theorem holds.   □

Corollary 10.1: *With input $\mu = \mu' = 0$, the steady state value of mood vector M^* obtained after n iteration is given by*

$$M^* = M(0)\ \{ \prod_{i=0}^{n-1} W(i)\ \}. \tag{10.23}$$

Proof: The proof follows directly from the theorem.

Corollary 10.2: *When the net positive influence $B\mu$ is equal to the net negative influence $C\mu'$, a steady state in mood vector can be attained at time $t = n$ where $W(n) = I$.*

It is evident from corollary 10.2 that $\mu = \mu' = 0$ is not mandatory for steady state. The steady-state mood vector can also be attained by setting total input to the system: $B\mu - C\mu'$ to be 0.

For controlling system states in Figure 10.11 towards happy state, we need to supply positive influences in the state diagram at any state (excluding surprised state). Similarly, for controlling state transition towards angry state from any state s (excluding surprised state), we submit negative influence at state s. The controllability of a given state vector to a desired state can be examined by the Theorem 10.3.

Theorem 10.3: *The necessary and sufficient condition for the state-transition system to be controllable is that the controllability matrix:*

$$\begin{bmatrix} B & WB & W^2B & \ldots\ldots & W^{n-1}B \end{bmatrix}$$

and $\begin{bmatrix} C & WC & W^2C \dots \dots W^{n-1}C \end{bmatrix}$

should have a rank equal to n.

Proof: Proof of the theorem directly follows from the test criterion of controllability of linear systems.

10.8 The Proportional Model of Human Mood Control

A brief outline to the proposed mood control scheme is presented in Figure 10.13.

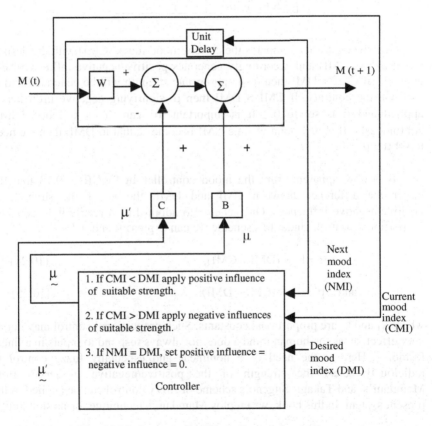

Fig. 10.13: A proportional logic for mood control.

One simple way to compare moods is to assign integer indices to the different moods. Table 10.1 provides the indices for the different moods. It is clear from Table 10.1 and Figure 10.13 that a positive influence causes an increase in mood indices, while a negative influence causes a decrease in mood indices. For example, suppose the current mood index is 3 (sad). A positive influence may change the mood index to 4 or 5, whereas a negative influence may force it to a lower value at 2 or 1. Such representation of mood indices supports the logic used in controller design in Figure 10.13.

Table 10.1: The mood index

Mood	Index
Happy	5
Anxious	4
Sad	3
Angry	2
Surprised	1

Here the controller compares the current mood index (CMI) with the desired mood index (DMI) and accordingly generates positive/negative influences. For example, if CMI < DMI then μ of appropriate strength will be generated and μ' = 0. On the contrary, if CMI > DMI then μ' signifying negative influence is applied and μ is set to 0. It is important to scan the next mood (NMI) continuously. If at any point of time NMI becomes equal to DMI, then we need to set $\mu = \mu' = 0$.

It is now apparent from the mood controller in FigURE 10.13 that the larger the difference between CMI and DMI, the larger the strength of positive/negative influences. One scheme to adopt this in reality is to consider proportional control. Under this scheme we can express μ and μ' by

$$\mu = k_1 \times (DMI - CMI), \tag{10.24}$$

$$\text{and} \quad \mu' = k_2 \times (CMI - DMI), \tag{10.25}$$

where k_1 and k_2 are proportional constants. Such proportional control may not be very effective, as the human mind a does not always respond to inputs in a linear fashion. Here is the need of a nonlinear and especially fuzzy control to judiciously select the strength of the positive/negative influence. Both Mamdani's and Takagi- Sugeno's scheme of fuzzy control can be applied in the present system. In this book, we employ Mamdani's technique for its simplicity.

10.9 Mamdani's Model for Mood Control

In Mamdani's model [18] a fuzzy rule of the form:

If x is A then y is B

is represented by a relational matrix R (x, y), where

$$R (x, y) = Min \{\mu_A(x), \mu_B(y)\} \tag{10.26}$$

and $\mu_A(x)$ and $\mu_B(y)$ denote the membership of x and belonging to fuzzy sets A and B, respectively. Given a input distribution for the fact: x is A' ($\approx$A), we may utilize it to instantiate the rule and consequently we get the distribution of y is B',

$$\mu_B'(y) = \mu_A'(x) \; o \; R (x, y) \tag{10.27}$$

This simple scheme of Mamdani's control and be extended for computing μ and μ' in Figure 10.15. Let error be a fuzzy linguistic variable defined by

$$Error = DM - CM \tag{10.28}$$

Let AMPLITUDE and SIGN be two fuzzy universes of error, where

VERY-LARGE, SMALL, MODERATE $\subseteq$ AMPLITUDE

and   POSITIVE, NEGATIVE $\subseteq$ SIGN.

Similarly, let STRENGTH be a fuzzy universe of positive or negative influence where

VERY-LARGE, LARGE, MODERATE, SMALL $\subseteq$ STRENGTH.

Now we reconstruct the fuzzy rules as follows:

Rule 1: If error is SMALL & POSITIVE
    Then apply positive influence of SMALL STRENGTH.

Rule 2: If error is MODERATE & POSITIVE
    Then apply positive influence of MODERATE STRENGTH.

Rule 3: If error is LARGE & POSITIVE
    Then apply positive influence of LARGE STRENGTH.

Rule 4: If error is VERY LARGE & POSITIVE
    Then apply positive influence of VERY LARGE STRENGTH.

Rule 5: If error is SMALL & NEGATIVE
    Then apply negative influence of SMALL STRENGTH.

Rule 6: If error is MODERATE & NEGATIVE
           Then apply negative influence of MODERATE STRENGTH.

Rule 7: If error is LARGE & NEGATIVE
           Then apply negative influence of LARGE STRENGTH

Rule 8: If error is VERY LARGE & NEGATIVE
           Then apply negative influence of VERY LARGE STRENGTH.

The fuzzy membership functions, which we will use in the present context are given in Figure 10.14.

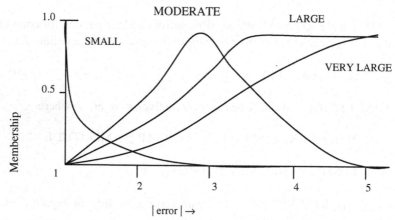

Fig. 10.14 (a): Membership distribution of error in fuzzy set SMALL, MODERATE, LARGE, and VERY LARGE.

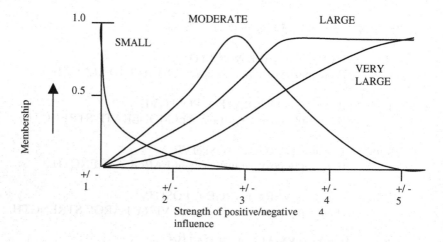

Fig. 10.14 (b): Membership distribution of the strength of positive/ negative influence to be SMALL, MODERATE, LARGE and VERY LARGE.

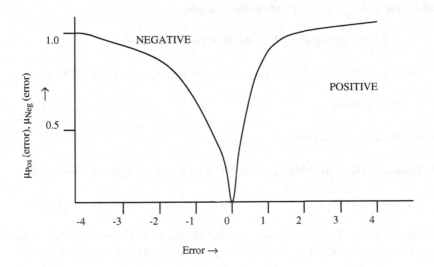

Fig. 10.14 (c): Membership distribution of error to be POSITIVE and NEGATIVE.

Let us now consider the rule PR_j,

PR$_j$: If error is SMALL & POSITIVE
   Then apply positive – influence of SMALL strength.

Also consider that we received a fact: error is MORE-OR-LESS-SMALL
and MORE-OR-LESS-POSITIVE.

  How can we infer about the positive influence? First we construct a fuzzy
relation R (error, error; pos-influence), where

$$R = \text{Min } [\text{Min } \{\mu_{SMALL} (error), \mu_{POSITIVE} (error)\}, \mu_{SMALL} (pos\text{-}in)]. \quad (10.29)$$

The outer Min in the last expression is due to Mamdani's implication rule. The
positive-influence for the j-th fired rule is given by

$$(\mu_{SMALL}{}' (pos\text{-}in) = (\mu_{MORE\text{ - }OR\text{ - }LESS\text{ -}SMALL} (error) \wedge \mu_{MORE\text{ - }OR\text{ - }LESS\text{ -}}$$

$$_{POSITIVE} (error)) \text{ o } R (error, error; pos\text{-}in), \quad (10.30)$$

where o denotes max-min composition operator

The generalized expression of $\mu_{STRENGTH}$ (pos-in) for the j-th rule is given below:

$$\mu_{STRENGTHj} (pos\text{-}in) = (\mu_{SIGNj} (error) \wedge \mu_{AMPLITUDEj} (error)) \text{ o } R (error, error;$$
$$pos\text{-}in) \quad (10.31)$$

where the suffix j corresponds to the j-th rule.

Now, taking aggregation of all the fired rules, we have:

$\mu_{STRENGTH}$ (pos-in) = Max {(μ_{SIGNj} (error) $\wedge$ $\mu_{AMPLITUDEj}$ (error)) o R (error,
$\quad\quad\quad\quad\quad\quad\quad$ $\forall j$

$\quad$ error; pos-in)}.   (10.32)

Similarly, we can compute

$\mu_{STRENGTH}$ (neg-in) = Max {(μ_{SIGNj} (error) $\wedge$ $\mu_{AMPLITUDEj}$ (error)) o R(error,
$\quad\quad\quad\quad\quad\quad\quad$ $\forall j$

$\quad$ error; neg-in)}.   (10.33)

Example 10.1: In this example we would like to illustrate the construction process of the relational matrix from the given distribution of error and pos-in.

$\mu_{POSITIVE}$ (error) = {1/0.2, 2/0.5, 3/0.6, 4/0.9}

and μ_{SMALL} (error) = {0/0.9, 1/0.1, 2/0.01, 3/0.005}

μ_{SMALL} (pos-in) = {10/09, 30/06, 60/0.21, 90/0.01}.

The relational matrix R (error, error; pos-in) now can be evaluated by Mamdani's implication function as follows:

Min [$\mu_{POSITIVE}$ (error), μ_{SMALL} (error)]

= {(1, 0)/(0.2 $\wedge$ 0.9), (1, 1)/(0.2 $\wedge$ 0.1), (1, 2)/(0.2 $\wedge$ 0.01), (1, 3)/(0.2 $\wedge$ 0.005);

$\quad$ (2, 0)/(0.5 $\wedge$ 0.9), (2, 1)/(0.5 $\wedge$ 0.1), (2, 2)/(0.5 $\wedge$ 0.01), (2, 3)/(0.5 $\wedge$ 0.005);

$\quad$ (3, 0)/(0.6 $\wedge$ 0.9), (3, 1)/(0.6 $\wedge$ 0.1), (3, 2)/(0.6 $\wedge$ 0.01), (3, 3)/(0.6 $\wedge$ 0.005);

$\quad$ (4, 0)/(0.9 $\wedge$ 0.9), (4, 1)/(0.9 $\wedge$ 0.1), (4, 2)/(0.9 $\wedge$ 0.01), (4, 3)/(0.9 $\wedge$ 0.005)}

= {(1, 0)/0.2, (1, 1)/0.1, (1, 2)/0.01, (1, 3)/0.005;

$\quad$ (2, 0)/0.5, (2, 1)/0.1, (2, 2)/0.01, (2, 3)/0.005;

$\quad$ (3, 0)/0.6, (3, 1)/0.1, (3,2)/0.01, (3,3)/0.005;

$\quad$ (4, 0)/0.9, (4, 1)/0.1, (4,2)/0.01, (4,3)/0.005}

The R (error, error; pos-in) matrix is now given below:

pos-in

error, error	10	30	60	90
1, 0	$(0.2 \wedge 0.9)$	$(0.2 \wedge 0.6)$	$(0.2 \wedge 0.2)$	$(0.2 \wedge 0.01)$
1, 1	$(0.1 \wedge 0.9)$	$(0.1 \wedge 0.6)$	$(0.1 \wedge 0.2)$	$(0.1 \wedge 0.01)$
1, 2	$(0.01 \wedge 0.9)$	$(0.01 \wedge 0.6)$	$(0.01 \wedge 0.2)$	$(0.01 \wedge 0.01)$
1, 3	$(0.005 \wedge 0.9)$	$(0.0 \wedge 0.6)$	$(0.0 \wedge 0.2)$	$(0.005 \wedge 0.01)$
2, 0	$(0.5 \wedge 0.9)$	$(0.5 \wedge 0.6)$	$(0.5 \wedge 0.2)$	$(0.5 \wedge 0.01)$
2, 1	$(0.1 \wedge 0.9)$	$(0.1 \wedge 0.6)$	$(0.1 \wedge 0.2)$	$(0.1 \wedge 0.01)$
2, 2	$(0.01 \wedge 0.9)$	$(0.01 \wedge 0.6)$	$(0.01 \wedge 0.2)$	$(0.01 \wedge 0.01)$
2, 3	$(0.005 \wedge 0.9)$	$(0.005 \wedge 0.6)$	$(0.005 \wedge 0.2)$	$(0.005 \wedge 0.01)$
3, 0	$(0.6 \wedge 0.9)$	$(0.6 \wedge 0.6)$	$(0.6 \wedge 0.2)$	$(0.6 \wedge 0.01)$
3, 1	$(0.1 \wedge 0.9)$	$(0.1 \wedge 0.6)$	$(0.1 \wedge 0.2)$	$(0.1 \wedge 0.01)$
3, 2	$(0.01 \wedge 0.90)$	$(0.01 \wedge 0.6)$	$(0.01 \wedge 0.2)$	$(0.01 \wedge 0.01)$
3, 3	$(0.005 \wedge 0.9)$	$(0.005 \wedge 0.6)$	$(0.005 \wedge 0.2)$	$(0.005 \wedge 0.01)$
4, 0	$(0.9 \wedge 0.9)$	$(0.9 \wedge 0.6)$	$(0.9 \wedge 0.2)$	$(0.9 \wedge 0.01)$
4, 1	$(0.1 \wedge 0.9)$	$(0.1 \wedge 0.6)$	$(0.1 \wedge 0.2)$	$(0.1 \wedge 0.01)$
4, 2	$(0.01 \wedge 0.9)$	$(0.01 \wedge 0.6)$	$(0.01 \wedge 0.2)$	$(0.01 \wedge 0.01)$
4, 3	$(0.005 \wedge 0.9)$	$(0.005 \wedge 0.6)$	$(0.005 \wedge 0.2)$	$(0.005 \wedge 0.01)$

$R=$

$$
=
\begin{pmatrix}
0.2 & 0.2 & 0.2 & 0.01 \\
0.1 & 0.1 & 0.1 & 0.01 \\
0.01 & 0.01 & 0.01 & 0.01 \\
0.005 & 0.005 & 0.005 & 0.005 \\
0.5 & 0.5 & 0.2 & 0.01 \\
0.1 & 0.1 & 0.1 & 0.01 \\
0.01 & 0.01 & 0.01 & 0.01 \\
0.005 & 0.005 & 0.005 & 0.005 \\
0.6 & 0.6 & 0.2 & 0.01 \\
0.1 & 0.1 & 0.1 & 0.01 \\
0.01 & 0.01 & 0.01 & 0.001 \\
0.005 & 0.005 & 0.005 & 0.005 \\
0.9 & 0.6 & 0.2 & 0.01 \\
0.1 & 0.1 & 0.1 & 0.01 \\
0.01 & 0.01 & 0.01 & 0.01 \\
0.005 & 0.005 & 0.005 & 0.005
\end{pmatrix}
$$

Let us now consider the observed membership distribution of error to be positive and small as follows.

$$\mu_{POSITIVE}'(error) = \{1/0.1, 2/0.5, 3/0.7, 4/0.4\}$$

and $\mu_{SMALL}'(error) = \{0/0.2, 1/0.1, 2/0.4, 3/0.5\}$

How should we compute the μ_{SMALL}' (pos-in)?

This can be computed by

$$\mu_{SMALL}'(pos\text{-}in) = Min\ [\mu_{POSITIVE}'(error), \mu_{SMALL}'(error)]\ o\ R\ (error, error;\ pos\text{-}in)$$

$$= [(1, 0)/0.1, (1, 1)/0.1, (1, 2)/0.1, (1, 3)/0.1,$$

$$(2, 0)/0.2, (2, 1)/0.1, (2, 2)/0.4, (2, 3)/0.5,$$

$$(3, 0)/0.2, (3, 1)/0.1, (3, 2)/0.4, (3, 3)/0.5,$$

$$(4, 0)/0.2, (4, 1)/0.1, (4, 2)/0.4, (4,3)/\ 0.4]\ o\ R(error,$$

$$error;\ pos\text{-}in)$$

$$= [0.1\ \ 0.1\ \ 0.1\ \ 0.1\ \ 0.2\ \ 0.1\ \ 0.4\ \ 0.5\ 0.2\ \ 0.1\ \ 0.4\ \ 0.5$$
$$0.2\ \ 0.1\ \ 0.4\ \ 0.4]\ o\ R\ (error,\ error;\ pos\text{-}in)$$

$$= [10/0.2\ \ \ 30/0.2\ \ \ 60/0.2\ \ \ 90/0.01].$$

For the selection of appropriate video or audio speech signals we need to defuzzify the control signals of positive strength. The method of centroid defuzzification may be employed here to absolutely determine the strength of pos-influence. Example 10.2 illustrates this phenomenon.

Example 10.2: The defuzzification of the control signal:

$$[10/0.2, 30/0.2, 60/0.2, 90/0.1]$$

by the center of gravity method is obtained as

$$\frac{10 \times 0.2 + 30 \times 0.2 + 60 \times 0.2 + 90 \times 0.1}{0.2 + 0.2 + 0.2 + 0.1}$$

$$= 14.75.$$

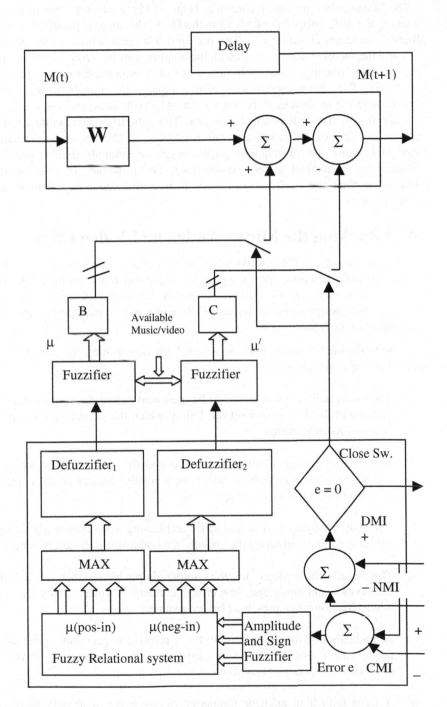

Fig. 10.15: The complete scheme of Mamdani type mood control.

The Mamdani's type Fuzzy controller (Fig. 10.15) senses the error in mood by taking the difference of the CMI from the DMI. This error is fuzzified in two different universes of fuzzy sets. The first universe deals with the sign of the error and the second universe deals with the amplitude of the error. Figure 10.14 (a) and (c) describe the membership functions to represent the error in respective fuzzy sets. The fuzzy relational system computes the membership of the positive/negative influences to be SMALL, MODERATE, LARGE, and VERY LARGE from the fuzzified membership of the error. The defuziffier units compute the absolute strength of the positive/negative influences. The appropriate music, video and audio clip of required positive/negative strength should then be presented to the subject under consideration. The principles of ranking the audio/video clips and music according to their positive/negative strength are outlined below.

10.10 Ranking the Music, Audio, and Video Clips

To provide the subject the positive/negative influence of appropriate strength, we need to rank the music, audio, and video sequences based on their relative strength. Since there are no clear-cut methods for such ranking, we in the present chapter design some strategies to efficiently assign appropriate ranks to these inputs of the human mind.

The following strategies may be adopted for determining the rank of the appropriate input signals.

1. The video/audio clips or music to be presented before the subject should match with his/her mental set-up, failing which the person may not pay attention to these events.

2. Signals that change in time and space and the content of which are conducive to the human mind will have a positive impact on the control of human mood towards the happy state.

3. The signals varying in time and space and having a pessimistic impact on the human mind will retard the growth in mood towards the angry state.

4. The small fluctuations in the amplitude/pitch/frequency including audio/video and music and slow changing theme of voice usually receive much attention when presented before a subject.

5. A gradually growing pitch usually has a positive impact and a gradually falling pitch usually has a negative impact as evidenced through observation on a large population.

6. A large fall-off in pitch or frequency causes anger or anxiety in most people.

7. For effective application of positive influences on the human mind, the rise in pitch or frequency of music should be of much longer duration compared with the time involvement in their respective fall-offs.

8. To control mood towards happy state, a wide-angle video of nature including sea, waterfalls, mountains, and green vegetation has been found to be most effective while experimenting with a large population. A steady growth in the wavelets on sea or river or an increased flow of water from a waterfall causes a positive impact on more than 50% of the experimented population.

9. An atmospheric storm, flood, war, famine, or other devastating forces that cause injury to the human beings usually generate negative impact on the human mind and consequently a sense of fear and anxiety appear in their face.

10. An encouraging voice which may (or may not) be contextual usually helps in the growth of the human mind to control their mood towards happy state.

11. A criticism and nonfavoring opinion may cause anger or anxiety in more than 60% of the experimental population.

For ranking the audio/video clips or music or voice message, we arranged to play those inputs before individuals of a large population and sought their opinion to assign a normalized strength that range from 0 to 1 to indicate their positive/negative effect on the human mind. The strength suggested by the population is then averaged and is labeled against the individual items. Such normalization by a large audience eliminates the scope of personal bias on the selection of the control inputs.

10.11 Experimental Results

We experimented with a large population of approximately 10,000 people of different age groups, sex, regions, and living standards. During the experiments we replaced the mind model part of Figure 10.15 with human subjects. The experiments reveal that people of lower age group (below 12) or people living in the rural regions directly respond to the selected control inputs in more than 86% cases. Approximately 67% of the urban population of age group 40 years and above reported to the control strategies suggested in this chapter. Women have been found to be more responsive than men in both rural and urban areas to the control policy for mood adaptation.

The mood control technique was also employed on psychologically impaired persons. It has been noted that for psychologically disabled people, controlling the mood towards the happy state needs softer music or more soothing videos in comparison to those required for normal people. A little annoying music or video can considerably change their mood towards anxious or angry.

10.12 Conclusions

The chapter presented a fuzzy relational scheme for mood detection from facial expressions and its control by judiciously selecting music, video/audio clips, and messages. Experimental results show good accuracy in both mood detection and control.

A fuzzy relational model has been used in both mood detection and control problem. For simplicity of Mamdani's model, we have employed his reasoning model to handle the present problem. Takagi-Sugeno's model can also be used in designing the mood control scheme.

In controlling human mood, emphasis is given on the selection of music, audio/video clips, and message. The defuzzifier units in the proposed fuzzy mood controller determine the strength of the necessary positive and negative influences. The available music, video, and audio messages are precalibrated with labels signifying their strength in mood control. It is indeed important to note that the labels suggested by the defuzzifier may not always be available in the databse of the respective items. Naturally, the music, video, and audio with strength-labels closest to that suggested by the defuzzifiers are selected for presentation. Much work still remains to use the proposed mood control scheme for clinical application.

Exercises

1. Develop a C/C++ program to read a .bmp or .tiff file as an array of characters and write the data part of the file in a 2-D array as integers. Further, construct a program for histogram plotting and segmentation of the hair region by thresholding.

2. Design a C/C++ program to implement the FCM clustering algorithm and use it for segmentation of the mouth region in a facial image.

3. Determine the mouth-opening by the method suggested by your own program.

4.   Consider an eyetemplate and use this to approximately locate the position of eyes in a facial image. Also determine eye-opening by the suggested scheme.

5.   Design a complete program for mood control using the suggested scheme.

6.   Prepare a database of music, video clips, and text messages and ask your friends to label them according to strength to control moods to a desired state. Use the labeling scheme suggested in the chapter to design the labels of your data items.

References

[1] Bezdek, J. C., *Fuzzy Mathematics in Pattern Classification*, Ph.D. Thesis, Applied Mathematics Center, Cornell University, New York Ithaca, 1973.

[2] Biswas, B. and Mukherjee, A. K., "Template matching with fuzzy descriptors," *J. of Inst. of Engineers*, 1997.

[3] Carroll, J. M. and Russel, J.A., "Do facial expression signal specific emotion? Judging emotion from the face in context," *Journal of Personality and Social Psychology*, vol. 70, no. 2, pp. 205-218, 1996.

[4] Ekman, P. and Friesen, W. V., *Unmasking the Face: A Guide to Recognizing Emotions from Facial Clues*, Prentice-Hall, Englewood Cliffs, NJ, 1975

[5] Fernandez-Dols, J. M., Wallbotl, H., and Sanhez, F., "Emotion category accessibility and the decoding of emotion from facial expression and context," *Journal of Nonverbal Behavior*, vol. 15, 1991.

[6] Izumitani, K., Mikami, T., and Inoue, K., "A model of expression grade for face graphs using fuzzy integral," *System and Control*, vol. 28, no. 10, pp. 590-596, 1984.

[7] Kawakami, F., Morishima, S., Yamada, H., and Harashima, H., "Construction of 3-D emotion space using neural network," *Proc. of the 3rd International Conference on Fuzzy Logic, Neural Nets and Soft Computing*, Iizuka, pp. 309-310, 1994.

[8] Klir, G. J. and Yuan, B., *Fuzzy Sets and Fuzzy Logic: Theory and Applications*, Prentice-Hall, Englwood Cliffs, NJ, 1995.

[9] Kobayashi, H. and Hara, F., "The recognition of basic facial expressions by neural network," *Trans. of the Society of Instrument and Control Engineers*, vol. 29, no.1, pp. 112-118, 1993.

[10] Kobayashi, H. and Hara, F., "Measurement of the strength of six basic facial expressions by neural network," *Trans. of the Japan Society of Mechanical Engineers*, vol. 59, no. 567, pp. 177-183, 1993.

[11] Kobayashi, H. and Hara, F., "Recognition of mixed facial expressions by neural network", *Trans of the Japan Society of Mechanical Engineers*, vol. 59, no. 567, pp. 184-189, 1993.

[12] Konar, A., *Computational Intelligence: Principles, Techniques and Applications*, Springer-Verlag, Heidelberg, 2005.

[13] Lowery, O., *Technical Report of Silent Sounds, Inc.*, Norcross, GA, 2004.

[14] Ueki, N., Morishima, S., Harashima, H., "Expression analysis/synthesis system based on emotion space constructed by multilayered neural network," *Systems and Computers in Japan,* vol. 25, no. 13, 1994.

[15] Uwechue, O. A. and Pandya, S. A., *Human Face Recognition Using Third-Order Synthetic Neural Networks*, Kluwer Academic Publishers, Boston, 1997.

[16] Vanger, P., Honlinger, R., and Haykin, H., "Applications of synnergetics in decoding facial expressions of emotions," *Proc. of Int. Workshop on Automatic Face and Gesture Recognition*, Zurich, pp. 24-29, 1995.

[17] Yamada, H., "Visual information for categorizing facial expression of emotion," *Applied Cognitive Psychology*, vol. 7, pp. 257-270, 1993.

[18] Zimmerman, H. J., *Fuzzy Set Theory and Its Applications*, Kluwer Academic, Dordrecht, The Netherlands, pp. 131-162, 1996.

Chapter 11

Distributed Planning and Multi-agent Coordination of Robots

This chapter provides an overview of multi-agent planning and coordination of mobile robots. It begins with single agent planning with special reference to the well-known blocks world problem. It then addresses the issues of task sharing in distributed planning, cooperation with/without communication, distributed versus centralized multi-agent planning, and homogeneous versus heterogeneous planning. The chapter next takes up multi agent transportation problem as a case study. The experimental details of two-agent cooperation in the said problem have been presented. The chapter finally analyzes the timing efficiency of two-agent and l-agent cooperation in the transportation problem. It concludes with a discussion on the scope of multi-agent planning in complex engineering systems.

11.1 Introduction

A robot is a controlled manipulator capable of performing complex tasks and decision making like the human beings [11]. A mobile robot can displace itself in its workspace through locomotion or flying. Mobile robots designed for factory automation usually have a mobile platform that runs on wheels [10]. A low-cost mobile robot [9], [12] contains ultrasonic/laser source-detector assemblies fixed around its physical structure to sense the location of objects in its world map [11]. These transducers work on the principle of time-of-flight (TOF) measurement [11].

The work presented in this chapter is based on Pioneer II robots, manufactured by *ActivMedia Robotics Corp. Amherst, NH, USA*. The model Pioneer II consists of eight ultrasonic source-detector pairs mounted around the structure of the robot. It also contains one video camera mounted on a pan-tilt platform and a frame grabber to hold the video frames for subsequent processing. The robot has two independently controlled motors to drive the side-wheels for movements in a given world map. It also has a caster wheel for mechanical balancing of the system. When the linear velocities of the side-wheels are equal in both magnitude and sign, the robot moves straight. When the linear velocity of the right (left) wheel is more than the left (right) wheel, the robot turns left (right). A desktop computer is connected with the robot in a radio-local area network (Radio-LAN) for TCP/IP communication of video and sonar packets of data. The robot usually receives the sensory data and sends them to the desktop machine for generating control commands for motion or camera control. The desktop machine in turn makes a control decision and sends the packets of data to the robot for execution of the control tasks.

This chapter is dedicated to multi-agent planning and cooperation. Here, we consider robots as hardwired agents. In multi-agent planning, agents together generate and execute a plan in an efficient manner. Unlike software agents that manage the main problems concerned with *information overload* [8], hardwired agents can perform actions on the environment. The load sharing in both plan-generation and its physical execution are equally important for hardwired agents. The most important issue of multi-agent systems is coordination among the agents. Coordination usually is of two common types: cooperation and competition. In box-pushing problems, agents cooperate with each other as they have a common goal to push the box towards the prescribed location. Competition comes into picture when the goals of the agents are mutually independent (or sometimes contradictory). For example, target-tracking problem realized with two robots, one as the tracker and the other as the moving target, is an example of coordination through competition. The tournament playing between two teams is also an example of competition among two groups of agents.

Readers may wonder, why we study agents in cognitive engineering. The science of cognition takes into account the detailed study of the psychological and behavioral processes that take place in the human brain. Agents in the real world are people who help us in solving a problem. Consider a real-world multi-agent system, such as a courier service, where a courier carries a parcel from India to the UK, the second courier from the UK to the USA, and the third local courier carries the parcel to the recipient. Similarly, in an intelligent multi-agent system, the agents participate jointly in solving a complex problem. How they share a complex task needs some analysis. The study of cognitive engineering this chapter in is very much needed as we have interest in modeling and realizing intelligent systems by emulating natural intelligence of agents.

The chapter is classified into seven sections. Section 11.2 illustrates single-agent planning using the classical blocks world problem. In Section 11.3, we introduce multi-agent planning with emphasis on task-sharing among the agents. The differences between centralized planning versus distributed planning and homogeneous versus heterogeneous planning are also outlined in this section. In Section 11.4, we take up a case study on multi-agent coordination in material transportation problem. Experimental results and observations for the transportation problem are outlined in Section 11.5. Timing analysis for the said problem with 2 and l-agents ($l > 2$) is undertaken in Section 11.6. Conclusions are listed in Section 11.7.

11.2 Single-Agent Planning

Robots like the human beings can perform complex tasks by generating a plan of schedules. The process of generating a plan of schedules is known as planning in texts on artificial intelligence [10]. To understand the planning process, let us consider the following example.

Example 11.1: In this example, we consider the problem of blocks world, where a number of blocks are to be stacked to a desired order from a given initial order. The initial and the goal states of the problem are shown in Figures 11.1 and 11.2, respectively. To solve this type of problem, we have to define a few operators using the if-add-delete structures, to be presented shortly.

The database corresponding to the initial and the goal states can be represented as follows.

The initial state:

>On (A,B)
>On (B, Table)
>On (C, Table)
>Clear (A)
>Clear (C)

The goal state:

>On (B, A)
>On (C, B)
>On (A, Table)
>Clear (C)

where On (X, Y) means the object X is on object Y and clear (X) means there is nothing on top of object X. The operators in the present context are given by the following if-add-delete rules.

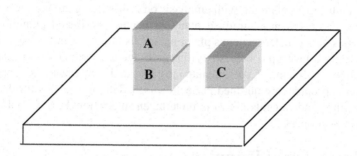

Fig. 11.1: The initial state of blocks world problem.

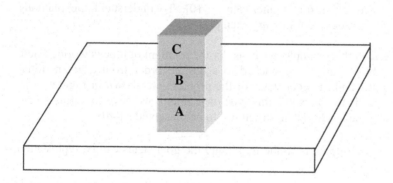

Fig. 11.2: The goal state of blocks world problem.

Rule 1: If On (X, Y), Clear (X), Clear (Z)
      Then Add-List: On (X, Z), Clear (Y),
      and Delete-List: On (X,Y), Clear (X).

Rule 2: If On (X,Y), Clear (X)
      Then Add-List: On (X, Table), Clear (Y),
      and Delete-List: On (X, Y).

Rule 3: If On (X, Table), Clear (X), Clear (Z)
      Then Add-List: On(X, Z),
      and Delete-List: Clear (Z), On (X, Table).

We try to solve the above problem by the following sequencing of operators. Rule 2 is applied to the initial problem state with an instantiation of X = A and Y = B to generate state S1 (Fig. 11.3). Then we apply Rule 3 with an instantiation of X = B and Z = A to generate state S2. Next, Rule 3 is applied once again to state S2 with an instantiation of X = C and Z = B to yield the goal state. There are many alternative approaches to handle the problem. Details of these are available in any textbook on artificial intelligence [10].

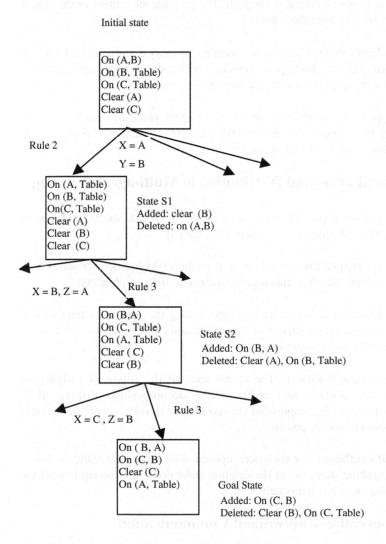

Initial state

On (A,B)
On (B, Table)
On (C, Table)
Clear (A)
Clear (C)

Rule 2   X = A

Y = B

On (A, Table)
On (B, Table)
On(C, Table)
Clear (A)
Clear (B)
Clear (C)

State S1
Added: clear (B)
Deleted: on (A,B)

X = B, Z = A   Rule 3

On (B,A)
On (C, Table)
On (A, Table)
Clear (C)
Clear (B)

State S2
Added: On (B, A)
Deleted: Clear (A), On (B, Table)

X = C , Z = B   Rule 3

On (B, A)
On (C, B)
Clear (C)
On (A, Table)

Goal State
Added: On (C, B)
Deleted: Clear (B), On (C, Table)

Fig. 11.3: The breadth-first search of the goal state.

11.3 Multi-agent Planning

In single-agent planning, the agent (robot) generates the complete plan itself. When the complexity of planning very high, a single agent requires significant amount of time to complete the plan. A group of agents is then employed to execute the plan. Some examples of multi-agent planning are outlined below.

Example 11.2: Consider the problem of carrying a large stick outside a room by multiple robotic agents. Here we need a leader robot, who shows the right path and the follower robot who traces the path. The plan for the motion of the robots is designed jointly by both the robots.

Example 11.3: In robot soccer, two teams of robots play a tournament. Here, we need to design plans for both goal protection (defensive actions) and attacking the opponent aggressively (offensive actions) [7].

In this chapter, we consider a similar multi-agent planning problem and its solution using neural networks. Before taking up the problem, we briefly outline some important features of mult-agent planning.

11.3.1 Task Sharing and Distribution in Multi-agent Planning

The most important aspect of multi-agent planning is task sharing. The main steps involved in task sharing are presented below [5].

1. **Task decomposition:** Decompose the total tasks into fragments or sub-tasks so that the fragments together represent the complete task.

2. **Task allocation:** Assign the sub-tasks among the agents, so that they can be executed in parallel. List the constraints that have to satisfied to allocate the tasks to the agents.

3. **Task accomplishment:** The agents execute the sub-tasks assigned to them. The sub-tasks may include further decomposition of tasks. If so happens, the decomposition is done recursively, until no further decompositions are possible.

4. **Result synthesis:** For each decomposed sub-task, perform the sub-tasks and combine them to get the original tasks done, and repeat it until the resulting action is generated.

11.3.2 Cooperation with/without Communication

In multi-agent systems, agents co-operate with/without communication [1-2]. Communications help the agents to share the sensory instances, thereby making them more knowledgeable about their world. Naturally, making decisions from more sensory instances becomes easier. In the absence of communication, the agents have to make decisions from approximate knowledge about their world, and occasionally the decisions suffer from inaccuracy. In the following examples, we illustrate the relative merits of communication in multi-agent systems.

Distributed Coordination in Box-Pushing: In the box-pushing problem, a box is shifted by the pushing actions of several agents. Here each agent generates a plan to move the box to the desired location. Usually, in the box-pushing problem, there exists no central coordinator.

Let θ_t denote the desired angle of box movement at time t. Suppose, there are n agents. Let the force applied by agent i be $F_i = A_i + jB_i$. Then in order to position the box correctly, we require:

$$\text{Angle of } \sum_{i=1}^{n} F_i = \theta_t \qquad (11.1)$$

$$\text{or,} \quad \tan^{-1} \left(\sum_{i=1}^{n} F_i \right) = \theta_t \qquad (11.2)$$

Substituting $F_i = A_i + jB_i$, we have:

$$\tan^{-1} \left(\sum_{i=1}^{n} B_i / \sum_{i=1}^{n} A_i \right) = \theta_t \qquad (11.3)$$

Since all the n-agents together move the box towards its next position, the contribution of the n-th agent is given by

$$\theta_t - \tan^{-1} \left(\sum_{i=1}^{n-1} B_i / \sum_{i=1}^{n-1} A_i \right). \qquad (11.4)$$

The n-th robot applies a force in a particular direction to reduce the above quantity to zero. The agents in the box-pushing problem learn online to make necessary corrections in their individual tasks. Any typical reinforcement learning algorithm such as Q-learning may be employed to enable the robot to make necessary corrections in their tasks.

As there is no central coordinator, but the agents themselves plan to move the box towards a predefined goal position, the planning for the individual movements of the agents is accomplished by the agents in a distributed manner.

Centralized Coordination of the Agents in Battlefields: Multi-agent robots designed for battlefields usually have a team leader, who makes the decisions about the movements of the armed-agents. Here the teammates of the leader, called followers, submit information about the local changes in the environment through a radio-network. The leader then processes the information from multiple followers, plans their actions and commands them to execute the plans [15].

11.3.3 Homogeneous and Heterogeneous Distributed Planning

The distributed planning of agents usually is of two common types: homogeneous and heterogeneous. In homogeneous planning, the role of each agent is identical. For example, in a box-pushing problem, the planning algorithms that the agents employ are identical, but there may be small difference in their actions.

In heterogeneous planning, the plans of action by the agents are different. Let P be the desired plan, accomplished by n-agents, and P_i and P_j be the sub-plans assigned to the respective agents i and j for execution, then $P_i \neq P_j$. For instance, consider the transportation problem of blocks by two robots (Fig. 11.4).

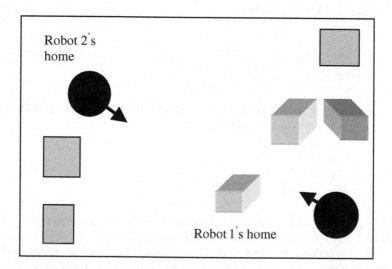

Fig. 11.4: The robots' world: Robot 1's home contains three blocks and robot 2's home is empty. The circles denote the robots and the rectangles denote obstacles in the space.

Here, robot 1 has a number of blocks in its home position. Robot 2's home is empty. The blocks need to be transported from robot 1's workspace to robot 2's workspace. Suppose, robot 1 is assigned a task to carry blocks, one at a time, towards robot 2's home, and on the way when it meets robot 2, the box is to be transferred to robot 2. The task assigned to robot 2 is to look for robot 1 in the path leading to robot 1's home and on meeting robot 1, it should receive the block from robot 1 and carry it to its home place. When robot 2 is carrying the box to its home, robot 1 is to return to its home place to fetch the next block. The similar steps are to be repeated to carry all the blocks to robot 2's home place.

It is important to note that the two robots' behavior and plans of action are different. In fact, the agents here perform a distributed heterogeneous planning.

11. 4 Vision-Based Transportation of Blocks by Two Robots

In this section, we present an implementation of the transportation problem by two robots, introduced in the last section. We used Pioneer II robots to handle the problem [14]. A brief introduction to Pioneer II robot is given in Section 11.1. The following behaviors of the robots have been synthesized to efficiently generate and execute the plans for the proposed multi-agent transportation problem.

    a)   Keen recognition by color segmentation [2] of opponent's image and localization, abbreviated as KR,

    b)   Block recognition by color segmentation and localization, abbreviated as BR,

    c)   Gripping (holding) a box, hereafter called GRIP,

    d)   Ungripping (releasing) a box, hereafter called UNGRIP,

    e)   Transferring a box to another agent, hereafter called TRANSFER,

    f)   Receiving a block from another agent, hereafter called RECEIVE,

    g)   Obstacle avoidance and path planning towards a given target position, hereafter called MOVE-TARGET-POSITION,

    h)   Return to home, hereafter called RH,

    i)   Camera zoom control, abbreviated as CZC, and

    j)   Camera pan and tilt control, abbreviated as CPTC.

Thus, the task assigned to robot 1 may be described by the following sequence of behaviors:

    BR, GRIP, MOVE-TARGET-POSITION, KR, TRANSFER, RH.

The camera zoom control is needed in synthesizing the TRANSFER behavior. The task assigned to robot 2 can also be represented by a sequence of behaviors, as presented below:

MOVE-TARGET-POSITION, KR, RECEIVE, RH, UNGRIP.

It is important to note that the target positions in the behavior: MOVE-TARGET-POSITION of the two robots are different. In the first case, the target position refers to robot 2's home position, whereas in the second case it refers to robot 1's home position. The principles used in designing the basic behaviors are outlined below.

Keen recognition: Since Pioneer II robots are dark red and there are no other red objects in the robots' world, the color segmentation is a good choice for keen recognition. We identified the regions on the image acquired by an agent, where the red pixel-intensity far exceeds the sum of the intensity for green and blue colors for the same pixels. Further, the selected pixels should form a region that can be separated from the background of the image. For localization of the keen's image, we determine the centroid position of the region, which is segmented as the keen.

Block recognition: The blocks used for our experiment are white-colored and can be easily segmented by histogram thresholding of the monochrome image. To determine the boundary (edges) of the block, we use a Sobel mask. Identification of the edges of the block and their localization is needed to hold the blocks by the grippers. The details of edge localization are beyond the scope of the chapter. This involves 3-D modeling of edges from multiple 2-D images using extended Kalman filtering [3], [13].

Gripping and **releasing** loads are simple behaviors, available in the software platform of the Pioneer II robots.

Transfer: To transfer a block from one agent to the other is a complex process. It includes tight co-operation of the agents. First the agents should understand that they are coming closer to each other by activating the ultrasonic transducers in the direction of the keen. The agents then should gradually decrease their velocity and set it to zero when they meet each other. The receiving agent opens its grippers, when it is at a safe distance from its partner. After the receiving agent grips the block, a signal is communicated to the transferring agent, and only then it releases the block. Each agent then moves back by a safe distance and turns 180° to come to their respective home positions. The last step is part of the RH behavior.

Move-target-position: This behavior is realized using back-propagation neural net. We have taken approximately 600 training instances of the robot's world map [4], [6]. Given the goal position of the robot in this map, the agent's direction of movement and speed are considered as output parameters. Thus, with eight sonar transducers, we form vector training instances of 12 attributes, out of which the first eight fields correspond to the obstacle positions in eight directions, separated by an interval of 45°. The 9th and the 10th fields denote the target (x, y) position, whereas the 11th and the 12th fields denote the heading direction (in angle) and the speed of the robot. After the agent is trained with the 600 instances, it can select its direction of movement and speed for a new reading of the sonar transducers.

The behavior **return to home** is simple and needs no explanation. It uses move-target-position as a sub-behavior, where the destination is its home station.

Camera zoom control: This behavior attempts to focus the kin or the white blocks by the camera so that it occupies the major regions in the grabbed image.

Camera pan and tilt control: Camera pan and tilt angle controls are needed to bring the desired object within the grabbed image. The camera tilt angle usually need not change much, but pan angle is gradually changed to examine the existence of a block or the keen within the grabbed image.

11.5 Experimental Results

An experiment on multi-agent coordination of two robots for the block transportation problems was performed in Artificial Intelligence Laboratory of Jadavpur University. Some interesting steps of the experiment are given in Figure 11.5. The following observations are noted from the experiment.

a) Quite an appreciable amount of time is consumed to execute the TRANSFER/RECEIVE behavior. On the PIONEER II platform, the transfer time/block is approximately 1 minute and 40 seconds.

b) BR and GRIP also consume significant time on the order of 1 minute 25 and seconds.

c) After an agent grips a white tall block, the frontal image of this agent is partitioned into two red regions separated by the white region of the block. We have taken special care in our implementation to handle this problem.

d) Approximately 38% of the total time required to transport the blocks by a single agent can be saved if two agents are employed to handle the same problem.

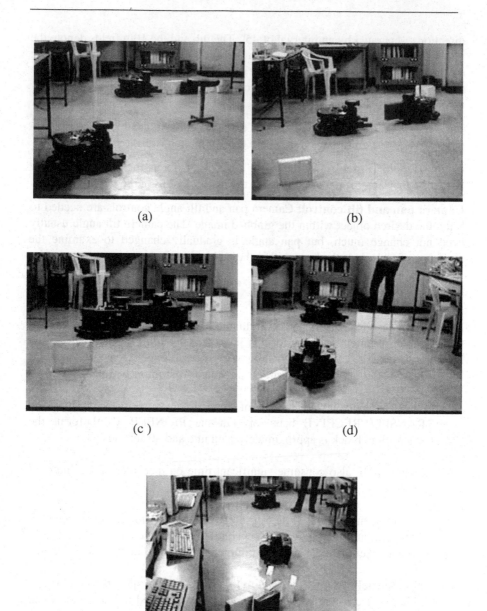

(a)   (b)

(c)   (d)

(e)

Fig. 11.5: Snapshots of cooperation by the agents: a) Initial position of the robots at their home stations, b) Robot 1 has the second block in its gripper and robot 2 has its gripper open, c) Transfer of block is taking place, d) Robot 2 is returning home with the second block, e) The transportation of the last block is just completed, robots returned to their respective homes.

11.6 Timing Analysis of the Transportation Problem

For simplicity in analysis, we first analyze the problem with two agents. Then we shall consider the analysis with l number of agents to handle the same problem

11.6.1 Analysis with Two Agents

Let

D	=	initial distance between the two robots,
v	=	velocity of the robot,
S_g	=	search time identify and grip an object,
S_{ungrip} =		time required to ungrip/ release a block,
S_r	=	search time to identify the companion robot,
n	=	number of blocks to be transferred,
t_r	=	transfer time,
T	=	total time needed by a robot to transfer n boxes,
d_i	=	distance traversed for transferring the i-th block,
T_1	=	time required by robot 1 to complete the task,
T_2	=	time required by robot 2 to complete the task.

T_1 = n (pick-up time for a block + search time to identify a robot + transfer

$$\text{time}) + 2 \sum_{i=1}^{n} \frac{d_i}{v} \qquad (11.5)$$

T_2 = n (release time for a box + search time to identify a robot + transfer

$$\text{time}) + 2 \sum_{i=1}^{n} \frac{D - d_i}{v} \qquad (11.6)$$

Since both the robots complete the sub-tasks at the same time, we have

$T_1 = T_2$, which implies

$$n (s_g + s_r + t_r) + 2 \sum_{i=1}^{n} \frac{d_i}{v} = n (s_{ungrip} + s_r + t_r) + 2 \sum_{i=1}^{n} \frac{D - d_i}{v}$$

$$\Rightarrow (n\, s_g - n\, s_{ungrip}) + 4 \sum_{i=1}^{n} \frac{d_i}{v} = \frac{2nD}{v}$$

$$\Rightarrow n \left\{ (s_g - s_{ungrip}) + \frac{2D}{v} \right\} = -4 \sum_{i=1}^{n} \frac{d_i}{v}$$

$$\Rightarrow \quad n\left\{\frac{2D}{v} - (s_g - s_{ungrip})\right\} = 4\sum_{i=1}^{n} \frac{d_i}{v}$$

$$\Rightarrow 2\sum_{i=1}^{n} \frac{d_i}{v} = n\left\{\frac{D}{v} - (s_g - s_{ungrip})/2\right\} \qquad (11.7)$$

$$T_1 = n(s_g + s_r + t_r) + 2\sum_{i=1}^{n} \frac{d_i}{V}$$

$$= n(s_g + s_r + t_r) + n\left\{\frac{D}{V} - (s_g - s_{ungrip})/2\right\}$$

$$= \left(\frac{n}{2} s_g + \frac{n}{2} s_{ungrip}\right) + n(s_r + t_r) + \frac{nD}{v}$$

$$= n\left(\frac{s_g + s_{ungrip}}{2}\right) + n(s_r + t_r) + \frac{nD}{v} \qquad (11.8)$$

For a single agent doing the same job,

$$T = n(s_g + s_{ungrip}) + 2n\frac{D}{v}$$

$$\text{Total saving in time} = \frac{T - T_1}{T} = \frac{nD/v + n/2(s_g + s_{ungrip}) - n(t_r + s_r)}{2nD/v + n(s_g + s_{ungrip})}$$

$$= \frac{1}{2} - \frac{(t_r + s_r)}{2D/v + (s_g + s_{ungrip})}. \qquad (11.9)$$

Thus, percentage saving in time $= \{(T - T_1)/T\} \times 100$

$$= [50 - (t_r + s_r)/\{2D/v + (s_g + s_{ungrip})\} \times 100]\%. \qquad (11.10)$$

The following observations follow from the last result.

i)   The percentage time saving is always less than 50%.

ii)   The smaller the transfer time (t_r) and search time to identify the companion robot (s_r), the greater the percentage saving.

11.6.2 Analysis with *l* Agents

In this section, we present an analysis with *l* agents.
Let

D	=	initial distance between the two robots,
v	=	velocity of the robot,
s_g	=	search time identify and grip an object,
s_{ungrip}	=	time required to ungrip/ release a block,
s_r	=	search time to identify the companion robot,
n	=	number of blocks to be transferred,
t_r	=	transfer time,
T	=	total time needed by a robot to transfer n boxes
d_{kj}	=	distance traversed by the k-th agent to transfer the j-th block,
T_j	=	total time required by the j-th agent,

$$T_1 = n\,(s_g + s_r + t_r) + 2 \sum_{j=1}^{n} \frac{d_{1j}}{v}, \tag{11.11}$$

$$T_2 = n\,(2s_g + 2t_r) + 2 \sum_{j=1}^{n} \frac{d_{2j}}{v}, \tag{11.12}$$

$$T_i = n\,(2s_r + 2t_r) + 2 \sum_{j=1}^{n} \frac{d_{jj}}{v}, \tag{11.13}$$

$$T_{last} = T_l = n\,(s_r + s_{ungrip} + t_r) + 2 \sum_{j=1}^{n} \frac{d_{lj}}{v}. \tag{11.14}$$

Again, $\displaystyle\sum_{k=1}^{l} d_{kj} = d_{1j} + d_{2j} + \ldots\ldots + d_{lj} = D.$ \hfill (11.15)

Now, $\displaystyle\sum_{k=1}^{l} T_k = n(s_g + s_{ungrip}) + \{n(s_r + t_r) + n(2s_r + 2t_r) + \ldots + n(2s_r + 2t_r) +$

$$n(s_r + t_r)\} + 2\,nD/v \tag{11.16}$$

Therefore, $\displaystyle\sum_{k=1}^{l} T_k$

$$= \frac{n}{l}\,(s_g + s_{ungrip}) + \frac{2n}{l}\,(s_r + t_r)\,(l - 1) + \frac{2nD}{lv}. \tag{11.17}$$

On the other hand, a single agent performing transfer of n-blocks requires a time

$$T = n\,(s_g + s_{ungrip}) + \frac{2nD}{v} \tag{11.18}$$

$\therefore$ total saving in time $= \dfrac{T - T_1}{T}$

$$= \frac{(n - \dfrac{n}{l})\,(s_g + s_{ungrip}) + \dfrac{2\,nD}{v}\,(1 - \dfrac{1}{l})\,\dfrac{2n}{l}\,(s_r + t_r)\,(l - 1)}{n\,(s_g + s_{ungrip}) + \dfrac{2nD}{v}}$$

$$= \frac{(l - 1)\,(s_g + s_{ungrip}) + \dfrac{2D}{v}\,(l - 1) - 2\,(s_r + t_r)\,(l - 1)}{l\,(s_g + s_{ungrip}) + \dfrac{2Dl}{v}}$$

$$= (\frac{l - 1}{l})\,[\{(s_g + s_{ungrip}) + \frac{2D}{v} - 2\,(s_r + t_r)\} / \{(s_g + s_{ungrip}) + \frac{2D}{v}\}] \tag{11.19}$$

Cases:

i) Marginal saving takes place when

$$(s_g + s_{ungrip}) + \frac{2D}{v} \geq (s_r + t_r). \tag{11.20}$$

ii) Maximum percentage saving takes place when l approaches infinity, i.e., number of agents is excessive.

iii) Percentage saving is zero when $l = 1$. This, of course, is a trivial case.

iv) Percentage saving is independent of n, the number of blocks.

11.7 Conclusions

This chapter presented an introduction to multi-agent planning and coordination of mobile robots. It is clear from the chapter that task distribution and assignment is one of the key factors in determining the time-efficiency of a multi-agent system. The design issues for a homogeneous distributed system are relatively simpler in comparison to those of a heterogeneous system. Proper synchronization is needed when the agents directly interact with one another. In the transportation problem, the coordination between the agents while transferring the block is most vital. It may be noted that after robot 2 receives the block from robot 1, it issues a signal to robot 1, indicating that the block now can be released by robot 1. Determination of the appropriate signals in the tight coordination phase remains a crucial problem.

Two important issues in distributed system are timing efficiency and resource utilization rate. Timing efficiency for the transportation problem has been analyzed in this chapter. In the transportation problem, resources, i.e., the agents are fully utilized. So the resource utilization rate for two-agent problem is 100%. However, with more than two agents, the resource utilization rate may decrease.

One interesting point to note is that high resource utilization rate in multi-agent system does not always ensure high speed-up. This is because some additional tasks, such as transfer of block in the transportation problem, that do not occur in a single agent system appear as additional problem in multi-agent systems.

Exercises

1. Consider the bricklayer's problem where there are two bricklayers and six assistants for mixing cement with sand and carrying bricks and the mixture to bricklayers for the construction of two walls concurrently. If the six assistants are replaced by six robots, how will you distribute their jobs? Think of two alternative job-distribution schedules and design a comparator to compare the relative merits of each schedule. Which schedule is better with respect to your comparator? [*open-ended problem*]

2. Suppose there are l agents in a transportation problem. Suppose that the agents are geographically distributed in a room. There are m locations in the room, each containing p paper waste boxes. The paper waste boxes are to be transported to a garbage-storage located at one end of the room. How will you assign tasks to the agents? Determine the time saving using l agents in comparison to the time required by a single agent to execute the job. [*open-ended problem*]

References

[1] Arkin, R. C., *Behavior-Based Robotics*, MIT Press, Cambridge, MA, 1998.

[2] Arkin, R. C. and Balch, T., "Cooperative multi-agent robotic systems," In *Artificial Intelligence and Mobile Robotics: Case Studies of Successful Robot Systems*, Kortenkamp, D., Bonasso, R. P. and Murphy, R. (Eds.), AAAI-MIT Press, Menlo Park, CA, 1998.

[3] Ayache, N., *Artificial Vision for Mobile Robots*, MIT Press, Cambridge, MA, 1991.

[4] Banerjee, R. K., Chowdhury, A. S., Chakraborty, I. and Konar, A., "Benchmark analysis for the path planning of a mobile robot using neural nets," *Proc. of Int. Conf. on Knowledge Based Computer Systems*, Bangalore, India, 2002.

[5] Durfee, E. H., "Distributed problem solving and planning," In *Multiagent Systems: A Modern Approach to Distributed Artificial Intelligence*, Weiss, G. (Ed.), MIT Press, Cambridge, MA, 1999.

[6] Jain. L. C. and Fukuda, T., *Soft Computing for Intelligent Robotic Systems*, Physica-Verlag, Heidelberg, 1998.

[7] Kim, J.-H., Kim, D.-H., Kim, Y.-J. and Seow, K.-T., *Soccer Robotics*, Springer-Verlag, New York, 2004.

[8] Klush, M. (Ed.), *Intelligent Information Agents: Agent-Based Discovery and Management in the Internet*, Springer-Verlag, Heidelberg, 1999.

[9] Liu, J. and Wu, J., *Multi-agent Robotic Systems*, CRC Press, Boca Raton, FL, 2001.

[10] Konar, A., *Artificial Intelligence and Soft Computing: Behavioral and Cognitive Modeling of the Human Brain*, CRC Press, Boca Raton, FL, 1999.

[11] Nehmzow, U., *Mobile Robotics: A Practical Introduction*, Springer-Verlag, London, 2000.

[12] Parker, L. E., Bekey, G. and Barhen, J. (Eds.), *Distributed Autonomous Robotic Systems*, Springer-Verlag, Hong Kong, 2000.

[13] Patnaik, S., Konar, A. and Mandal, A. K., "Building 3D visual perception of a mobile robot employing extended Kalman filter," *J. of Intelligent and Robotic Systems*, vol. 34, pp. 99-120, 2002.

[14] Roy, K., Shaw, A. and Konar, A., "Neuro-visual perception in multi-agent coordination of mobile robots for application in material transportation problem," *Proc. of National Conf. on Advances in Manufacturing Systems*, Kolkata, 2003.

[15] Ray, S., De, D. and Konar, A., "Secured transportation management in battlefield by multi-agent robotic systems," *Proc. of Int. Conf. on Intelligent Signal Processing and Robotics*, Allahabad, India, 2004.

Index